199 Terranova

DELEUZIAN INTERSECTIONS

DELEUZIAN INTERSECTIONS

Science, Technology, Anthropology

Edited by
Casper Bruun Jensen and Kjetil Rödje

Berghahn Books
New York • Oxford

Published in 2010 by
Berghahn Books
www.berghahnbooks.com

Library of Congress Cataloging-in-Publication Data
Deleuzian intersections : science, technology, and anthropology / edited by Casper Bruun Jensen and Kjetil Rødje.
p. cm.
Includes bibliographical references and index.
ISBN 978-1-84545-614-6 (alk. paper)
1. Deleuze, Gilles, 1925-1995--Criticism and interpretation. 2. Technology--Anthropological aspects. 3. Technology and civilization. 4. Science and civilization. 5. Culture--Semiotic models. I. Jensen, Casper Bruun. II. Rødje, Kjetil.
B2430.D454D485 2009
194--dc22

2009045266

British Library Cataloguing in Publication Data
A catalogue record for this book is available from the British Library

Printed in the United States on acid-free paper.

ISBN: 978–1–84545–614–6 (hardback)

Contents

Acknowledgements

The idea for this book was conceived during conversations at the *Description and Creativity: Approaches to Collaboration and Value from Anthropology, Art, Science and Technology* conference at King's College, Cambridge, in July 2005 between Casper Bruun Jensen and Eduardo Viveiros de Castro. Later conversations with numerous scholars – about the interest and viability of putting together a volume that would put Deleuzian thought in the context of science and technology studies and social anthropology – generated invariable enthusiasm for the project; several are now contributors.

The editors' Deleuzian inclinations are, of course, themselves the result of intellectual engagements and conversations over a longer period, in Casper's case not least with Geoff Bowker, Steve Brown, Manuel de Landa, Adrian Mackenzie, Evan Selinger and Isabelle Stengers.

Versions of the introduction have been presented at the 4S conference in Vancouver, November 2006, at the PhD workshop *'Ethnography and Technology in Relation to Gilles Deleuze's Philosophy'* held at the IT University in Copenhagen, May 2007, and at the International Seminar on Symmetrical Anthropology held at the Museu Nacional – Universidade Federal do Rio de Janeiro. We would like to thank the audience for their stimulating feedback and the organizers of the latter two events for letting us present our material.

We are very grateful for the constructive criticism of the introduction offered by Martin Holbraad, Morten Nielsen, Eduardo Viveiros de Castro, Marcio Goldman and especially Morten Pedersen, who read it carefully in several versions.

Finally, we want to express our appreciation to Marion Berghahn of Berghahn Books, for her continued support of this project.

Introduction

Casper Bruun Jensen and Kjetil Rödje

Gilles Deleuze was a thinker whose main concern was creation and differentiation, and according to whom new assemblages constantly emerge, reconfiguring reality in the process. Rather than accepting already established philosophical categories and distinctions he reassembled thought in new and inventive ways, thereby producing conceptual hybrids with unusual qualities and different potentials. The basic elements in Deleuzian thought are not static but entities in becoming. Consequently, the question to be asked is not what something is, but rather what it is turning into, or might be capable of turning into. Practice, knowledge, politics, culture and agency are seen as continually produced in heterogeneous processes without definite control mechanisms. Further, such processes traverse modern distinctions including the human and non-human, the material and ideal and the theoretical and practical. This volume raises the question of what a Deleuzian approach might entail for social anthropology and for science and technology studies (STS).[1]

While ideas related to and inspired by Deleuzian themes have emerged in fields/areas such as actor–network theory and non-humanist theory, there has been little sustained exploration of the specific challenges and possibilities that Deleuzian thought could bring to STS.[2] And while Deleuze and Guattari made use of anthropology 'in free variation' relatively few anthropologists have made use of their work in turn.[3] The distinction between STS and anthropology evoked here is somewhat elusive. Indeed, several Deleuze-inspired anthropologists are, precisely, anthropologists of science. However, it is not our ambition to attempt to disentangle these complicated relations. Rather, our general argument is that Deleuzian analysis offers many opportunities for rethinking important issues *both in and among* social anthropology and STS. It offers, we suggest, new insights into methodology, epistemology and ontology in these fields. It facilitates an arguably increasingly important rethinking of the relations between science, technology, culture and politics. And it suggests different ways of conceiving the links between these fields and the practices they study.

There are several reasons why it is both an enriching and a challenging task to tease out the implications of Deleuzian thought for anthropology and STS. Deleuze was a prolific writer both alone and with Felix Guattari.[4] He was also a notoriously difficult writer and the interpretation of his *oeuvre* is a minor industry in contemporary continental philosophy and cultural theory.[5] In these discussions, Deleuze and

Guattari's work is often presented as especially radical, both by their advocates and opponents, a description with which we largely concur. Yet it is far from obvious what kind of radicality can be claimed for their philosophy. Commenting on their use of social anthropology, for example, Christopher Miller notes that the ethnographic material Deleuze and Guattari make use of bears little resemblance to what he would consider 'a truly radical, transformed ethnography'. He estimates that in comparison with the stylistic requirements of such a transformed ethnography '*A Thousand Plateaus* appear to be quite old-fashioned' (Miller 1993: 13). However, it might appear so because Deleuze and Guattari's radicality has little to do with achieving an appropriately reflexive writing style.[6] If, indeed, Miller's truly radical ethnography is 'defined neither by form nor by relation to external objects' this description is certainly a far cry from Deleuzian thought, which emphasises interactions between multiple material forces. Deleuze's analyses moved across fields, drawing for instance on mathematics, ethnology, biology, political theory and anthropology. Yet Deleuze was never faithful to any of these disciplines. Rather he used their materials for what he viewed as the unique philosophical task of producing new concepts (see contributions by Brown and Stengers). As we shall see below, his idiosyncratic use of ethnographic material makes his relation to anthropology unusually complicated.

Deleuze's interest in the sciences also indirectly intersects with STS. However, even though Deleuze shares some analytical concerns with non-humanist STS approaches such as Donna Haraway's cyborg analyses (Haraway 1997) and Annemarie Mol's empirical philosophy (Mol 2002) and has directly inspired others such as actor-network theory (ANT; see Latour 2005) and Andrew Pickering's *The Mangle of Practice* (Pickering 1995)[7] his approach differs from such analyses both in scope and in the manner in which it treats scientific material. Whereas STS is often characterized by close empirical studies of scientific practice and discourse, and, at least in the guise of social constructivism, has had the aim of redefining science from a rational truth-seeking endeavour to a product of social interest and negotiation, Deleuze drew freely on physical, mathematical and biological concepts while paying little attention to the social processes through which these were generated (see, for example, de Landa 2002). Thus, while Deleuze has offered a plethora of interesting concepts that anthropology and STS might use to rethink their own vocabularies, these latter fields and his own philosophical pursuits appear incongruent in certain ways (see Brown, this volume).

How to relate to this tension? We can indicate immediately that our own approach to this issue is as pragmatic as was Deleuze's use of anthropology. He connected pragmatism to the notion of forming a rhizome and argued that, indeed, the term pragmatics 'has no other meaning: Make a rhizome. But you don't know what you can make a rhizome with, you don't know which subterranean stem is effectively going to make a rhizome, or enter a becoming, people your desert. So experiment' (Deleuze and Guattari 1987: 251). This pragmatism provides a rationale for working with anthropological material in what Deleuze and Guattari called 'free variation', certainly freed from the intentions or motivations of the original anthropologists or their informants. Instead Deleuzian pragmatism entails what we can call a strategy of the 'and', which aims to carry 'enough force to shake and uproot the verb "to be"' (Deleuze and Guattari 1987: 25).

In his paper 'And', Eduardo Viveiros de Castro suggests that the 'little magic word "and" … is to the universe of relations as the notion of *mana* is to the universe of substances. "And" is a kind of zero-relator, a relational *mana* of sorts – the floating signifier of the class of connectives – whose function is to oppose the absence of relation, but without specifying any relation in particular' (Viveiros de Castro 2003: 1–2, italics in original). This lack of preconceived specificity ('without specifying any relation in particular') points in the direction of an experimental attitude because it demands of research that it investigates just what is connected with what in each specific situation. Thus, also, it is through *experiments with* different materials from anthropology and STS that Deleuzian intersections in these fields can be explored. The task of this volume, then, is to explore potentials located within the materials traversed by these fields. Consequently we have encouraged contributors to *do new things* with Deleuzian modes of thought.

Viveiros de Castro draws a more wide-ranging conclusion with respect to the entailments of the strategy of the *and*. He proposes that following this strategy inevitably leads one towards a choice between two different images of anthropology as such (and by extension social science in general). The first views research as 'the outcome of applying concepts extrinsic to its object: we know beforehand what social relations are, or cognition, kinship, religion, politics and so on, and our aim is to see how these entities take shape in this or that ethnographic context' (Viveiros de Castro 2003: 7–8).

In contrast, the alternative encouraged by the strategy of the *and* is to note that the 'procedures characterizing the investigation are *conceptually* of the same kind as those to be investigated' (ibid., italics in original). In the term of Bruno Latour, they are symmetrical (Latour 1993). Yet similarity at this level immediately gives way to a potentially radical dissimilarity at all other levels. For:

> While the first conception of anthropology imagines each culture or society as the embodiment of a specific solution to a generic problem – as the specification of a universal form (the anthropological concept) with a particular content (the indigenous representation) – the second by contrast imagines that the problems themselves are radically distinct. More than this: it starts out from the principle that the anthropologist cannot know beforehand what these problems may be. (Viveiros de Castro 2003: 8)

It is thus *because* a symmetrical anthropology views the 'problems' of actors as *a priori* radically undetermined that it is enabled to see how these actors – whoever they are and whatever problems they may have – build up their worlds. Indeed, observing, documenting and analysing such world-building activities become a methodological requirement – perhaps the only requirement. This specification, in turn, allows us to view ethnography as an empirical method for dealing with Deleuze's Spinozist dictum (and problem) that 'we do not yet know what a body can do', or, indeed, what a society or cosmology can do (Spinoza 1959; Deleuze 1990a; Viveiros de Castro this volume). It also offers an entry point for understanding Deleuze and Guattari's so-called 'magic formula' PLURALISM = MONISM (Deleuze and Guattari 1987: 20–21) in relation to Latourian actor–network ontology. According to this formula the world *should* be approached in the mode of monism, that is, as an undifferentiated mass of forces, energies and actants, because only this starting point allows one to subsequently appreciate its rhizomatic plurality and to see the proliferating connections between entities that traditional distinctions would keep separated.[8]

As we see it, to approach anthropology and STS in the mode of the *and* has serious consequences. Most prominently, it entails a fundamental change away from the idea that anthropology deals in more or less adequate representations of 'the native'. As Viveiros de Castro states, following Vassos Argyrou, this means that anthropology's true problems are not epistemological but ontological; we are not operating in a situation of multiculturalism based on one common nature, but in a situation of multinaturalism, in which different natures may or may not be aligned through pragmatic efforts (Viveiros de Castro 2005).

Notions such as 'free variation' or the 'and', with their emphasis on the creativity of connection making, also make vivid the fact that anthropologists and sociologists, like everyone else, participate in worlds and actively contribute to their maintenance or transformation. This is the case whether social scientists are aware of the fact or not. By default we thus engage with, rather than simply observe or interpret, the complex and heterogeneous realities under study. It follows from this that the fruitfulness of the connections and affiliations proposed by researchers cannot be evaluated solely in terms of representational adequacy. Indeed, representations are themselves created through what Andrew Pickering has called 'complex representational chains' (Pickering 1995: 99, emphasis removed).[9] In turn, this means that 'the conceptual stratum of scientific culture is itself multiple rather than monolithic' so that 'disparate layers … typically have to be linked together in bringing experience into relation with the higher levels of theory' (ibid.). We might question the metaphor of 'higher levels' evoked here. But the important thing to note is that methods and theories in this view are implicated in, rather than external to, the complex chains through which anthropological representations emerge because such representations are precisely the condensed end-points of multiple interactions among implicated and enchained actors. This suggests that a significantly expanded notion of relevance must be the criteria for determining the adequacy of a representation, a concept, an analysis or an experiment. Indeed, to Isabelle Stengers it suggests that any relevant notion of truth is directly related to a duly expanded conception of interest, for as she argues:

> the demands of the 'politics of reason' and those of the city, in a more classical sense, intersect, and it is in this sense that I have been able to employ the double qualifier, used rather infrequently, 'antidemocratic, that is, irrational'. In fact, as

> soon as one puts aside the classical division of responsibilities, which gives the sciences and their experts the task of 'informing' politics, of telling it 'what it is' and deciding what it 'must be', one comes face to face with the *inseparability of principle* between the 'democratic' quality of the process of political decision and the 'rational' quality of the expert ... This double quality depends on the way in which the production of expertise will be provoked on the part of all those, scientific or not, who are or could be interested in a decision. (Stengers 2000: 160)

It follows from this consideration that relevance can be reduced neither to utility nor to a question of motivation and volition since one inevitably acts with and against many disparate actors and practices as part of establishing the complex chains of ethnography. It also follows that the aim for productive relevance can in no way be reduced to a call for position-taking, precisely because the positions that one might become able to take themselves emerge from unpredictable processes. Needless to say, this performative approach to social science is far from generally accepted. On the contrary, it is a highly controversial proposition, which relates directly to issues of anthropological politics.

Nomadic Thought and Anthropological Representation

To highlight the controversy it engenders, we start out by addressing an unlikely argument about the relation between Deleuzian thinking and ethnography. In 1993, literary scholar Christopher L. Miller published an article entitled 'The Post-identitarian Predicament in the Footnotes of *A Thousand Plateaus*: Nomadology, Anthropology and Authority' in *Diacritics*. Reprinted as a book chapter, this analysis was harshly criticized in a review by the well-known Deleuze scholar Eugene W. Holland (2003).[10] Miller's concern is with how Deleuze and Guattari appropriate anthropological concepts and, sometimes, empirical illustrations in order to develop their 'post-identitarian' nomad thought, a thought, however, that remains 'overwhelmingly academic' (Miller 1993: 9). Nomad thought, in Miller's reading, claims to be non-authoritarian, labile and flexible but in fact relies on a rhetorical trick that offers to its authors authority without corresponding responsibility towards the actors from which their notions are derived, for example, the real nomads who are at issue in the anthropological texts on which Deleuze and Guattari draw. This is the 'epistemological paradox of nomadology: nomads don't represent themselves in writing, they must be represented' (ibid.). Asking 'what if anything does this project of nomadology have to do with real *and* "actual" nomads' (ibid: italics in original) Miller himself answers that 'nothing' would be a compelling answer: 'the only nomads to deal with would be Deleuze and Guattari themselves and their intellectual fellow travellers'.

Real nomadism in this reading is translated into intellectual nomadism and this enables Deleuze and Guattari to move between Kandinsky paintings and Mongolian nomadic motifs without interruption (ibid.: with reference to Deleuze and Guattari 1987: 575n38; see also Pedersen 2007). But what, Miller asks, 'do DG [Deleuze and Guattari] know about Mongolian or other nomadic motifs, and where did they learn it? The short answer is obvious: they read some books and articles, from which they borrowed information and authority on their subject' (Miller 1993: 12). It is this 'semi-assimilation' of anthropological discourse that allows Deleuze and Guattari to make their nomadic connections in the first place.

Quite to the contrary of what is claimed for nomadic thinking, however, 'anthropology has been nothing if not representational', and this is the epistemological paradox entailed by Deleuze's use of anthropology. As Miller sees it, this sliding between scales of realism and reference and the free drawing of connections offer Deleuze and Guattari a 'nomadological immunity' (Miller 1993: 21). According to Miller, they could have developed a pure concept of the nomad by remaining 'pure philosophers', 'but they want to have it both ways: to propose a "pure idea" of nomads mixed with "actual" information' (ibid.: 25), drawn from a hotchpotch of accepted and dubious ideas from the literature. As he further argues, Deleuze and Guattari 'have great faith in the anthropologists they quote, even if they sometimes correct their interpretation of data' (Miller 1993: 19). It is this acceptance of anthropological representations that subsequently enables them to pronounce on nomads, sorcery and many other issues in a manner claimed to be non-representational.

In response, Holland considers Miller's analysis to 'more or less completely miss ... its mark' (Holland 2003: 160), even referring to it as a 'maniacal attack' (ibid.: 162). Holland notes that 'the crux of the matter, as Miller recognizes, is the *representational status* of their concept of nomadism' (ibid.: 163, italics in original). As Holland complains, 'He [Miller] claims it is representational; they insist it is not. He claims they are doing ethnography and representing people, i.e., actual nomads; they insist they are doing philosophy, and creating specifically philosophical concepts ... not (social) scientific ones' (Holland 2003: 163). The difference is that 'unlike science, which is representative, philosophy is creative, serving as a kind of relay between one practical orientation to the world and another, new (and hopefully improved) one' (ibid.). Citing *A Thousand Plateaus*, Holland reminds us that 'the book is not an image of the world. It forms a rhizome with the world' (ibid., referring to Deleuze and Guattari 1987: 11), and that it is 'entirely oriented towards an experimentation in contact with the real' (ibid., referring to Deleuze and Guattari 1987: 13). Consequently, 'we shouldn't judge DG on the "accuracy" of their concept of nomads' (ibid.: 164), since, as with their other 'conceptual personae', 'they are embodiments of different invented philosophical modes of thought' (ibid.).

In response Miller argues that, if the philosophy of Deleuze and Guattari is oriented towards experimentation in contact with the real, the interface of this contact remains the problem, the way in which a 'creative virtuality is "extracted" from an *actual* state of affairs' (Miller 2003: 133, emphasis in original):

> My essay was concerned with the implications of that act of extraction, with the refractions of meaning that took place at the 'points of contact' between *A Thousand Plateaus* and its sources. At those points, strange things happen: ethnographic authority is exerted even as it is denied; the real comes and goes; the shadow of colonialism fades in and out. (ibid.)

This debate clearly shows how challenging it is to articulate the implications of the strategy of the *and*. But, while usefully highlighting the challenges involved in this change of analytical topos, the debate is also disappointingly familiar. Because of the incompatible presuppositions of the discussants, a point of mutual engagement and contrast never emerges. Neither do they evoke any particularly interesting or constructive modes of thought. Miller is highly critical of the ways in which fervent Deleuze followers accept his arguments as authoritative. As he concludes: 'It is significant that no taint from colonialist sources troubles Holland's reading of *A Thousand Plateaus;* Deleuze and Guattari are above all that, because they do not represent. And how do we know that they do not represent? *Because they say so,* and because Holland repeats their claim' (Miller 2003: 134, italics in original). Yet, because Miller's critical effects are generated by means of his staunchly representational viewpoint, he never pauses to consider (much less to appreciate) what the purposes or, indeed, virtues, in intellectual and political terms, attending nomadic thought might be. Meanwhile, Holland, in spite of his insistence that the book forms a rhizome with the world rather than representing it, fails to see that this also holds with respect to the uptake of Deleuze and Guattari's own work. Indeed, it is because he fails to recognize this implication that he falls prey to the argument from authority, identified by Miller. Significantly his rebuttal of Miller's article is entitled 'Representation and Misrepresentation in Postcolonial Literature and Theory', and one of his most damning critiques is that 'for someone who claims elsewhere to be concerned with the politics of representation, the degree of blatant misrepresentation in this essay is truly astounding' (Holland 2003: 165). Curiously this critique mirrors the one provided by Miller of Deleuze and Guattari's use of anthropology: that the analysis is not faithful to the intentions of the original authors. Yet, precisely from a Deleuzian point of view, which sees the text as forming a rhizome with the world, the accusation of blatant misrepresentation seems less than pertinent. For both Deleuzians and their detractors the discussion thus leaves something to be desired in terms of productive outcomes.

It should be emphasized once again that a pragmatic and performative approach to anthropology is quite different from rejecting the existence of a reality located outside academic texts (as in the worry that 'everything is discursive' or 'everything is a social construction'). Nor does Deleuze provide a carte blanche for making anything whatsoever out of ethnographic material (as in the worry that 'anything goes').

So where do we locate Deleuzian radicalism? Although Deleuzianism is sometimes said to demand across-the-board reconfigurations of aesthetics and ethics, knowledge and politics, our wager is that such calls for generalized change, while likely to sound radical, are bound to be of limited consequence and remain marginal to anthropology and STS.

A quite different entry point is to consider that Deleuzian pragmatism supports an extended conception of what counts as the ethnographic, since it allows us to treat 'theories' as ingredients in the empirical (see Riles 2001; Maurer 2005). This enables us to view the work of ethnography as having to do with putting in conjunction heterogeneous sets of material irrespective of their usual compartmentalization into the 'theoretical framework' and the 'ethnographic description', with standardized roles ascribed for each component. But if the relation between the explanation and explained is destabilized and rendered flexible, then one's ambition cannot be to achieve a more or less adequate 'matching' of the two. Instead the aspiration must be to create associations that mutually enrich and reciprocally transform each part of the material.

In an evocative discussion of the problem of painting, Deleuze makes a similar observation, which deserves a long quote:

> It is a mistake to think that the painter works on a white surface. The figurative belief follows from this mistake. If the painter were before a white surface, he – or she – could reproduce on it an external object functioning as a model. But such is not the case. The painter has many things in his head, or around him, or in his studio. Now everything he has in his head or around him is already in the canvas, more or less virtually, more or less actually, before he begins his work. They are all present in the canvas as so many images, actual or virtual, so that the painter does not have to cover a blank surface, but rather would have to empty it, clear it, clean it. He does not paint in order to reproduce on the canvas an object functioning as a model; he paints on images that are already there, in order to produce a canvas whose functioning will reverse the relations between the model and the copy. In short, what we have to define are all these 'givens' [*données*] that are on the canvas before the painter's work begins, and determine, among these givens, which are an obstacle, which are a help, or even the effects of a preparatory work. (Deleuze 2003: 86–87)

Let us consider carefully what is at stake in this argument. The first part will seem well known to the reflexive ethnographer. One never approaches the world as a tabula rasa, with a white canvas to paint on, or as a blank slate, on which the empirical can imprint itself. There can be no question of a pure representation, since numerous heterogeneous resources are always already part of the composition. The positivist might attempt to find a scientific method to empty the canvas. However, this is not at all what Deleuze suggests. Rather, he proposes that one is required to orient oneself among available actual and virtual resources. Following Henri Bergson, Deleuze sees perception as a process generating a double, yet complementary, system of images, where the perception-images belong to different systems of reference from those of the images being perceived (Deleuze 1986: 62–63). Every 'thing' exists through its relations to innumerable other 'things', both virtual and actual, which can never be completely captured in perception.

It follows from this that the relevant question is not how to rid oneself of (theoretical) prejudice, but how to pick from among one's 'prejudices' those that will help in the articulation of particular problems and how to remove those that are likely to turn into obstacles. Transferred to ethnography, the problem is not as such that the anthropologist draws on Western theory or introduces external categories to articulate indigenous practice. Indeed, this is a pseudo-problem that retains the positivist presumption that it would be possible to return to an empty ethnographic canvas. However, since such a return is impossible, it follows that general worries about the imposition of theory on ethnographic data are entirely misconceived.

Instead the specific relations established in any given case become crucial. This engages the researcher in a process of simultaneous conceptual and empirical selection and combination that finds no generic solution. Thus, if there is a problem with theory in anthropology (and social science in general), it has to do not with choosing between more or less theoretical or more or less empirical approaches. Instead it has to do with the many uses of theory by ethnographers claiming to merely represent; that is, it has to do not with using wrong theory but with being *used unwittingly by theories or categories* (for example about kinship, gender, power or culture, for example shaped by Kantian, Marxian or phenomenological assumptions) that are so thoroughly taken for granted that their analytical/prejudicial effects remain invisible.

This consideration indicates why general calls for a change in anthropological theory are far less interesting to us than specific explorations of the multiple *concrete interfaces* at which the previously mentioned experimentation with the real takes place. Further it suggests why it is crucial to do so without returning either to the mode of criticism and referentiality expounded by Miller, or the alternative proffered by Holland, according to which Deleuzian concepts (theory) are detached from, rather than an integral part of, the empirical. In other words, this poses the challenge of moving the thought of Deleuze beyond philosophy, into the domains of science, politics, culture and society – into territories where he did not venture. To get both the challenges and opportunities of Deleuzian thought better into view, we proceed by elaborating on Deleuze and Guattari's conception of an engagement, first, with science, and, secondly, with anthropology.

Deleuze and Science

In *What is Philosophy?* Deleuze and Guattari distinguish between philosophy and science, and locate their own activity squarely within the former. They suggest that philosophy and science can both be creative practices, but, whereas philosophy carries the potential to create concepts, science creates propositions. Philosophical concepts relate to, and are constituted by, other concepts, philosophical planes and problems, but they do not refer back to objects in the external world. The object of philosophy is rather thought itself, and the operation of thought generates concepts (Deleuze and Guattari 1994: 18–22). What is generated by philosophy, then, are not objects of reference, but rather, self-referentially, the very territories on which philosophy operates. Thought is perennially drawn towards the construction of new concepts, enabling conceptual stabilizations and destabilizations and shaping ideas into productive or dangerous tools for action. Concepts may facilitate resistance to present realities, as well as the construction of alternative realities. For this reason, they can never be disconnected from reality (e.g. as ideal) but must be viewed rather as operating in a distinct way within the real, namely as providing virtual tools for engaging with practice.

According to Deleuze and Guattari, scientific propositions are different from philosophical concepts, as they refer to states of affairs, existing outside the propositions themselves (Deleuze and Guattari 1994: 22–23). This does not mean that science neutrally and passively describes reality. The object of science is rather to create functions that relate to the outside world and derive their power by referencing states of affairs, things or other propositions (ibid.: 138). Whereas propositions constitute discursive formations, according to Deleuze and Guattari, this is not the case with concepts (ibid.: 22, 50). Scientific discourse, linking propositions with external referents, is always about some object of study. While the generative power of science lies in the movements of thought enabling the constitution or modification of states of affairs or bodies (ibid.: 138), philosophical concepts make it possible to think what was previously unthinkable.

Yet, as Steve Brown notes (see Brown this volume), the human and social sciences are left out of the classification of *What is Philosophy?*. In an interview, Deleuze was confronted with this omission. In a somewhat evasive answer he contrasted history with philosophy, arguing that becoming is the business only of the latter (Deleuze 2006: 377–78). He argued that, although history relates to events and the functions according to which they come to pass, it stays within the limits of the existing events and does not enter the terrain of becoming. Deleuze never indicated whether he thinks this is the case with the human and social sciences in general but in any case this answer leaves much to be desired from the point of view of both STS and anthropology.[11]

As noted, the approach of Deleuze and Guattari involves a philosophical engagement with the sciences. Rather than treating science as a social practice among others, Deleuze and Guattari seek to localize what is special and unique about it. Their interest is not with science in the making, but rather with the potential outcomes of science already made – bearded science, in Isabelle Stengers's evocative phrase.[12] Rather than taking science apart, in order to expose underlying social powers and structures, Deleuze and Guattari look at the results of science and attempt to understand their actual and virtual capacities. They treat science as a distant and somewhat foreign friend whose tools and problems are different from their own, but who, precisely for that reason, can teach important lessons. Deleuze and Guattari's approach is thus constructive rather than deconstructive, a disposition they share with figures such as Isabelle Stengers (2000) and Bruno Latour (2005).

Manuel de Landa has made a strong argument that a Deleuzian ontology is realist (de Landa 2002: 2; Escobar and Osterweil, this volume). By this he means, among other things, that it is not concerned with social interests and interpretations, but rather with the production of reality as actual states of affairs. Like ANT, the Deleuzian approach is then ontological rather than epistemological, or, to be more precise, the separation between epistemology and ontology breaks down. However, if that is the case the distinction between realism and constructivism also breaks down.[13] What is under construction is, indeed, reality, not just meaning and interpretation. As Latour puts it: the more constructed, the more real (see Latour 2005: 88–93). Yet this obviously entails the recognition that humans are not the sole constructors, and thus that sciences cannot be reduced to cultural or social representation. Indeed, the social can be given no special status, but must be seen as produced in permutational networks consisting of heterogeneous elements.

Deleuze and Guattari locate constructive potentials within science, not unconditionally and as a purveyor of truth, but as a provider of difference – of other kinds of creation. Like philosophy and art, science may carry virtual political implications, that is, implications to be determined, rather than built-in outcomes. Although differently constituted, philosophical concepts and scientific propositions both carry the potential to alter practices and modes of thought (see Mackenzie, this volume). We might then consider what the potential contributions of science are. The answer again refers back to the strategy of the *and*. Science carries the potential of adding to and diversifying the world. By providing alternatives, other solutions, realities, worlds and forms of life, science is a political force to reckon with. By staking out the immanent potentials of reality, science may bring forth strategies for what Deleuze and Guattari in *What is Philosophy?* call resistance to the present (1994: 108). What science can tell us is that the world need not be as it seems, but that it carries a potential for engendering other states of affairs and modes of life.

This may all sound overly idealistic and romantic. Surely science does not always work in such a creative, subversive and emancipating manner. Not being completely ignorant of the world beyond their writing desks, Deleuze and Guattari are of course well aware of this fact. They propose that creativity is indeed a rare ingredient of scientific practice. As they see it, science most of the time does not add anything new to the world, but rather strengthens and reproduces existing states of affairs. However, they insist that alternatives are possible and that some may already exist. Indeed, Deleuze and Guattari stress the specific reality of the possible (Deleuze and Guattari 1994: 17). At best science can open up and elaborate on virtual capacities – enlarging rather than reducing the world.

Now, as Deleuze and Guattari see it, this creative potential is largely incompatible with capitalist modes of production. Like art and philosophy, science only functions creatively if separated from commercial imperatives, because these aim to shape the results of research in the form of commodities and profits, thereby only supporting the reproduction of the market (ibid.: 106). Just as Deleuze in an interview denounced commercial art as a contradiction in terms (Deleuze 2006: 288), it seems clear that for Deleuze and Guattari science cannot easily be commercialized without going in accord with its productive potentials.

Yet many studies in STS have argued that scientific activity is not, and hardly ever has been, free of political and financial interests (see, for example, Shapin and Schaffer 1985; Biagioli 1993). Instead it can be suggested that science is a contested territory with a multitude of different and contradicting forces at work – among them capitalist imperatives. Commercial logics may certainly imply a reterritorializing of scientific activities and outcomes. In such cases, which Stengers qualifies as instances of 'mimicking' science, invention would only be encouraged to the extent that it can be reterritorialized directly into profit. Yet capitalism is itself no all-encompassing entity (Callon 1998; Mackenzie 2006; de Landa n.d.). In practice, holes and possibilities of breakthrough emerge through which new pathways and territories can be generated – even if only temporarily. It would appear to be an important task for both social anthropology and STS to provide empirical and conceptual analysis of such events and their dangers as well as inventive potentials.

This brings us to the question of how to distinguish between kinds of science, for, as we have discussed, Deleuze and Guattari claim that science can function in both a reterritorializing and a deterritorializing manner. Whereas the royal or major sciences operate within and strengthen existing knowledge and states of affairs, the nomad or minor sciences provide creative potentials for exploring new pathways and territories (see Pickering, this volume). What Deleuze and Guattari characterize as minor sciences is science disentangled from, while still relating to, the major sciences: 'Major science has a perpetual need for the inspiration of the minor; but the minor would be nothing if it did not confront and conform to the highest scientific requirements' (Deleuze and Guattari 1987: 486). The science promoted by Deleuze and Guattari is thus one of becoming and heterogeneity. But, contrary to the argument that science should be secluded from society, 'good' science, in a Deleuzian view, would always be ready to explore new spaces from which to extract creative possibility.

Minor sciences are also described by Deleuze and Guattari as a 'becoming smooth' of science. Not smooth as in the workings of well-oiled machinery, but smooth as open, slippery and evasive. In the vocabulary of Deleuze and Guattari, smooth space is opposed to striated space, although this is not an exclusive opposition, as the two tend to exist in a mixture (1987: 474). The conceptual difference between smooth and striated space concerns not so much scale as position and relation to the field of study (see also Escobar and Osterweil, this volume). Striated space is the space of an observer whose view is premised on distance and detachment, not unlike Donna Haraway's description of the God's-eye view (Haraway 1997). Striated space is affiliated by Deleuze and Guattari with the reproducing sciences, which are oriented towards constants and universals and seek the conditions for the reproduction of the same. This implies a permanent and fixed point of view external to what it sees (Deleuze and Guattari 1987: 372). The outcomes of such scientific procedures are stable patterns operating according to law-like conditions, allowing for and accommodating predictable phenomena and activities.

In contrast the science of smooth space is not reproductive but *following;* it is implicated and flows from the middle (ibid.: 372–73). The following sciences are concerned with *singularities* – precisely non-reproduceable phenomena. Rather than extracting constants from variables, one sets out to engage in continuous variation (ibid.: 372). Hence Deleuze and Guattari's fascination with chaos theory and sciences exploring complexity, as these concern the potentials of science to open up towards unknown territories.[14]

This appraisal of the sciences brings us to the recurrent worry within STS about the political stance of the analyst. Deleuze and Guattari's engagement with the sciences has a political and ethical agenda as they bring to bear on science a Nietzsche-inspired evaluation, according to which merit depends on the capacity to provide virtual and actual difference (e.g. Jensen and Selinger 2003).

As should be clear, this evaluative component does not at all align Deleuze and Guattari with those who criticize STS studies for a deficiency of normativity (e.g. Winner 1993; Woodhouse et al. 2002; Feenberg 2003). Often such voices aim to bring into analytical focus the macro-features of scientific and cultural production, for example by considering institutional and economic structures, with the purpose of offering critiques of science. In spite of their calls for radicalism, however, the analytical result of bringing in 'the macro' is often prosaic, as it relies on the use of stable social, economic or political categories, supposed to be capable of structuring science and technology from the outside. Here science and technology become dependent variables, shaped by economic power and ideological structures. Yet precisely this way of thinking, according to which the world is already catalogued and prepared for evaluation by the critical theorist, is suspended by the Deleuzian strategy of the *and,* which instead aims to draw nearer to the transformative undercurrents of scientific and technical processes in order to understand their immanent consequences.

Although a Deleuzian approach to science shares little with such critical perspectives, neither does it align with the *Politics of Nature* proposed by Bruno Latour (Latour 2004; Jensen 2006; Fraser, this volume). Unlike Latour, Deleuze and Guattari do not seek a political model operating according to standards of representative democracy. Whereas Latour seeks to join forces and incorporate actors with established networks, Deleuze and Guattari always seek what can be termed minor potentials. While Latour seeks ways of rearranging existing networks, Deleuze and Guattari seek to escape from dominating territories and constellations. Their work is imbued with scepticism towards anything grand, solid and major. Their preference for the minor and fluid is on the one hand an anti-utopian rejection of enduring institutions working for the common good and on the other hand a positive belief in the ever-present existence of potentials for change.

For the same reason, and contrary to currently popular interpretations of science as increasingly use-oriented and transdisciplinary, epitomized by the notion of mode-2 science (Nowotny et al. 2001; see also Jensen and Zuiderent-Jerak 2007), Deleuze and Guattari reject communication, discussion, and consensual solutions as adequate answers to problems of concern and relevance. Dismissing discussion as 'an exercise in narcissism where everyone takes turns showing off', and describing the phrase 'let's discuss it' as an 'act of terror' (Deleuze 2006: 380), Deleuze calls for a focus on the need to pose problems rather than to discuss solutions to what is already agreed upon. Rather than engaging in a continuous dialogue in search of compromises, common terminology and standardized practises, the Deleuzian strategy seems to be to: pose a problem – step back – pose a problem – step back, ad infinitum.

Such an avowedly disentangled practice seems relatively easier to maintain for philosophy than for either the natural or the social scientist. The hermit scientist is not a likely figure and probably never has been.[15] But in reality the situation is also far more ambiguous as scientists are often heterogeneous figures who transverse a number of positions in various networks as part of doing science (Latour 1987; Law 1987). Neither the disentangled hermit nor the enmeshed researcher can be seen as static and unitary figures since both operate through complex series of connections internal and external to their practices. Yet, if a completely autonomous position is impossible, this might be precisely the condition for the production of other kinds of 'links and knots' (see Stengers, this volume).

If one conceives of STS itself as a following science (indeed, a science that is precisely following science), its potentials for 'critical intervention' might go in the direction of experimental studies of and with scientific practices, rather than science criticism (see contributions by Bowker and Vann). Following this line of thought, the problem becomes how to approach the sciences (including STS and anthropology) as world-building practices, so as to tease out potentials for new modes of action and thought provided by different scientific adventures (see contributions by Pickering and Schienke), as well as adventures at their intersections.

Anthropologists in Deleuze, Deleuze in Anthropology

At the time when Deleuze and Guattari were writing together STS was still in its infancy. This was by no means the case with anthropology. Indeed, they refer to an enormous body of anthropological work. A non-exhaustive list from *Anti-Oedipus* and *A Thousand Plateaus* includes Gregory Bateson, Pierre Clastres, Meyer Fortes, Maurice Godelier, Marcel Griaule, Edmund Leach, André Leroi-Gourhan, Claude Lévi-Strauss, Bronislaw Malinowski, Marcel Mauss, Marshall Sahlins and Victor Turner. As noted, the way Deleuze and Guattari use anthropological ideas and concepts is similar to how they make use of other scientific and artistic fields; that is, with a pragmatic lack of fidelity to the intentions and ambitions of the authors as well as to disciplinary concerns. This is a situation that makes simple synopsis – unequivocally determining that 'this is what they say' – complicated. It is likewise one that makes critical evaluation – deciding whether what they do say is appropriate in anthropological terms – problematic. When Deleuze and Guattari use anthropological concepts these are usually transformed and detached from their formative context in order to link them up or exchange them with other terms and notions formulated elsewhere, often for different purposes.

It is obviously impossible to offer a comprehensive analysis of Deleuze and Guattari's use of anthropology. Instead our approach is to move selectively through what we consider some 'highlights'. Our discussion moves – pseudo-chronologically – from the 'archaic' towards the contemporary. We first discuss Deleuze and Guattari's analysis of debt and exchange, which leads to the conclusion that Nietzsche's *On the Genealogy of Morals*, rather than Marcel Mauss' *The Gift* (1967), ought to be seen as the 'great book of modern ethnology' (Deleuze and Guattari 1983: 190).[16] We then consider in more detail how this argument links up with their analysis of 'capitalism and schizophrenia'. Here, the extent to which Deleuze and Guattari rely on the work of Gregory Bateson is especially striking. We show how several central notions, including the plateau, schizophrenia and even the rhizome, have been adapted from Bateson's work among the Iatmul, the Balinese and the North Americans (in psychotherapeutic settings).

Finally, we use this analysis as an entry point for considering how Marilyn Strathern's classical post-Maussian work on *The Gender of the Gift* might be characterized as Deleuzian in a certain sense. Indeed, our analysis indicates how a rhizomatic, 'subterranean stem' might link Deleuze and Guattari's work to Marilyn Strathern, with Gregory Bateson's work as a 'dark precursor' (Deleuze 1994: 119).

Debt and Exchange in Nietzsche and Mauss

Deleuze and Guattari trace back to Marcel Mauss the question of whether debt is primary to exchange or an aspect thereof. They claim that Mauss himself left this question open, but that Lévi-Strauss closed off the question, in order to claim the primacy of exchange – of which debt was to be seen merely as a superstructure (Deleuze and Guattari 1983: 185).

As a consequence of his analysis of inscriptions on the social and biological body, Nietzsche, however, insisted on the primacy of debt. For this reason, according to Deleuze and Guattari, 'the great book of modern ethnology is not so much Mauss' *The Gift* as Nietzsche's *On the Genealogy of Morals*. At least it should be' (Deleuze and Guattati 1983: 190). Although Deleuze and Guattari acknowledge that Nietzsche has only 'a meagre set of tools at his disposal – some ancient Germanic law, a little Hindu law' (ibid.), they argue that his analysis goes beyond Mauss. Although Deleuze stresses the importance of evaluating Nietzsche's *On the Genealogy of Morals* in relation to later ethnographic texts, especially studies of the potlatch (Deleuze 1998: 199n.6), it is clear that he evaluates the greatness of Nietzsche not in terms of detailed empirical knowledge but rather with a view to the creative potential that can be extracted from his analysis. Deleuze and Guattari's concern with Nietzsche's argument consequently has to do less with anthropology as the study of other cultures than with its genealogical implications for political economy and philosophy. By eroding the ground under exchange as a fundamental and universal principle for social organization, Deleuze and Guattari expose social relations and the constitution of society as founded not upon a social contract but upon force. Nietzsche's greatness, then, lies 'in having shown, without any hesitation, that *the creditor-debtor relation was primary in relation to all exchange*' (Deleuze 1998: 127, italics in original).

Following Nietzsche, Deleuze and Guattari see debt as a direct result of territorial and corporeal inscription (Deleuze and Guattari 1983: 190). A body, or territory, is inscribed and marked by dominating forces, and the debtor is thereby subordinated to the creditor. In this view, relationships of exchange, far from being essentially contractual, are founded upon domination and enforced regulation of desire and production. As desire, according to Deleuze and Guattari is produced in abundance and always threatening to overflow, the fundamental task in building and maintaining a social formation is the management of these flows. Correspondingly, it becomes an important conceptual task to analyse how desire is produced and recorded upon the social body. Deleuze and Guattari here distinguish between three modes of social production and recording: the primitive territorial machine; the barbarian despotic machine; and the civilized capitalist machine. Each machine is characterized by its specific ways of producing, organizing and recording flows of desire.

The primitive territorial machine is concerned with inscription of the earth (in its widest social and material sense). Inscribed on the earth are alliances (which organize politics and the economy) and filiations (which organize administration and hierarchies) (Deleuze and Guattari 1983: 146). This practice of inscription produces a social assemblage linking bodies and flows of desire. Social order is thus constantly (re)produced and recorded through a process in which everybody is prescribed a given social location.

The barbarian despotic machine implies new forms of alliances, with a strict hierarchical organization where a despot in direct filiation with the deity is located above the people (ibid.: 192). The territorial inscriptions of alliances and filiations, from the primitive social machinery are kept intact in the despotic machinery, but they are at the same time connected to new alliances and filiations, by means of which they are located in a hierarchy underneath the despot (ibid.: 196–98). The despotic machinery is viewed by Deleuze and Guattari as paranoid, always in need of controlling productive flows. Thus, while this despotic machinery deterritorializes and decodes the flows that constituted the territorial machinery, at the same time it overcodes pre-existing recordings and inscriptions in order to make them fit into the hierarchical despotic machinery (ibid.: 198–99). Entities of the territorial machinery are thus disconnected from their previous location in the social field, only to be relocated and given new functions within the despotic social machinery. Tribes, lineages and communities are still present, but are now subordinated to a despotic rule, which reaps the benefits of flows and productions. In this mode, desire is thus invested not only in the local alliances and filiations, but also in flows and connections linking these to the despot.

According to Deleuze and Guattari, deterritorializations taking place within the despotic social machinery are accelerated with the coming of capitalism. Capitalism is distinguished as a social machinery based upon decoded flows. Capitalism decodes desire and its products in order to recode them into abstract quantities in form of money and commodities (Deleuze and Guattari 1983: 139–40, 239). This decoding is met by a recoding by the despotic state (ibid.: 222–23). The state thus maintains an important role within capitalism in maintaining and regulating the flows (ibid.: 253). So-called civilized modern societies deterritorialize with one hand and reterritorialize with the other (ibid.: 257).

In this analysis, the economy functions not according to monetary exchange and the free and voluntary flow of goods, but rather through the management and domination of production and flows of desire. According to Deleuze and Guattari, the fundamental anthropological concepts should therefore not be scarcity and reciprocity, exchange and contractual relations, but ought rather to include debt, abundance and domination. The formations described by Deleuze and Guattari all centre on regulation or, indeed, domination of flows in order to maintain social coherence. The problem of the social, they therefore argue, should be seen in terms not of distribution and the maintenance of contractual relations but of the opening and closing of flows. Correspondingly, scarcity is viewed not as a natural situation, but rather as produced in order to close off certain kinds of flows while keeping others open. Extrapolating from Nietzsche, Deleuze and Guattari argue that, 'The fundamental problem of the primitive socius, which is the problem of inscription, of coding, of marking, has never been raised in such an incisive fashion. Man must constitute himself through the repression of the intense germinal influx, the great biocosmic memory that threatens to deluge every attempt at collectivity' (Deleuze and Guattari 1983: 190).

According to Deleuze and Guattari the problem with the anthropological turn taken by Mauss, and to an even larger extent Lévi-Strauss (Deleuze and Guattari 1983: 185) lies in appending debt to a system of exchange, rather than questioning the contractual basis of the analysis of exchange. Whereas a system of debt is dynamic, installing and feeding upon relations of inequality, a system of exchange tends towards equilibrium: goods and money circulate and are distributed according to relations founded upon equality. In their view, the concept of exchange thus functions as a reterritorializing mechanism that encapsulates social relations and modes of production and makes them appear static and governed by universal rules. For Deleuze and Guattari it is this fundamentally structural-functionalist perception of a society in equilibrium, where all relations are parts of a contractual system of exchange, which is opened up by Nietzsche's refusal of the primacy of exchange, as this destabilizes cultural formations and turns social relations dynamic. 'One begins by promising, and becomes indebted not to a god but to a partner, depending on the forces that pass between the parties, which provoke a change of state and create something new in them: an affect' (Deleuze 1998: 127). Change and affect – not equilibrium – are the primary outcome of the forces making up social relations.

Deleuze and Guattari's critical reappraisal of debt and exchange begin with Mauss and moves subsequently to Lévi-Strauss and, by implication, anthropology in general. Aside from the question of whether this is a fair characterization of Mauss and Lévi-Strauss, a general problem with their rather sweeping critique is that accepting it wholesale would obviate the need to consider whether later anthropological analyses of debt and exchange might resonate with their analysis. In fact, several newer analyses are available which are certainly dynamic as well as attentive to power, if not Nietzschean (consider Sahlins 1972, Munn 1986, or Strathern 1988). Indeed, as we shall argue in a later section, Marilyn Strathern's *The Gender of the Gift* offers an analysis of exchange and debt that is at once more conceptually nuanced than Mauss and far more empirically detailed than anything one might get from Nietzsche's reading of 'a little Hindu law'. This particular exemplar will also allow us to indicate how Deleuzian thought intersects with the 'post-social' anthropology developed by Strathern through the shared precursor of Gregory Bateson. First, though, we turn to a discussion of how precisely Bateson's work informed Deleuze and Guattari.

Plateaus, Rhizomes, Schizophrenia: Bateson in Deleuze

If any particular anthropologist were to be singled out for his importance for Deleuze and Guattari's *oeuvre*, it would have to be Gregory Bateson. In his Spinoza lectures, Deleuze offhandedly refers to 'Bateson, who is a genius'.[17] Indeed, Deleuze and Guattari take Bateson's own life as an illustration both of the creative lines of flight enabled by anthropology and of their uneasy relations to the state and capitalist machines:

> Let us consider the more striking example of a career *à l'américaine*, with abrupt mutations, just as we imagined such a career to be: Gregory Bateson begins by fleeing the civilized world, by becoming an ethnologist and following the primitive codes and savage flows; then he turns in the direction of flows that are more and more decoded, those of schizophrenia, from which he extracts an interesting psychoanalytic theory; then, still in search of a beyond, of another wall to break through, he turns to dolphins, to the language of dolphins, to flows that are even stranger and more deterritorrialized. But where does the

> dolphin flux end, if not with the basic research projects of the American army, which brings us back to preparations for war and to the absorption of surplus value. (Deleuze and Guattari 1983: 236, italics in original)

This is a beautiful picture of the struggles of the rhizomatic, countercultural ethnographer, who increasingly deterritorrializes his work and interests but finds himself, at the maximum point of alterity – funded by the army – in the belly of the beast, as Donna Haraway might have put it. Deleuze and Guattari's interest in Bateson, however, was not solely illustrative and biographical. On the contrary, several of their most well known concepts were inspired by Bateson's work in anthropology and psychotherapy. Prominent among these is the notion of the plateau, which made it into the title of Deleuze and Guattari's most famous work.

Bateson had observed what he termed the 'schismogenic' behaviour among the Iatmul of Papua New Guinea, where boasting contests and 'buffoonery' progressively destabilize social integration to the point where special ceremonies were required to relieve the mounting tension. Arriving subsequently at Bali, Bateson came to use the notion of the plateau to describe 'Balinese character', in which 'climactic sequences' like those observed among the Iatmul appeared to be entirely absent. Noting how mother–child interactions were never allowed to escalate into confrontation, but were always defused by the parent, Bateson argued that, in Bali, 'The perhaps basically human tendency towards cumulative personal interaction is thus muted' (Bateson 1972: 112–13; see also Geertz 1975: 403).

Instead there appeared a 'continuing plateau of intensity' (Bateson 1972: 113). Reinterpreting the passage, Deleuze and Guattari argue that Bateson 'uses the term *plateau* for continuous regions of intensity constituted in such a way that they do not allow themselves to be interrupted by any external termination, any more than they allow themselves to build toward a climax … A plateau is a piece of immanence' (Deleuze and Guattari 1987: 158, emphasis in original).

In their introduction to rhizomatic analysis in *A Thousand Plateaus*, Deleuze and Guattari explain that 'A plateau is always in the middle, not at the beginning or the end' (ibid.: 21–22). They suggest that the plateau 'designate[s] something very special: a continuous, self-vibrating region of intensities whose development avoids any orientation toward a culmination point or external end' (ibid.: 22). Precisely because the plateau is not oriented towards pre-given intentions or ultimate ends it constitutes a 'piece of immanence'. In this it contrasts with the 'regrettable characteristic of the Western mind to relate expressions and actions to exterior or transcendent ends, instead of evaluating them on the basis of their intrinsic value' (ibid.: 22), intrinsic here meaning not 'essential', but rather immanent to the composition of the plateau in and through which action and expression take place.

Although the anticlimactic orientation characteristic of Balinese culture was especially striking compared with Iatmul interactions, Bateson made the more general observation that social organization in Bali 'differs very markedly from our own, from that of the Iatmul, from those systems of social opposition which Radcliffe-Brown has analyzed and from any social structure postulated by Marxian analysis' (Bateson 1972: 115). It was this basic difference that he aimed to capture with the notion of the plateau. As Deleuze and Guattari might say, the Balinese study seemed to illustrate a societal organization outside the scope of both 'capitalism and schizophrenia'.

Bateson, however, did not remain on Bali. Instead he returned to the US, where he studied family-therapeutic situations and the construction of schizophrenia. To account for the formation of this disorder, Bateson proposed the concept of the double bind, which, according to Deleuze and Guattari designates 'the simultaneous transmission of two kinds of messages, one of which contradicts the other, as for example the father who says to his son: go ahead, criticize me, but strongly hints that all effective criticism – at least a certain type of criticism – will be very unwelcome' (Deleuze and Guattari 1983: 79). For Bateson, inability to creatively resolve the question of how to respond to this paradoxical communication forms the basis for a schizophrenic development. Deleuze and Guattari disagree, however, arguing that there is nothing in 'two contradictory injunctions ... with which to make a schizophrenic' (ibid.: 360).

They did not, however, draw the conclusion that this made the double bind a useless conception. Rather, they complained that Bateson had been too restrictive in proposing the concept of the double bind only in the limited context of familial communication: 'if there is a veritable impasse, a veritable contradiction, it is the one into which the researcher himself is led, when he claims to assign schizophrenogenic social mechanisms, and at the same time discover them within the order of the family, which both social production and the schizophrenic process escapes' (ibid.: 360). Rather than believing in the 'Oedipal fiction' according to which every significant event is related to family matters, we should recognize that 'social production and the schizophrenic process' inevitably escape the narrow confines of the family.

The production of the schizophrenic should thus not be approached psychologically and focused on the question of family communication and interaction. Instead, it should be approached with special attention to the role that the material and cultural conditions of the patient – the assemblage he inhabits – has in shaping his mental state. In this sense, Deleuze and Guattari advocated a transformed version of Bateson's broadly cultural schismogenetic analysis against his more narrowly confined analysis of familial interaction.

Yet Deleuze and Guattari were not satisfied with playing out one transformed version of Bateson against another. On the contrary, their analysis exhibits a kind of crossbreeding of Batesonian themes. For just as the anticlimactic Balinese plateaus were seen by Bateson as a manner of avoiding schismogenesis, so the schizophrenic emerging from the violent productive forces of the capitalist machine reacts by making nomadic connections and by withdrawing to the tension-free, anticlimactic zone of the body without organs (Deleuze and Guattari 1987: 80). Witnessing this fractal criss crossing of the repertoires of the micro (family interaction, psychotheraphy) and the macro (the organization of the capitalist assemblage in general) makes it obvious that we are not here in the realm of unilinear causation, but are entering a realm of emergent and hybrid processes that move the researcher across usually separated domains of enquiry.

What has happened to the plateau in this process? It appears that the plateau is now, among other things, a piece of immanence that a schizophrenic within capitalism attempts to compose and inhabit. But, then, what has happened to the question of Balinese character that Bateson originally had in mind when proposing the concept?

As noted, Bateson viewed the 'nonschismogenic' Balinese system as an especially effective way of achieving the 'steady state' – the cybernetic term for a system in equilibrium (Bateson 1972: 126).[18] Indeed, the striking first impression of the plateau is one of balance and harmony. Contrary to the climactic and schismogenic Iatmul, the Balinese seemed to have perfected a way of interacting that was relieved of animosity and tension; an efficient auto-poietic machine. However, as we have seen in the discussion of debt and exchange, such holism is a far cry from Deleuze and Guattari's Nietzschean conception of violently interacting human and non-human forces.

Accordingly, in their reformulation, stability is assumed to be at most a surface manifestation generated by and responding to heterogeneous material and cultural forces. To be sure this reading constitutes a leap out of bounds from the point of view of Bateson's own analysis of the plateau. Yet by moving once again elsewhere in Bateson's work Deleuze and Guattari are capable of forging a link by way of the rhizome, which enable them to imagine how to connect disparate elements: unlike with unlike.

With respect to the Iatmul, Bateson noted that these: 'natives see their community, not as a closed system, but as an infinitely proliferating and ramifying stock ... The idea that a community is closed is probably incompatible with the idea of it as something which continually divides and sends out offspring "like the rhizome of a lotus"' (Bateson 1958: 249).[19] Nowhere does Bateson pick up on this notion. He certainly does not connect it with his Balinese studies, characterized by harmonic plateaus of intensity. Yet we find in his *Naven* an almost verbatim formulation of the rhizome as Deleuze and Guattari will come to use the term.

It is striking to see how this notion, apparently from a Iatmul informant, is turned into a key concept by Deleuze and Guattari, who subsequently recombine it with the entirely different notion of the plateau, in order to analyse the making and unmaking of schizophrenic relations under the capitalist machine. Even more, Deleuze and Guattari used this thoroughly Bateson-entangled analysis to critique the same author's analysis of how schizophrenia is generated in North American family interaction for failing to notice that rhizomatic undercurrents always infuse the seemingly stable plateau of the family with other elements, which 'ceaselessly establishes connections between semiotic chains, organizations of power, and circumstances relative to the arts, sciences and social struggles' (Deleuze and Guattari 1987: 7).

It would be easy to criticize this uninhibited bricolage for failing to take into account original aims and contexts. Yet who can deny that something new and different is being established through these analyses? It is hard to see Bateson, himself an intellectual maverick, in much disagreement with the procedure, arguing with respect to his own analysis of Balinese culture that 'a tool or a method can scarcely be proven false. It can only be shown to be not useful' (Bateson 1972: 108). Indeed, if we momentarily go along with Roy Wagner's claim that 'what passes in the literature for the steadfastness or self-consistency of a cultural pattern endlessly reproducing itself may simply be a reflection of the fact that Melanesian anthropology has not advanced one jot beyond the "schismogenesis" model proposed by Gregory Bateson in the 1930s' (Wagner 2005: 221), it might well be that an infusion of Batesonian anthropology as refigured by Deleuze and Guattari could provide tools for such further conceptual advance. Indeed, we should look forward with interest to the future where this might happen.

In the meantime, however, it is also possible to question whether Wagner's characterization is entirely fair. Indeed, a different and yet quite obvious idea might be to revisit Marilyn Strathern's *The Gender of the Gift*, which appears to do nothing if not advance Melanesian anthropology 'one jot' not only 'beyond the "schismogenesis" model' proposed by Bateson, but also beyond the Maussian theory of the gift so harshly criticized by Deleuze and Guattari. Making our own leap out of bounds, we therefore propose to view the analysis provided in *The Gender of the Gift* as one analytical format through which social anthropology and Deleuzian ideas might intersect – or, indeed, as an example of how it has already 'subterraneously' done so.

From Mauss to Strathern

One obvious reason for this proposal is precisely that Strathern's *The Gender of the Gift* (1988) breaks with the boundary between the social and the pre-social by arguing that this distinction is based on a Western metaphysical understanding that cannot be universalized. We emphasize this point since Strathern has in fact been criticized by Annette Weiner, for exactly the same reason that Deleuze and Guattari criticized anthropology at large – that is, for being too Lévi-Straussian and universalistic.[20] Weiner reads Strathern as merely presuming 'to move beyond positivist theory by beginning with indigenous notions of personal identity and then showing the types of social relations that are their concomitants' (Weiner 1992: 14). She continues by disqualifying this supposed presumption by arguing that, 'being grounded in the a priori essentialism of the norm of reciprocity, Strathern's argument cannot account for the temporal aspects of the movements of persons and possessions and the cultural configurations that limit or expand the reproduction or dissipation of social and political relationships through time' (ibid.: 14–15). Weiner views Strathern's effort as one of finding indigenous ideas that would correspond to the Lévi-Straussian notion of the norm of reciprocity, which she has thereby mistakenly universalized. The critique is curious, as the very first sentence of *The Gender of the Gift* argues against using indigenous concepts of personal identity to respond directly to problems in Western social theory: 'it might sound absurd for a social anthropologist to suggest he or she could imagine people having no society. Yet the argument of this book is that however useful the concept of society may be to analysis, we are not going to justify its use by appealing to indigenous counterparts' (Strathern 1988: 3). Contrary to Weiner's suggestion, this formulation indicates that Strathern's endeavour resonates with Viveiros de Castro's call to move away from a conception of anthropology that views every culture as offering specific means of dealing with the same set of universal problems. It is precisely because Melanesians do not have Western problems that Strathern proposes that '[t]he task is not to imagine one can replace exogenous concepts by indigenous counterparts; rather the task is to convey the complexity of the indigenous concepts in reference' (ibid.: 6).

Hence, far from starting with the 'a priori essentialism of the norm of reciprocity', Strathern argues that '[o]ne must, in fact, get away altogether from the model of a "model" that takes symbolic representation as ordered reflection. Domaining is better taken as an activity, the creation/implementation of difference as a social act' (Strathern 1988: 96). On the one hand, this is a move away from general theories of society and representation towards articulation of the ethno- or infra-theories that can encompass potentials for difference and innovation immanent in the processes under study. However, on the other hand, this is not a call to find a truly indigenous point of view, since Strathern's 'ethnographic canvas' is clearly filled with many actual and virtual resources that have been nourished so as to enable her articulation of Melanesian relationships.

Strathern argues that, in Melanesia, 'there is no indigenous supposition of a society that lies over or above or is inclusive of individual acts and unique events. There is no domain that represents a condensation of social forces controlling elements inferior or in resistance to it' (ibid.: 102). However, this does not imply that the Melanesians lack, or have failed to discover, these concepts. Rather, analytical categories such as 'society' are themselves products of Western metaphysical root metaphors that may be quite inadequate for describing or understanding the life of Melanesians.

Likewise, the Western concept of the gift, as well as its antonym commodity, is derived from a commodity economy foreign to Melanesian cultures where such conceptual distinctions are not found (ibid.: 136). This argument, then, is directed against analyses that claim to find commodity-based forms of domination in Papua New Guinea. However, Strathern's alternative is not to rule out the existence of unequal relations. It is, rather, to open for the analysis of other forms of relation and domination than those imaginable through the lens of Western political economy. Making use of a distinction between mediated and unmediated relations, Strathern argues as follows:

> Through mediated relations, items flow between persons, creating their mutual enchainment. The items carry the influence that one partner may hope to have on another. Through unmediated relations, one person directly affects the disposition of another towards him or her, or that person's health and growth,

> of which the work that spouses do for another or the mother's capability to grow a child within her are examples. Or one may think of the contrast between the circulation of blows in warfare, and the harm that one body can do to another because of its very nature, as evinced in so-called pollution beliefs. Despite the absence of mediating objects, these latter interactions have the form of an 'exchange' in so far as each party is affected by the other; for instance, a mother is held to 'grow' a child because the child, so to speak, also 'grows' her. (Strathern 1988: 178–79)

The point is precisely to show that some Melanesian exchanges remain entirely unrecognizable from a theoretical point of view that claims to know in advance the characteristics of gifts and exchanges. For, as Strathern comments, 'whereas the first mode of symbolization is recognizable to us through Mauss' description of gift exchange – persons transfer parts of themselves – the second would not normally be recognized as a type of exchange at all' (ibid.: 179). This is not only a problem because the latter form of exchanges remain invisible but also because it impedes understanding of even those relations that appear as recognizable Western-style exchanges: 'One cannot, out of the workings of Melanesian social action, extract one set of relations as typical of "gift relations" and another as typical of non-gift relations: the unmediated mode takes its force from the presence of the mediated mode and vice versa' (ibid.).

Through this analysis Strathern frees the gift economy from its bonds to commodity exchange, and thus from the notion of gifts as the exchange of property items, at least in the Melanesian case. Rather than exchange of disposable items, Strathern speaks of enchainment as a condition of gift relations (1988: 161). Enchainment involves a series of transformational relations. Gift exchange here becomes an open system, or assemblage, where people enter relations and undergo transformations through productive activities. Domination is not ruled out, but it cannot be conceptually predetermined and it is certainly not bound to take forms easily recognizable to the critical anthropologist. Rather, domination in a gift economy becomes a matter of determining 'the connections and disconnections created by the circulation of objects' (Strathern 1988: 167, see also 326–27).

In summary, Strathern makes the notion of the gift into a tool for describing the transformative capacities of not necessarily commodified actions and relations. In her own words: 'Transactions ... appear to make relations afresh, for the instruments of mediation – the gifts – appear to be creating the relationship. Transactions are thus experienced as holding out the possibility of relationships ever newly invented, as though there could be an infinite expansion of social connections' (ibid.: 180) – not unlike the rhizomatic and 'infinitely proliferating and ramifying stock' remarked on by Bateson's Iatmul informant.

We might sum up by noting that Strathern's ethnographic mode of theoretical thinking breaks with Deleuze and Guattari's distinction between gift and debt, while implicitly taking a Deleuzian understanding of ethnography further than Deleuze himself ever went, although drawing on some of the same 'virtual and actual' resources. In the next section we remark on some of these resources, namely the inspirations from chaos theory, which Strathern has developed into her idea of *partial connections* and which Deleuze and Guattari made use of in conceptualizing reproductive and following sciences.

Rhizomes and Fractals: A Deleuzian Anthropology?

In their Bateson-inspired vocabulary, Deleuze and Guattari explained that 'we call a "plateau" any multiplicity connected to other multiplicities by superficial underground stems in such a way as to form or extend a rhizome' (Deleuze and Guattari 1987: 22). In this description plateaus appear as relatively stable formations that, if examined more attentively, turn out to be constituted by a set of heterogeneous elements, through each of which connections to other places and times proliferate. As this is the case, plateaus are not understandable in terms of any one type of 'domain' knowledge, such as technical, cultural, political or biological. Indeed, this is why they must be approached with the purpose of elucidating the links that connect seemingly disparate entities.

Discussing Viveiros de Castro's notion of multinaturalism, José Antonio Kelly notes that in Amerindian societies bodies are understood not as the material dimension of the human being, 'but rather as "bundles of affects"' organizing its capacities (Viveiros de Castro 1998: 478; Kelly 2005: 112). The example highlights the affiliation between the Deleuze-inspired notions of multinaturalism and fractal analyses as promoted by Marilyn Strathern and Roy Wagner, because it makes the question of 'level' of analysis unanswerable in principle and forces the analyst to reconsider the very notion of 'relation' (Strathern 1991). Where are we to locate bundles of affects, for example? In the bodies of individual Amerindians? In technical artefacts such as clothes and painting? In cultural systems? In cosmology? These are moot questions, because bodies, cultures, cosmologies and techniques are constituted mutually in the same process, rather than determined at any clearly demarcated level (Jensen 2007). Such a fractal analysis might be seen as related to Deleuze and Guattari's 'smooth space', which is characterized by not having a 'dimension higher than that which moves through it or is inscribed in it' (Deleuze and Guattari 1987: 488). At a first estimation, this may appear a surprising suggestion, since the notion of the fractal suggests that the social is constituted through self-similar processes taking place simultaneously at multiple scales, whereas 'smooth space' is defined as flattened and one-dimensional. Yet we would suggest that smooth and fractal space simply entail the adoption of a viewpoint from opposite sides of the PLURALISM = MONISM equation. The 'amorphous' smooth space of Deleuze and Guattari is space as seen from the side of monism, which is to say outside the bounds of clear and a priori categorization. It is precisely because of its 'smoothness' that this space is in practice capable of giving rise to 'zones of indiscernibility', where diverse elements (clothes, affects, cosmology), that from a traditional point of view ought not to be mixed, are nevertheless seen to be capable of entering into relations through 'proximity' and form 'new bodies'. However, through this process of actualization emerges a non-smooth space, whose elements are linked in partial connection and through which unstable hierarchies may take form. This would be the social space of Strathernian ethnography, which renders visible a fractal topology (see also Pedersen 2007: 314–15).

Indeed, we suggest that the formula PLURALISM = MONISM facilitates analysis in a mode where neither hierarchies nor stable relations can be taken for granted. But it is for this very reason that such analyses enable the ethnographer to grapple with issues of becoming, for example of new cultural formations, political forces or subjects, without prejudging the qualities that such becoming must have. If that is the case, however, we can go even further and propose that it is by accepting this ontological diversity and fractality and developing analytical tools for dealing with them, that social anthropology and STS might become 'smooth', 'following', minor sciences, rather than 'striated', 'reproductive' and major ones.

This suggestion, however, presumes that a negative answer is given to the question of whether anthropology and STS are not already following, smooth sciences. Is the very definition of good ethnography not precisely that it attempts to take into account all aspects, no matter how surprising, of the cultures and peoples it studies?

Consider, for example, one of Deleuze and Guattari's own examples of good ethnography. Surprising as it may seem, this is Victor Turner's study of the Ndembu (Turner 1967).[21] Turner, like Bateson, was seen by Deleuze as Guattari as an exceptionally creative 'smooth' thinker because of his capacity for analysing rhizomatic multiplicities, involving mutually constitutive relations between seemingly separated domains, such as language, symbolism, material culture and social organization.

We may now ask whether Bateson or Turner, Strathern or Viveiros de Castro stand as true illustrations of a Deleuzian anthropology. Obviously, given their many differences, answering this question in the affirmative might be close to vacuous. In addition, it would be quite possible to pick apart the presuppositions of Turner's liminal anthropology, just as it would be possible to deconstruct Bateson's cybernetic and psychotherapeutic discourse. Alternatively, more insidiously one might argue that, if ethnography by definition aims to deal simultaneously with multiple domains of human existence, then Deleuze and Guattari provide at most an elaborated theoretical affirmation of what everyone already knows to be the case.

Of course, we believe that following the strategy of the *and* offers something more consequential and far more interesting than an overtheorized version of an already well-known anthropology. On the one hand, we can follow Strathern, who proposes that 'anthropological exegesis must be taken for what it is: an effort to create a world parallel to the perceived world in an expressive medium (writing) that sets down its own conditions of intelligibility' (Strathern 1988: 17). Yet, with a performative disposition, taking anthropological exegesis for what it is, is in fact quite a challenge, since it has precisely never been quite what has so often been claimed: the best available representation of savage, native or indigenous culture. We have argued instead that anthropological representation is the outcome of linking together disparate agencies and materials – virtual and actual – in performative chains. Ethnography may therefore be seen as a matter of inventively rearranging already overflowing canvases in order to articulate both 'theories' and 'practices' without assuming any particular relation between them.

This, of course, is also why we would need to qualify the question of 'how to do Deleuzian anthropology' as itself relying on a representational strategy. Obviously, searching for the correct answer to this question will lead to disappointment. But then this is as we would expect from Deleuze and Guattari. For, as we have seen, their pragmatic approach is not to reproduce the meanings or agendas of those from whom they learn, but to extract their concepts and graft them on to new concerns, placing them in proximity with other issues.

If this is not a representational strategy, neither is it a purely discursive one. Rather it is performative and fractal: a strategy of the *and*; evaluated according to its ability to move between disciplinary domains and hierarchies and thereby generate new capacities for thinking and acting. Yet, if this is a *strategy*, it is also a peculiar one, precisely because it offers no general, methodical or theoretical recipe for anthropology or STS. It also has no particular stylistic or aesthetic requirements. And it certainly does not demand the regular invocation of any particular 'order-word' – even Deleuzian terms such as 'deterritorrialization', 'nomad' or 'rhizome'. It does not demand any of these, which is precisely why styles, methods and guiding concepts can be used at the intersection of multiple and heterogeneous situations, contexts and people, that is, at the sites of risk, where ethnographical moments and events may occur.

The Book

The volume has three parts, loosely defined by the contributors' overall approach to the theme of Deleuzian intersections in science, technology and anthropology. The first part, entitled 'Deleuzian Sciences?', contains four chapters, which deal in broadly philosophical and theoretical terms with the question of what a Deleuzian approach to social science might entail.

In Chapter I, 'Experimenting with *What is Philosophy?*', philosopher of science Isabelle Stengers confronts the perplexity with which Deleuze and Guattari's last book was met by many readers, a perplexity related to its sobriety and apparent conservatism in comparison with the wildness of *Anti-Oedipus and A Thousand Plateaus*. Stengers proposes a new reading, identifying in Deleuze and Guattari's authorship an acritical turn, which aims neither to celebrate nor to criticize science but rather to offer resistances to the present. Specifically, Stengers here confronts the challenge of how to resist the contemporary appropriation – and indeed destruction – of the sciences in the name of utility. Accordingly, science is to be empowered, with the purpose of enabling a creative evolution rather than producing an objectivized picture of the world. Still, Stengers admonishes us to approach the scientific path of risky invention with caution.

In Chapter 2 'Facts, Ethics and Event', Mariam Fraser explores the reality of facts as understood by a series of constructivist thinkers that she links together: Bruno Latour, Isabelle Stengers, Alfred North Whitehead and finally Gilles Deleuze. Specifically, Fraser examines the concept of the event and how this relates to scientific facts and to ethics. Key in this discussion is the idea of potentiality as developed by Whitehead, and taken up by Deleuze: an idea following which ethics is not solely related to the actual – what already exists – but also to virtual, presently unrealized, potentialities. The Deleuzian take on the event is explored as posing problems that enter the virtual and reach beyond what is actually accessible. Fraser thus calls for an ethics of the event informed by an openness towards the virtual.

In Chapter 3 'Irony and Humour, Toward a Deleuzian Science Studies', Katie Vann enquires into the categories of irony and humour in Deleuze, following an invitation to engage in 'humorous' analyses in Isabelle Stengers' *Invention of Modern Sciences.* Vann turns to Deleuze's underexplored work 'Coldness and Cruelty' from *Masochism,* in order to consider the history of these categories in Deleuze's work. Carefully unravelling these changing Deleuzian conceptions, from 'Coldness and Cruelty' to *Difference and Repetition,* Vann suggests that a significant difference – between the classical and modern conceptions of law – frames Deleuze's treatment of these terms.Vann considers the implications of this structuring for the prospects of crafting humorous Deleuzian science studies.

Finally, in Chapter 4, 'Between the Planes: Deleuze and Social Science', Steve Brown notes that the typology developed in *What is Philosophy?,* with its three 'planes' of philosophy, science and art, leaves no space for the social sciences. The consequence is that social sciences can be safely relegated by Deleuzians to a status of second-hand philosophy (which means that philosophers must help to ground those sciences properly) or of sloppy imitators of natural science (which means that natural scientists must help to formalize them properly). Critically engaging with Manuel de Landa's assemblage theory, Brown proposes that alternative Deleuzian strategies might be made available for constructing the plane of social science.

Part Two of the volume is named 'Sociotechnial Becomings' and has three chapters, all of which approach science and technology studies from a Deleuzian angle. Returning to Deleuze's work on Leibniz, Geoffrey C. Bowker argues in Chapter 5 that the notion of the fold could enable science and technology studies to dissolve – or, rather, re-knot – multiple binaries that have traditionally haunted such studies. Exploring in particular the dualism between foreground and background, inside and outside and identity and multiplicity, and drawing on a diverse set of examples, especially from biogeography, 'A Plea for Pleats' ends by encouraging further explorations of pleated thinking in science, technology and anthropology.

If Bowker presents the reader with a breadth of examples, Chapter 6, 'Every Thing

Thinks: Subrepresentative Differences in Digital Video Codecs', by Adrian Mackenzie, on the other hand, works in depth with a specific case. Raising the question of what it would mean to take seriously Deleuze's claim that 'every thing thinks' (see also Henare et al. 2007), Mackenzie takes a close look at how codecs are simultaneously involved in processes of difference and repetition at the level of complex calculations relating to audio-visual imagecompression and at the level of cultural consumption. Through his analysis, Mackenzie seeks to get an analytical purchase on this sociotechnical becoming by means of the notion of the centre of envelopment. The differences enveloped in the codecs provide for differences across scales and modalities, impacting patterns of making, viewing and moving images, thus facilitating new ecologies of spectatorship, consumption and citizenship.

Chapter 7, Andrew Pickering's 'Cybernetics as Nomad Science', ends the second part of the book. In this chapter, Pickering takes a closer look at the distinction made in *A Thousand Plateaus* between royal and nomad science. Rather than seeing these as two phases of science, Pickering proposes instead to consider the royal and the nomad as two distinct kinds of science. With this typology in mind, he explores cybernetics as an example of such a nomad science, engendering its own peculiar and nomadic becomings. Pickering traces a line through the institutional links – or rather lack of links – and the haphazard modes of transmission and development of English cybernetics. He views cybernetics as an exemplar of a radically different kind of science; one that existed outside the circuits of royal science. Pickering ends by calling for more attention to the interplay between the royal and the nomad, for example through analysis of how nomadic sciences feed back into the institutions and practices of royal science.

The three chapters of the final section of the book – 'Minor Assemblages' – engage with Deleuzian thought with a set of distinctly anthropological concerns in mind. In Chapter 8, 'Cinematics of Scientific Images: Ecological Movement-Images', Erich W. Schienke draws on cultural anthropology and STS to explore the analytical potential in Deleuze's work on cinema for studying scientific images. Like the cinema, scientific images are framed and edited forms of communication, where numerous activities – cultural and political – are involved in determining what actually ends up within the frame. Schienke explores Deleuze's concept of the movement-image as a device for categorizing and analysing the contemporary production of scientific images and applies Deleuze's four categories of the movement-image – perception-images, affection-images, action-images and relation-images – to ecological sets, in order to study how these scientific images are framed and closed. Of particular interest here is the notion of scale in the production of ecological images, and how seemingly miniscule decisions on the scale of an image can have far-reaching effects for governance as well as for science.

In Chapter 9, 'Social Movements and the Politics of the Virtual: Deleuzian Strategies', Arturo Escobar and Michal Osterweil explore what they see as a move towards ontological questions in recent social theory. In this light, they approach the issue of how to rethink social movements on the basis of network theories and Deleuzian philosophy, especially in the interpretation given by Manuel de Landa. Following de Landa's distinction between hierarchies and self-organizing meshworks, and not unlike Pickering's distinction between modes of science, Escobar and Osterweil propose to see these two kinds of networks as defining alternative approaches to political and social action. Eschewing the centralized control structures and military connotations of the former, they suggest that the rhizomatic meshwork is a more adequate organizational format for the new movements, based as it is on heterogeneity and diversity. Even within the organizing principle of meshworks, however, a further distinction can be drawn between 'dominant actor-networks' and 'subaltern actor-networks', where only the latter can be seen as a fully appropriate model for thinking about the new social movements. This turn away from hierarchies and central organization implies a step towards what Escobar and Osterweil describe as 'flat alternatives' for political stragies and movements, entailing an openness towards the virtual.

The volume ends with Chapter 10, 'Intensive Filiation and Demonic Alliance', in which Eduardo Viveiros de Castro embarks on a wide-ranging theoretical and empirical voyage, in order to explore the potentials for anthropological thought of concepts and ideas in *Anti-Oedipus* and *A Thousand Plateaus*. The chapter sets the agenda by elucidating Deleuze and Guattari's theory of multiplicities and its apparently contradictory tendency to contrast concepts in terms of dualities. In the second part of the chapter, an exploration of the distinction between the intensive and the extensive allows Viveiros de Castro to reinterpret the notions of alliance and filiation, central to kinship theory, by reworking them in terms of difference and multiplicity. Through a reading that contrasts *Anti-Oedipus* with *A Thousand Plateaus*, Viveiros de Castro then argues for a re-evaluation of alliance in terms of becoming and calls for a new concept of exchange, based not on contractual relations or dialectical synthesis but rather on ongoing change and variation.

Notes

1. Although for disciplinary reasons it might appear strange to connect discussions of Deleuze's philosophy with both STS and social anthropology we think there are even better reasons for treating these intersections in a shared volume. Prominent among these is the very significant ethnographic influence on STS and the ongoing cross-fertilization between the two fields. Among the many ethnographers who works at the interface between STS and social/cultural anthropology are Traweek (1988); Heath (1998); Strathern (1999b); Fortun (2001); Fischer (2003); Hayden (2003); Battaglia (2005); Tsing (2005).
2. For some exceptions, see Pickering (1995); Brown and Capdevila (1999); Latour (1999). More sustained efforts are Wise (1997) and Mackenzie (2002). Stengers (1997, 2000) and Despret (2004a,b) provide insightful discussions, which, if not strictly STS, are relevant to the discussions in the field.
3. Noteworthy exceptions include Appadurai (1990); Escobar (1994); Martin (1996); Viveiros de Castro (2004, this volume); Maurer (2005); Ong and Collier (2005); Henare *et al* (2007); Pedersen (2007); Nielsen (2008).

4. This is not the place to provide a comprehensive bibliography, but central works include: Deleuze (1983; 1990a, b; 1991; 1993; 1994); Deleuze and Guattari (1983; 1987; 1994).
5. See for example Hardt (1983); Massumi (1996); Rodowick (1997); Ansell-Pearson (1999); Olkowski (1999); Gatens (2000); Hansen (2000); Patton (2000); Rajchman (2000); Protevi (2001); de Landa (2002); Smith (2003).
6. He refers to Stephen Tyler (e.g. Tyler 1984).
7. In addition, these authors draw upon several other sources whose thinking is related to Deleuze's, such as Michel Serres (e.g. 1982); Ilya Prigogine and Isabelle Stengers (1984); Michel Foucault (e.g. 2001) and pragmatism, especially Alfred North Whitehead (e.g. 1929) and William James (e.g. 1996). See Jensen (2004) for a general discussion.
8. It is noteworthy that Gregory Bateson, one of Deleuze's main sources of anthropological inspiration, identified the same equivalence in the thought of the Iatmul, who adhered simultaneously to: '(a) A sense of pluralism: of the multiplicity of the objects, people, and spiritual things in the world. (b) A sense of monism; that everything is fundamentally one or at least derived from one origin' (Bateson 1958: 235).
9. See further Pickering and Guzik (2008).
10. For his Deleuze scholarship see Holland (1999).
11. Consider Philip Goodchild's gloss, according to which science 'is capable of sufficiently managing the behaviour of the objective world as to allow the conformity between theory and observation'. The humanities, on the other hand, 'may attempt to model themselves on the physical sciences, but they have a more difficult task: much human thought and behaviour is conditioned by a wide variety of mutually dependent variables which cannot be easily individuated and isolated' (Goodchild 1996: 20). In this excerpt the categorization of science and the humanities into distinct compartments with radically different functions corresponding to ontological domains is accepted without question. As we see it, however, the strategy of the *and* rather encourages an exploration of how sciences, humanities and the world with which they engage interlink and commingle and an interest in how sciences construct new fields of investigation through such processes. See Smith (2005) for a discussion of the compartmentalization of knowledge production into science and humanities.
12. See Stengers, this volume. Stengers here plays upon the two Janus faces of science presented in Bruno Latour's *Science in Action* (1987).
13. For a similar ANT rejection of the binarisms of ontology–epistemology and realism–constructivism, see Latour (1999).
14. See Mosko and Damon (2005) for analyses of chaos theory in social anthropology.
15. For a fascinating exploration of the functions, limits and possibilities of the hermit in academic practice, see the conversations between Michel Serres (whose position here closely resembles Deleuze and Guattari) and Bruno Latour (Serres and Latour 1995).
16. See Mauss (1967) and Nietzsche (1999).
17. See www.webdeleuze.com (lecture of 17 February 1981).
18. In contrast, schismogenesis is an example of what in cybernetic terms would be called a positive feedback loop.
19. Apparently Bateson is citing a native informant here.
20. The question of whether and how Lévi-Strauss himself was quite as 'Lévi-Straussian' and universalistic as often assumed has recently been explored by Viveiros de Castro (2008).
21. Turner's well-known proposal was to develop an anthropology of the liminal. His attention would go to: 'multivocal symbols and metaphors – each susceptible to many meanings, but with the core meanings linked analogically to basic human problems of the epoch which may be pictured in biological, or mechanistic, or some other terms – these multivocals will yield to the action of the thought technicians who clear intellectual jungles, and organized systems of univocal concepts and signs will replace them' (Turner 1974: 28).

Bibliography

Ansell-Pearson, K. (1999). *Germinal Life: the Difference and Repetition of Deleuze*. London, Routledge.

Appadurai, A. (1990). 'Disjuncture and Difference in the Global Cultural Economy', in M. Featherstone (ed.) *Global Culture: Nationalism, Globalization and Modernity*. London, Thousand Oaks, and New Delhi, Sage, pp. 295–311.

Bateson, G. (1958). *Naven: A Survey of the Problems Suggested by a Composite Picture of the Culture of a New Guinea Tribe Drawn from Three Points of View*. Stanford, CA, Stanford University Press.

———. (1972). *Steps to an Ecology of Mind – A Revolutionary Approach to Man's Understanding of Himself*. New York, Ballantine Books.

Battaglia, D. (ed.). (2005). *E.T. Culture: Anthropology in Outerspaces*. Durham, NC, Duke University Press.

Biagioli, M. (1993). *Galileo, Courtier – The Practice of Science in the Culture of Absolutism*. Chicago and London, University of Chicago Press.

Brown, S.D. and R. Capdevila. (1999). 'Perpetuum Mobile: Substance, Force and the Sociology of Translation', in J. Law and J. Hassard (eds) *Actor Network Theory and After*. Oxford, Blackwell Publishers, pp. 26–51.

Callon, M. (ed.). (1998). *The Laws of the Market*. Oxford, Blackwell Publishers.

de Landa, M. (2002). *Intensive Science and Virtual Philosophy*. New York, Continuum Press.

———. (n.d.) 'Markets and Anti-markets in the World Economy'. Available at http://www.t0.or.at/delanda/a-market.htm

Deleuze, G. (1983). *Nietzsche and Philosophy*, London, Athlone Press.

———. (1986). *Cinema 1: The Movement-Image*. Minneapolis, University of Minnesota Press.

———. (1990a). *Expressionism in Philosophy: Spinoza*. New York, Zone Books.

———. (1990b). *The Logic of Sense*. New York, Columbia University Press.

———. (1991). *Bergsonism*. New York, Zone Books.

———. (1993). *The Fold – Leibniz and the Baroque*. London, Athlone Press.

———. (1994). *Difference and Repetition*. New York, Columbia University Press.

———. (1994). *Difference and Repetition*. New York, Columbia University Press.

———. (1998). *Essays Critical and Clinical*, London, Verso.

———. (2003). *Francis Bacon: The Logic of Sensation*. New York and London, Continuum Press.

———. (2006). *Two Regimes of Madness: Texts and Interviews 1975–1995*. New York and Los Angeles, Semiotext(e).

Deleuze, G. and F. Guattari. (1983). *Anti-Oedipus: Capitalism and Schizophrenia*. London, Athlone Press.

———. (1987). *A Thousand Plateaus: Capitalism and Schizophrenia*. Minneapolis and London, University of Minnesota Press.

———. (1994). *What is Philosophy?* New York, Columbia University Press.

Despret, V. (2004a). 'The Body We Care For: Figures of Anthropo-zoo-genesis', *Body and Society* 10 (2–3), 111–34.

———. (2004b). *Our Emotional Make-Up: Ethnopsychology and Selfhood*. Other Press, New York.

Escobar, A. (1994). Welcome to Cyberia: Notes on the Anthropology of Cyberculture', *Current Anthropology* 35 (3): 211–31.

Feenberg, A. (2003). 'Modernity Theory and Technology Studies: Reflections on Bridging the Gap', in T. Misa, P. Brey and A. Feenberg (eds) *Modernity and Technology*. Cambridge, MA, MIT Press, pp. 73–104.

Fischer, M.J. (2003). *Emergent Forms of Life and the Anthropological Voice*. Durham, Duke University Press.

Fortun, K. (2001). *Advocacy after Bhopal: Environmentalism, Disaster, New Global Orders*. Chicago, University of Chicago Press.

Foucault, M. (2001). *Power*. New York, New Press.

Gatens, M. (2000). 'Feminism as "Password": Re-thinking the "Possible" with Spinoza and Deleuze', *Hypatia* 15 (2), 59–75.

Geertz, C. (1975). *The Interpretation of Cultures: Selected Essays*. New York, Basic Books.

Goodchild, P. (1996). *Deleuze and Guattari: An Introduction to the Politics of Desire*. London, Sage.

Hansen, M. (2000). 'Becoming as Creative Involution: Contextualizing Deleuze and Guattari's Biophilosophy', *Postmodern Culture* 11 (1). Available at muse.jhu.edu/journals/pmc/v011/11.1hansen.html

Haraway, D. (1997). *Modest_Witness@Second_Millennium.FemaleMan©_Meets_OncoMouse™ – Feminism and Technoscience*. New York, Routledge.

Hardt, M. (1983). *Gilles Deleuze – An Apprenticeship in Philosophy*. London, UCL Press.

Hayden, C.P. (2003). *When Nature Goes Public: The Making and Unmaking of Bio-prospecting in Mexico*. Princeton, NJ, Princeton University Press.

Heath, D. (1998). 'Bodies, Antibodies and Modest Interventions', in G.L. Downey, J. Dumit and S. Traweek (eds) *Cyborgs and Citadels – Anthropological Interventions in Emerging Sciences and Technologies*. Santa Fe, NM, School of American Research Press, pp. 67–83.

Henare, A., M. Holbraad and S. Wastel. (2007) *Thinking Through Things: Theorising Artefacts Ethnographically*. New York, Routledge.

Holland, E.W. (1999). *Deleuze and Guattari's Anti-Oedipus: Introduction to Schizoanalysis*. New York and London, Routledge.

———. (2003). 'Representation and Misrepresentation in Postcolonial Literature and Theory', *Research in African Literatures* 34 (1), 159–73.

James, W. (1996). *A Pluralistic Universe*. Lincoln and London, University of Nebraska Press.

Jensen, C.B. (2004). 'A Non-humanist Disposition: On Performativity, Practical Ontology, and Intervention', *Configurations* 12, 229–61.

———. (2006). 'Experimenting with Political Ecology – Review Essay: Latour, Bruno, 2004, *Politics of Nature: How to Bring the Sciences into Democracy* (Cambridge, MA and London: Harvard University Press)', *Human Studies* 29 (1), 107–22.

———. (2007). 'Infrastructural Fractals: Revisiting the Micro–Macro Distinction in Social Theory', *Environment and Planning D: Society and Space* 25 (5), 832–50.

Jensen, C.B. and E.M. Selinger. (2003). 'Distance and Alignment: Haraway's and Latour's Nietzschean Legacies', in D. Ihde and E. M. Selinger (eds) *Chasing Technoscience: Matrix for Materiality*. Indianapolis, IN, University of Indiana Press, pp. 195–213.

Jensen, C.B. and T. Zuiderent-Jerak. (2007). 'Unpacking "Intervention" in Science and Technology Studies', *Science as Culture* (special issue) 16 (3).

Kelly, J.A. (2005). 'Fractality and the Exchange of Perspectives', in M.S. Mosko and F.H. Damon (eds) *On the Order of Chaos: Social Anthropology and the Science of Chaos*. New York, Berghahn Books, pp. 108–36.

Latour, B. (1987). *Science in Action: How to Follow Scientists and Engineers through Society*. Cambridge, MA, Harvard University Press.

———. (1993). *We Have Never Been Modern*. New York, Harvester-Wheatsheaf.

———. (1999). *Pandora's Hope – Essays on the Reality of Science Studies*. Cambridge, MA, Harvard University Press.

———. (2004). *Politics of Nature: How to Bring the Sciences into Democracy*, Cambridge, MA, and London, Harvard University Press.

———. (2005). *Reassembling the Social: an Introduction to Actor–Network Theory*. Oxford, Oxford University Press.

Law, J. (1987). 'Technology and Heterogeneous Engineering: the Case of Portuguese Expansion', in W. Bijker, T. Hughes and T. Pinch (eds) *The Social Construction of Technological Systems*. Cambridge, MA, MIT Press, pp. 111–34.

Mackenzie, A. (2002). *Transductions: Bodies and Machines at Speed*. London, Continuum.

Mackenzie, D. (2006). *An Engine, Not a Camera: How Financial Models Shape Markets*. Cambridge, MA, MIT Press.

Martin, E. (1996). 'Citadels, Rhizomes, and String Figures', in S. Aronowitz, B. Martinsons and M. Menser (eds) *Technoscience and Cyberculture*. New York and London, Routledge, pp. 97–111.

Massumi, B. (1996). *A User's Guide to Capitalism and Schizophrenia*. Cambridge, MIT Press.

Maurer, B. (2005). *Mutual Life, Limited: Islamic Banking, Alternative Currencies, Lateral Reason*. Princeton, NJ, Princeton University Press.

Mauss, M. (1967). *The Gift: Forms and Functions of Exchange in Archaic Societies*. New York and London, W.W. Norton.
Miller, C.L. (1993). 'The Post-identitarian Predicament in the Footnotes of *A Thousand Plateaus*: Nomadology, Anthropology, and Authority', *Diacritics* 23 (3), 6–35.
———. (2003). '"We shouldn't judge Deleuze and Guattari": a response to Eugene Holland', *Research in African Literatures* 34 (3), 129–41.
Mol, A. (2002). *The Body Multiple: Ontology in Medical Practice*. Durham, Duke University Press.
Mosko, M.S. and F.H. Damon (eds). (2005). *On the Order of Chaos: Social Anthropology and the Science of Chaos*. New York, Berghahn Books.
Munn, N.D. (1986). *The Fame of Gawa: A Symbolic Study of Value Transformation in a Massim (Papua New Guinea) Society*. Cambridge, Cambridge University Press.
Nielsen, M. (2008). 'Constructing a House – Building a Life. Peri-urban Housing Strategies in Mozambique'. PhD Dissertation, University of Copenhagen.
Nietzsche, F. (1999). *On the Genealogy of Morals*. Cambridge, Cambridge University Press.
Nowotny, H., P. Scott and M. Gibbons. (2001). *Re-thinking Science: Knowledge and the Public in an Age of Uncertainty*. London, Polity Press.
Olkowski, D. (1999). *Gilles Deleuze and the Ruin of Representation*. Berkeley, Los Angeles and London, University of California Press.
Ong, A. and S.J. Collier. (2005). *Global Assemblages: Technology, Politics, and Ethics as Anthropological Problems*. Malden, MA, Blackwell Publishing.
Patton, P. (2000). *Deleuze and the Political*. London, Routledge.
Pedersen, M.A. (2007). 'Multiplicity without Myth: Theorising Darhad Perspectivism', *Inner Asia* 9 (2): 311–28.
Pickering, A. (1995). *The Mangle of Practice – Time, Agency, and Science*. Chicago, University of Chicago Press.
Pickering, A. and K. Guzik (eds). (2008). *The Mangle in Practice: Science, Society and Becoming*. Durham, NC, Duke University Press.
Prigogine, I. and I. Stengers. (1984). *Order Out of Chaos: Man's New Dialogue With Nature*. New York, Bantam Books.
Protevi, J. (2001). *Political Physics: Deleuze, Derrida and the Body Politic*. London, Athlone.
Rajchman, J. (2000). *The Deleuze Connections*. Cambridge, MA, MIT Press.
Riles, A. (2001). *The Network Inside Out*. Ann Arbor, University of Michigan Press.
Rodowick, D.N. (1997). *Gilles Deleuze's Time Machine*. Durham and London, Duke University Press.
Sahlins, M. (1972). *Stone Age Economics*. Chicago, Aldine-Atherton.
Serres, M. (1982). *Hermes: Literature, Science, Philosophy*. Baltimore, Johns Hopkins University Press.
Serres, M. and B. Latour. (1995). *Conversations on Science, Culture, and Time*. Ann Arbor, University of Michigan Press.
Shapin, S. and S. Schaffer (1985). *Leviathan and the Air-pump – Hobbes, Boyle, and the Experimental Life*. Princeton, Princeton University Press.
Smith, D.W. (2003). 'Deleuze and the Liberal Tradition: Normativity, Freedom and Judgment', *Economy and Society* 2, 299–324.
Smith, B.H. (2005). *Scandalous Knowledge: Science, Truth and the Human*. Edinburgh, Edinburgh University Press.
Spinoza, B. de. (1959). *Ethics*. London, J.M. Dent.
Stengers, I. (1997). *Powers and Invention*. Minneapolis, University of Minnesota Press.
———. (2000). *The Invention of Modern Science*. Minneapolis and London, University of Minnesota Press.
Strathern, M. (1988). *The Gender of the Gift: Problems with Women and Problems with Society in Melanesia*. Berkeley, Los Angeles and London, University of California Press.
———. (1991). *Partial Connections*. Lanham, MD, Rowman and Littlefield.
———. (1999b). *Property, Substance and Effect: Anthropological Essays on Persons and Things*. London, Continuum Press.
Traweek, S. (1988). *Beamtimes and Lifetimes: The World of High Energy Physicists*. Cambridge and London, Harvard University Press.

Tsing, A.L. (2005). *Friction: An Ethnography of Global Connection*. Princeton, NJ, Princeton University Press.

Turner, V. (1967), 'Themes in the Symbolism of Ndembu Hunting Ritual', in J. Middleton (ed.) *Myth and Cosmos: Readings in Mythology and Symbolism*. Garden City, NY, Natural History Press, pp. 249–69.

———. (1974). *Dramas, Fields, and Metaphors: Symbolic Action in Human Society*. Ithaca, Cornell University Press.

Tyler, S. (1984). 'Post-modern Ethnography: From Document of the Occult to Occult Document', in J. Clifford and G.E. Marcus (eds) *Writing Culture – The Poetics and Politics of Ethnography*. Berkeley, Los Angeles and London, University of California Press, pp. 122–41.

Viveiros de Castro, E. (1998). 'Cosmological Deixis and Amerindian Perspectivism', *Journal of the Royal Anthropological Institute* 4 (3), 469–88.

———. (2003). 'And'. After-dinner Speech Given at "Anthropology and science", 5th Decennial Conference of the Association of Social Anthropologists Association', *Manchester Papers in Social Anthropology* 7.

———. (2004) 'Exchanging Perspectives: the Transformation of Objects into Subjects in Amerindian Ontologies'. *Common Knowledge* 10 (3): 463–84.

———. (2005) 'From Multiculturalism to Multinaturalism', in M. Ohanian and J.-C. Royoux (eds) *Cosmograms*. New York, Sternberg Press, 137–57.

———. (2008) 'Xamanismo Transversal: Lévi-Strauss e a Cosmopolitica Amazônica', in R.C. de Queiroz and R. Freire Nobre (eds) *Lévi-Strauss: Leituras Brasileiras*. Belo Horizonte, Editora da UFMG, pp. 79–124. Presented in English as 'Transversal Shamanism: Form and Force in Amazonian Cosmopolitics'. Lecture to the Department of Social Anthropology, University of Cambridge, 26 October 2007.

Wagner, R. (2005). 'Afterword: Order is What Happens When Chaos Loses its Temper', in M.S. Mosko and F.H. Damon (eds) *On the Order of Chaos: Social Anthropology and the Science of Chaos*. New York, Berghahn Books, pp. 206–49.

Weiner, A.B. (1992). *Inalienable Possessions: The Paradox of Keeping-While-Giving*, Berkeley. University of California Press.

Whitehead, A.N. (1929). *Process and Reality: An Essay in Cosmology. Gifford Lectures Delivered in the University of Edinburgh During the Session 1927–28*. New York, Macmillan Press.

Winner, L. (1993). 'Upon Opening the Blackbox and Finding It Empty – Social Constructivism and the Philosophy of Science', *Science, Technology and Human Values* 18 (3), 362–78.

Wise, J.M. (1997). *Exploring Technology and Social Space*. London, Sage.

Woodhouse, E., D. Hess, S. Breyman and B. Martin. (2002) 'Science Studies and Activism: Possibilities and Problems for Reconstructivist Agendas', *Social Studies of Science* 32 (2): 297–319.

Part I
Deleuzian Sciences?

Chapter 1
Experimenting with *What is Philosophy?*
Isabelle Stengers

Perplexities

'It is in their full maturity, and not in the process of their constitution, that concepts and functions necessarily intersect, each being created by their own specific means' (Deleuze and Guattari 1994: 161). In other words, scientists do not need philosophers, and philosophers should not intervene when scientists are at work, or are facing new troubling questions, even if it may seem obvious that the elucidation of philosophical presuppositions could play a role, and even if it seems quite desirable that scientists experiment with new philosophical possibilities. Functions, as associated with scientific creation, and concepts, as associated with philosophical creation, indeed intersect but only after they have achieved their own specific process of self-fulfilment, that is, also after they have fully unfolded their requisites and consequences and, as such, do not entail that philosophers and scientists share a common concern about the questions arising from this unfolding. 'Philosophy can speak of science only by allusion, and science can speak of philosophy only as of a cloud. If the two lines are inseparable it is in their respective sufficiency, and philosophical concepts act no more in the constitution of scientific functions than do functions in the constitutions of concepts' (ibid.).

Taking a position that sounds like a biblical prohibition, 'Thou shall not mix' immature creations, Deleuze and Guattari seem to turn their backs against all those who had promoted them as the thinkers of productive connections, the creation of deterritorializing processes escaping fixed identities, transgressing boundaries and static classifications. The 'sufficiency' of the philosophical and scientific lines has also been a matter of disappointment for those who took for granted that Deleuze and Guattari would be allies in the debunking of the self-proclaimed autonomy of science and philosophy, underlining the open character of the process of the constitution of scientific enunciations, as well as its undetermined boundaries with politics, economics and cultural imperialism. They anticipated a joyful celebration of experimentations that subvert its very identity, that undermine the very persona of the philosopher. Instead, they got exemplifications from so-called 'great philosophers', Plato, Descartes, even Kant. As if, when the question 'What is philosophy?' was directly at stake, Deleuze had chosen to side with his great forerunners and forget his allies in deterritorialization.

As if philosophy itself, as the work of Dead White Males, was suddenly innocent of any connection with power, gender, imperialism, and so on.

To those matters of perplexity and disappointment, I would add my own, which concerned the very 'modern' character of the tripartition between philosophy, art and science, whose complementary lines seem together to define the notion 'creation'. Why this privileged connection between creation and modernity?

In producing such questions, *What is Philosophy?* confronts its reader with an alternative between two lines of thought. One may try to understand what felt like a betrayal in the terms of the book. Alternatively, one may experiment with 'outside' ingredients, which, if the experimentation is not a failure, should connect different aspects of the book that would otherwise appear to be mutually independent. In any case, what matters is to follow Deleuze's own advice: we should be interested in tools for thinking, not in an exegesis of ideas. An idea is always engaged in what he called a matter; always a specific one. An idea needs to be engaged in this way in order to enable a process of articulation of how and why this idea indeed matters and what kind of difference it makes: the process that Deleuze calls 'actualization' or 'effectuation'.

Before following this second line of thought, I shall describe what is entailed by thinking in terms of the first line. I start by recalling that for Deleuze and Guattari the question 'What is Philosophy?' is not a general one. It is a question they posed at 'that twilight hour when one distrusts even the friend' (Deleuze and Guattari 1994: 2). And it is a question about a threatened practice, the beginning of which they associate with 'contingent reason' and the end of which may also be contingent: no 'natural death' but a destruction. In other words Deleuze and Guattari do not define philosophy as a transcultural, transepochal feature of humanity as such (a Chinese philosophy, an African philosophy ...). When they talk about the need for a 'pedagogy of concept', we must understand that 'concepts' are irreducible to 'expressions' of thought, that they are something you need to encounter and experience in order to understand that very particular adventure of thought that is called philosophy. And that nobody would 'miss' philosophy if ever the conditions for this encounter disappeared.

Obviously pedagogy is not, in this case, a matter of faithful transmission; rather, it is a matter of relays. Relay transmission is always contingent as it implies both a taking over and a handing over. The taking over is always a creation, but the act of handing over also requires a creation. As Deleuze recalled in his *Abécédaire*,[1] it was the event of encountering concepts that produced him as a philosopher. To create the concept of concept, as distinct from science's functions and art's blocs of sensation, is to create and hand over what makes up the particular necessity of philosophy.

However, the point is not, or not only, the survival of philosophy. The point of *What is Philosophy?* is our 'lack of resistance to the present' (Deleuze and Guattari 1994: 108), a lack of resistance that science and probably art also share, entailing the very strong likelihood that they also may well be destroyed. Learning how to resist is a task that tolerates no economy. No great masterword (*mot d'ordre*) designating a common enemy may spare those who belong to a threatened practice from asking what kind of specific vulnerability this enemy is exploiting since he (or it) does not need to use violent, repressive means. In order to become the witness for their threatened practice, philosophers must speak concretely, that is, situate their practice among other threatened practices, each from the point of view of its own weakness and capacity to resist. When calling for the seemingly modest task of a 'pedagogy of the concept', but also when speaking about art and science, Deleuze and Guattari speak concretely. They speak about each practice's own specific 'bad will', which forces the practitioner to think and create, as opposed to 'good will', which facilitates thinking through consensual evidence (even the consensual evidence according to which our time would demand a general subversion of identities: to 'distrust even the friend').

At the twilight hour when it was written, at a time when relays can no longer be taken for granted, *What is Philosophy?* asks us to consider what Deleuze's favourite 'conceptual persona', the idiot, keeps saying while others hurry towards consensual goals: 'there may be something more important'. 'Something more important' does not mean something that would transcend our disagreements and reconcile us around a sacred cause, for instance the survival of philosophy. The idiot is unable to mobilize or convince; but perhaps he can slow down the mobilization and make some mobilized certainties stutter.

The idiot will never acknowledge that somebody has correctly understood *what* was more important. It is the task for everyone to learn and feel where and how his or her own slowing down and stuttering actually happen. The experimental reading I shall now propose stems from the feeling that Deleuze and Guattari were addressing their epoch, that is, their friends, while distrusting them, thereby asking us to think with the epochal fact that bad will as such can no longer be taken for granted. Continuing Deleuze's seemingly 'acritical' turn, his apparent forgetting that he had been thinking all his life 'against' the image of thought associated with the 'great philosophers' whom he portrays as creators in *What is Philosophy?*, my reading will take the reader towards a still more acritical position. It may well be that what we vitally need now is to honour what forces us to escape goodwill and consensual thought; to honour that which indeed causes us to think, each with diverging means. And it may well be that what we have to honour will designate us as survivors, having to disentangle ourselves from all the words that would ratify that survival as 'normal', and in particular from those 'critical words' forged in order to sever any relation between the survivors and those practices that 'deserved' to be destroyed.

Science with a Beard

I shall first address Deleuze and Guattari's characterization of science as a creation of functions as referring to matter of facts,[2] things and bodies. I shall not, however, dwell upon this distinction, even if it is an important one. Indeed, it offers a line of escape from the Great Sad Problem of scientific reductionism and its poisoning consequence for philosophy, with philosophers seeing it as their sacred task to defend human values, experience or responsibility against their reduction to scientific 'objectivity'. When a scientist affirms that experience should be and will be naturalized, explained (away) in 'scientific terms', as derivable from the 'state of the central nervous system', we recognize the inexorable advance of scientific knowledge and usually forget to ask about the possibility to define the brain as a 'state', with well-defined variables. I shall rather consentrate on the striking contrast with *A Thousand Plateaus* and its opposition between Royal Science, that is, unquestionably, the science of functions, and nomad sciences. Why do nomad sciences appear nowhere in *What is Philosophy?*, why is scientific creation enclosed in a definition that limits its relevance to what can be framed in terms of functions? A reader of *A Thousand Plateaus* can only be astonished. I was such a reader, and it is this astonishment that led me to my present reading of *What is Philosophy?*

I need now to complicate the problem by recalling another aspect of the situation, namely Deleuze and Guattari's suggestion that it is only in their full maturity that philosophy and science may intersect. Again the idea of 'maturity' is relevant for Royal Science only, but Deleuze and Guattari seem also to create a privilege for what is usually called 'science made' against the vivid, open, risky construction of 'science in the making', which most contemporary studies take as the relevant access to science. In order to dramatize this choice, I shall refer to the contrast between 'science made' and 'science in the making' as characterized by Bruno Latour in his well-known *Science in Action* (1987) by a double, Janus-like figure. One face is that of a beardless youth describing the risky production of scientific facts and their social constructive dimensions. This production requires the coming together of people whose interest must be gained and who participate in the very definition of the meaning and importance of the scientific facts. The other face is that of an old bearded man explaining the robustness of science by its truth, by its objectivity, by its respect of settled matters of fact, and so on.

This Janus-like figure may be sufficient to explain why Deleuze and Guattari claimed that philosophers should refrain from intervening in the collective construction of 'science in the making', even if the young beardless scientist is quite ready to welcome them, to quote them and to gain their interest. They should resist the temptation, resist being seduced by the openness of 'science in the making', and also resist believing in the promises of so-called 'new sciences' contradicting the closed, dogmatic character of 'science made'. Indeed the two faces offer no contradiction but a contrast, a contrasted unity. The kind of science that the youth has learned is the bearded one. He is speaking about a problem in construction but he knows that, if he is to succeed, if the story of the construction is to be told as the story of a scientific achievement, it will be told in the terms of the bearded old man. In other words, the dreams of the youth, his ambitions, are bearded ones. If he succeeds and gets the beard of his dreams, philosophers will be left outside because the successful, stable, 'mature' definition of matters of fact will be related to science's own specific means.

Deleuze and Guattari thus ask philosophers to resist understanding a description such as Bruno Latour's as a denunciation, understanding the claim of the bearded old man as lies, which hide the truth that 'matters of fact' are really just socially stabilized states of affairs. They ask us to resist the temptation to state that philosophers are left outside 'mature science' because the 'mature' scientists have acquired the social power to claim that their results are 'purely scientific'. When Deleuze and Guattari defined the 'creation of scientific function by science's own specific means' they certainly did not agree with the old bearded explanation, since this explanation denies creation. However, they nevertheless asked us to relate science as creation to the 'specific means', associated with the possibility for a scientist to get a beard.

Of course, the temptation to denounce the bearded old face is strong. I would even add that I cannot resist it when the contrast between the two faces is not alive – some so-called sciences do indeed seem to be born with a beard. This is why I shall concentrate on experimental sciences; in their case we may certainly imagine, wish for and struggle for a less dissociated or amnesic personality than the Janus figure, for a bearded old man who would remember and celebrate the adventurous, intricate constructive processes that any scientific achievement entails, instead of describing the achieved result as the direct consequence of a normal, rational method. This may indeed appear as the most important challenge, in political terms, because the price paid for the reduction of experimental achievement to a normal, rational operation is the general authority attributed to such an operation. This is the definition of science and scientific expertise as what I would call the thinking head of mankind. It is because of its resistance to this figure that social constructivism may so easily be identified with political emancipation against the authority of science.

Bearded dreams may well entail important problems of political power, but following my reading of the original 'idiotic' political stance of *What is Philosophy?* there is something that is still more important. Indeed, the result of the denial that science would have 'specific means' is that, whatever the scientific proposition, we know how to resist. This is why many readers will have identified Bruno Latour's problematic contrast between the two faces of the scientist as 'social constructivism': they 'recognized' the possibility of deriding the old bearded face, of claiming that behind any scientific (matter of) fact there is a state of affairs dressed with the social power to parade as authorizing scientific claims. Yet this is precisely the path Deleuze and Guattari refused to take when they chose to celebrate the mature scientific functions (and matters of fact) as a creation (science is a creation of functions).

A socially stabilized state of affairs, having acquired consensual authority, allowing scientists to feel that they know what they are saying or that they can define what they are observing, is the very characterization of what Deleuze and Guattari proposed to name 'functions of the lived' (*fonctions du vécu*): functions whose arguments are consensual perceptions and affections. For those functions, there is no creation, only recognition. Deleuze and Guattari were unsure whether all the human sciences should be included in this category: however sophisticatedly presented or statistically verified, they would merely constitute scientific opinion. But they did not hesitate with regard to logic as it came to dominate the philosophy that followed the route marked out by Frege and Russell (Deleuze and Guattari 1994: 135): logicism heralds the very triumph of goodwill functions depending on states of affairs, functions whose argument depends on consensual recognition. For Deleuze and Guattari, identifying science or whatever other practice as a question of 'states of affairs only' is not a 'lucid statement' but is the denial of their relation with 'creation'. There is certainly a strong appeal in debunking science by relating it to 'states of affairs'. But it is a consensual appeal[3] that offers no possibility of resisting the possibility that 'functions of the lived' may eventually come to define everything. Such a 'lucid' identification has a powerful taste of truth, but this is the poisoning taste of resentment: it means telling scientists: 'Wake up, you are just like everybody else'. And, as is always the case with resentment, it participates in the destruction of what is more important: namely the capacity to resist.

I shall certainly not take as a confirmation of my thesis the dreadful historical irony that social constructivism may be described as unwittingly collaborating in the destruction of those very aspects of science that it derided. As we know, scientists are now asked more and more insistently to renounce their dreams and to deserve the money they get. And in order to deserve money they are asked to forget about the distinction between scientific matters of fact and states of affairs. For instance, biotechnology, and the possibility to insert new genes in a genome, in no way means that biology would be able to define the function articulating the so-called genes with the features they are meant to explain. But this does not matter: what matters is that the production of genetically modified organisms (GMOs) has a very interesting strategic meaning for industrial states of affairs. This is not a confirmation that implies that the social constructivist analysis would be 'objectively' guilty, as Stalinists would put it. But it confirms that we live in a dangerous world and that, when one takes the path of denouncing and deconstructing the dreams of others, there is always the danger of discovering that one has strange bedfellows.

Resisting Social Constructivism

'There is something more important', 'we lack resistance to the present'. 'We', here, means all of us, whatever our (good-) will. Each practice is weakened by its own poisons, is infected by its own lack of resistance. The dissociated personality of the Janus-like scientists, the way the bearded old man describes as a matter of general methodology the scientific definition of a matter of fact, may indeed poison scientists and make them unable to resist the reduction of their practice to technoscience. But we are also poisoned, lacking resistance to the present, when we denounce the bearded dreams of science without creating the means to resist being joined to others who aim to destroy the dreamer. Resistance is a matter of creation, not of sincerity. Deleuze and

Guattari were not 'sincere' when celebrating mature science; they did not participate in a consensual belief in the 'autonomy of science'. By celebrating mature science as creation, they endeavoured to create philosophical means with which to tell another story: to escape the critical deconstruction of the claims of a bearded science. It is this process of creation I wish to continue.

Let us begin again, explicitly taking into account what Deleuze and Guattari seem to ignore, namely the social constructive activity of science in the making. How do we characterize (mature) scientific functions and their matters of fact as creation, that is, how do we resist viewing science as a simple case of social construction? When a scientific function has been created, how do we recognize that something new has entered the world?

Again, Bruno Latour helps us here. In *Pandora's Hope* (Latour 1999: 99–108), he describes the kind of 'state of affairs' the 'young' unbearded scientist must organize in order to succeed in meeting the demands of a 'science in the making'. This is a complex task, as it includes four kinds of ongoing, distinct and correlated processes of social construction. In order for his work to be possible, to gain importance and to achieve consideration, an innovative scientist has to form *alliances* with state or industrial powers, so as to get them to decide that they indeed need the kinds of results he is working to obtain. He has to achieve *academic recognition*, meaning that the academic demands and criteria for the new innovative field will be relevant ones, as autonomously produced and discussed by colleagues in this new field. He has to succeed in *mobilizing* the world, which involves getting the resources needed (i.e. the relevant instruments but also ambitious, innovative and competent students with interest in and loyalty to the new field). Finally, he has to produce a *public representation* of this field; to have it accepted as scientifically legitimate, as rationally answering important questions and/or as promising positive consequences for human development.

None of these four processes of construction are by themselves specific. Their correlation is somewhat more specific, because if one of them fails then the others will also tend to be problematized. But what makes them properly scientific is a fifth ongoing activity, which Latour characterizes as the making of *Links and Knots*. This designates the kind of activity most scientists would define as what truly matters, the actual production of those very matters of fact that the bearded old scientist will later claim are *sufficient* to explain the scientific achievement.

We could describe the first four activities, as Latour characterizes them, as constructing a 'state of affairs', creating a relation with an outside that must be both actively interested and also located at a distance (with a different kind of distance in a different kind of space for each one). But the point of this 'state of affairs', what designates science's 'own specific means', is precisely that this construction should not be reduced to a mere 'social construction'. Indeed, in all cases the distance means that whatever the success of those four processes of construction, it must be possible to present and describe them as nourishing the making of Links and Knots, as conditions for the scientific achievement, not the explanation of this achievement. If one claims that this is a matter of presentation only, Latour's description will be reduced to simple social constructivism and lose any relevance for reading *What is Philosophy?*

Avoiding this claim does not mean accepting the kind of easy separation that the art of distances is meant to promote. But it means addressing the making of Links and Knots as the point of 'idiocy' proper to science, as what matters for scientists, what they have to protect, what makes the specificity of a scientific achievement.

What does it mean to achieve Links and Knots? Scientists are certainly not linked together by their common bearded submission to rationality or objectivity, or by some good-will that explains how they are able to listen to each other and respect the rules of rational discussion. As I shall try to show, the link is not among humans as such, but exists only because some humans confer on the creation of reliable knots (i.e. the creation of reliable references between function and matter of fact) the power to link them, to force them to interact and entertain the kind of agonistic cooperation on which this achievement depends. When Galileo wrote that one man will win against a thousand rhetoricians, whatever their gift for persuasion or the authority of their references, if this one man has the facts on his side, we usually recognize it as some kind of positivist statement. And indeed Galileo was in the process of building the first public representation of experimental science, producing a state of affairs where experimental facts claim the power to silence both philosophers and theologians. But we should not forget that the Galileo who was writing was himself the product of the first experimental achievement, the first experimental knot.

In *The Invention of Modern Science*, I characterized this achievement as the ability to short-circuit the sceptical argument that refers any general statement to the power of fiction. I described the event of the experimental invention that produced Galileo as its spokesperson as 'the invention of the power to confer on things the power of conferring on the experimenter the power to speak in their name' (Stengers 2000: 88). Power intervenes three times in this description, each time, as I shall now show, with a different meaning.

'Invention of the power to confer on …' refers to Galileo in the very process of discovering the power of the first experimental device, the inclined plane. This device gave him the power to transform a usual state of affairs, a consensual function of the lived perception of falling bodies, into a scientific matter of fact correlated to a mathematical function. Indeed, the power of the inclined plane was to transform the fact that heavy bodies are perceived to fall into an articulated fact, defined in terms of independent variables, variables whose value can be modified at will, and the articulation of which produces a functional (state) description. In other words, the reference of a mathematical function to an experimental matter of fact is neither some kind of right belonging to scientific reason nor an enigma, but actually the very meaning of an experimental achievement.

'… a thing having the power of conferring on a human …' is the very definition of what the experimental 'knot' achieves when tying a matter of fact to a scientific function. 'Knot' is a very happy word and this achievement is a case of what Deleuze and Guattari called a 'marriage against nature'. The idea that science would 'naturalize' anything is complete nonsense. It would mean attributing to nature what we must describe as an event, the local entanglement between two lines that have nothing general in common. Before the event, falling bodies and the problem of the kind of knowledge we can gain about things were mutually indifferent. There was only a discursive reference, among others, to the perceptual fact that bodies do fall down. Such a falling down had in itself no power to force thinking. All heavy bodies fall down. This apple is a heavy body. Thus this apple will fall down if I open my hand. After Galileo, you no longer have apples, or a tree crashing to the ground, or a man falling from a window. You have something new: a Galilean body, a body that can exist nowhere but in the lab or in the sky, since its motion must approximate a frictionless one in order for the function to have any power of definition. Correlatively, the line of human discursive argumentation about the definition of valid knowledge as opposed to interpretation and fiction has bifurcated. One particular interpretation has received a reliable witness. Galileo no longer has any need to argue; he is able to turn his back on his human brothers, to cut any kind of intersubjective debate. The experimental device gave the Galilean body the power to allow Galileo to remain mute, to show just the facts.

'. . . the power to speak in their name'. It is because Galileo can present himself as essentially mute that he can claim that he has the power to speak in the name of the falling bodies. He is just representing the thing. The so-called objective scientific representation is an event because it may claim to be authorized by what is represented, while what is represented has no human voice. But who will be interested in this claim? Who will celebrate as an event the fact that Galileo is able to represent the way in which a body falls in a frictionless environment? Galileo needs colleagues who will take as primordial the verification that a knot has indeed been created, that none of their objections can defeat it. He needs colleagues who will accept being linked not to him, not because of his persuasive power, but by the production of the knot. In order for the specific character of his achievement to be verified, he needs not the goodwill of colleagues agreeing with him but the specific bad will of colleagues for whom what matters first is to test the reliability of the witness he claims to have produced.

In order to celebrate the successful knot, scientists will sometimes announce that 'Nature has spoken.' This triumphal statement is obviously misleading, but the way in which it is misleading is important. Indeed, the point is not the traditional one that things are mute and only humans speak. The point is that if nature had indeed spoken, the event would concern all humans, while the knot, as a 'marriage against nature' is always a local, selective event, the making of a spokesperson not for nature but for an experimentally framed aspect of what we call nature. The only ones concerned are those who belong to the two lines, those non-humans that can effectively, that is experimentally, be defined as reliable witnesses, and those humans – those I call competent colleagues – who will consider it as crucial to their own active practice to verify that a colleague was indeed authorized to claim the triple power of the achievement. Those and only those will be linked by the event as a matter of collective concern, exploring the consequences, testing the 'if ... then maybe ...' that may follow as eventually entailed by the event.

A 'marriage against nature' is never between 'man' and 'nature', which would mean a convergence, the access to a finally adequate knowledge. The marriage knot produces a divergence: it links these kinds of humans, endorsing very strong specific obligations, and those kinds of phenomena, or framed aspects of 'nature' verifying very selective requirements.

Science, Philosophy and the Public

The production of Links and Knots as the fabrication of an actively diverging adventure echoes Deleuze's claim that there is no relativity of truth but there is truth only of what is relative (Deleuze and Guattari 1994: 130). There is a scientific experimental kind of truth because science is relative to the adventure of the creation of Links and Knots, to the creation of knots and the production of links as what scientists explore together. The important point here is that there are many kinds of adventures, and each has its own truth and its own kind of loyalty, as it affirms its own diverging value. We may think here of the adventure of making a movie, of writing a text, and also of an alpinist's careful and risky climb, or of a mathematician in the process of producing a demonstration, or even of a judge hesitating about his or her judgment. The important point is that none of these adventures needs to belittle the other ones in order to affirm itself. Each of them is by definition a minority adventure, characterized by Deleuze and Guattari as that which does not dream of becoming a majority. And it is precisely because a minority collectively produces a divergence without a dream of convergence, of representing a future majority or consensus, that transversal connections are possible; a writer can understand something about an alpinist's discipline, or a mathematician about the judge's selective and creative processing of a case.

This is how I understand Deleuze and Guattari's proposition about the complementary lines of science and philosophy (1994: 159). Science would actualize and effectuate the event of the created knot, through its entangled processes of construction as characterized by Latour. And philosophy would counter-effectuate the event and isolate, that is create, its concept: Not reflecting on science but diverging from science. Indeed, such a philosophical 'counter-effectuation' would create by its own means what busy scientists so easily forget, namely the 'dignity of the event' that makes them busy. This is a case of vital communication between diverging adventures. It is also a demanding and selective issue for the sciences, as the question of the event they are in the process of effectuating would leave so much of today's so-called science 'born with a beard' aghast and speechless.

Returning to Latour's characterization of 'science in the making', it is crucial to note that the four other correlated constructive processes he distinguishes are marked by a contrast. A scientist will never tell a colleague, a demanding ally or a provider of mobilizing resources that 'Nature has spoken' and that they will have to listen. This claim is addressed only to those whose interests do not have to be won over, that is to 'the public'. It is for the public only that 'scientific rationality' as such matters, as its only chance to escape irrational belief and blind interests. And it is here that philosophers play a rather sorry role when they put their own means, whatever they may be, at the service of the image of thought that science requires. Scientists may well deride the means of philosophers (Stephen Weinberg wrote that 'to tell a physicist that the laws of nature are not explanations of natural phenomena is like telling a tiger stalking prey that all flesh is grass' (1993: 21–22)), but they usually welcome the possibility to identify science as valid knowledge, as opposed to subjective, irrational opinion, thus turning the scientific adventure into an epic, through which the scientist becomes the thinking head of mankind.

It may be interesting to approach the situation I have just described with the concepts that Deleuze and Guattari used in *A Thousand Plateaus* in order to describe how the war machine was 'encasted' (1992: 425) by the state. Indeed, when science is presented as an epic, scientists come to be respected as a 'caste' in the famous ivory tower. But, as a caste, they are captured, bound into the service of the (modern) state by the identification with the rational, apolitical aim they come to embody. The public representation of science is no mere ideology; it is the bond that makes it possible to transform the diverging creations of science into the kinds of convergent values the state needs for its own production. We can then go further and describe the present-day situation as the transition from being encasted to being appropriated. Read Donna Haraway's *Modest Witness* (1997), for instance, to feel what is now happening. The kind of adventure I have just characterized may be relevant for Galileo Galilei, Robert Boyle, Louis Pasteur and Frédéric Joliot-Curie, but emphatically not for contemporary biomedicine or biotechnology. Indeed, in those cases we can say that the art of distance required by demanding construction of Links and Knots has been swept away. It is no longer possible to tell the tale of those developments without having as leading protagonists those powerful allies who are quite ready to accept nominal definitions provided it means new patents and possibilities for industrial innovation.

The old bearded face in Latour's Janus-like figure, the face unable to celebrate a scientific achievement in words other than those that present the general, consensual triumph of rationality over opinion, is threatened with destruction. There could remain only one face, but it will not be the youthful face of the enterprising young scientist but the sad or cynical face of the one for whom the adventure of Links and Knots is a thing of the past. In other words, Deleuze and Guattari's characterization of science as creation indeed designated a threatened practice. And it may well be that the source of vulnerability for both science and philosophy is similar: in both cases the 'idiots', for whom there is something more important than public claims for rationality and authority, were unable to affirm their own specific 'bad will' that made them philosophers and scientists, and were unable to escape being bound by the majority claims they endorsed.

Can scientists resist the destruction that threatens them? Deleuze and Guattari ask that we think '*par le milieu*', meaning without going either to the root or to the final aim of a question, but rather taking into account the environment that this question both requires and creates. It may be that the 'milieu', which means the 'public' that is asked to accept science as authority, is the unknown of this question. While citizens are a consensual political fiction, the public is much more interesting because it poses the problem of a potential bad will, facilitating resistance to the present.

There may be a small, precarious possibility, which is part of our epoch, that a new kind of public is emerging, and that such a public may be able to make another kind of difference. We can see the stuttering emergence of new publics that have features of what Deleuze and Guattari called 'minorities', creating diverging lines of escape, through their own specific ways and means. These may be called 'objecting minorities', minorities who, not as their defining aim but in the very process of their emergence, produce the power to object and to intervene in matters they discover to be their concern. I would define the emergence of such divergent, problem-creating, multiple, 'empowered' minorities as an unknown of our epoch, one that does not concern the sciences specifically. The emergence of such minorities is the only possibility I can envisage against the probability of the scientific adventure being appropriated – that is, being destroyed.

The emergence of empowered minorities already produces a very interesting transformation in the 'milieu' of science, completely foreign to the 'fundamental questions' about the grounds or aims of knowledge, but effectively modifying the scientists' bearded dreams, which bound them to the state and capitalism. In recent years, the fact that public minorities were able to object has been a source of great surprise for concerned scientists. Among competent colleagues, objections certainly do matter. But the public was identified as something that was either 'against science' or 'supporting science'. Some scientists now begin to wonder whether the powerful allies who claimed to protect them against an irrational public were not instead using the public representation of science as opposed to 'mere opinion' in order to silence objections they are discovering to be important. They begin to feel the blind divide into the so-called scientific, objective definition of a problem, on the one hand, and ignorant irrational beliefs and traditional values, on the other, as a bond that now strangles them.

Celebrating Conventions?

The slight possibility of new emergent empowered minorities giving a new 'milieu' to science, providing them with a new environment that demands relevance against authority, also provides an opportunity for 'nomad sciences' to follow problems rather than framing them. This possibility, however, does not depend on philosophers. It must rather be produced against the philosopher's encasted role as the one who has to identify with scientific rationality. What then of philosophy? Are there specific, philosophical, means for philosophers to escape the encasted role that dooms philosophy to lack resistance to the present?

Here we may make a brief return to the problem of social constructivism. The fact that social constructivists entered the scene as those who were betraying the role that the public representation of science attributed to philosophers is something that requires attention. Indeed, you cannot betray a role without first claiming it, which in this case involves claiming it as your job to test the great opposition between science and mere opinion. But social constructivists did not see themselves as philosophers. It is thus possible to take up a role analogous to that of philosophers while claiming to belong to fields such as anthropology, sociology or cultural studies. It is sufficient to turn the generalities philosophers have produced about science and rationality into social and cultural generalities about human conventions, as a matter of the 'functions of the lived', as reducible to opinions, related as such to habits, settled interests and the balance of power.

The opposition between 'mere opinion' and what transcends opinion was a poison that modern science, since Galileo, inherited from philosophy and used to produce its own public representation. But when we come to philosophy it is not a matter of public representation only, it is the common-sense consensual conviction most philosophers inherit before they begin to think. To create philosophical means to resist this opposition between what would be reducible to human conventions only and what would transcend such reduction may well be what philosophy vitally needs today. And it is precisely such a creation that Deleuze and Guattari celebrate in the vibrant homage they pay to pragmatism as associated with the Anglo-American tradition of adventurous – non-analytical – empiricism:

> The English nomadize over the old Greek earth, broken up, fractalized, and extended to the entire universe. We cannot even say that they have concepts like the French and the Germans; but they acquire them, they only believe in what is acquired – not because everything comes from the senses but because a concept is acquired by inhabiting, by pitching one's tent, by contracting a habit ... Wherever there are habits there are concepts, and habits are developed and given up on the plane of immanence of radical experience: they are 'conventions'. That is why English philosophy is a free and wild creation of concepts. To what convention is a given proposition due; what is the habit that constitutes its concept? This is the question of pragmatism. (Deleuze and Guattari 1994: 105–6)

With each new convention, something new has entered the world, a new tent has been pitched, and not on a settled ground that would explain its stability but in a process of fractalization: the new ground is produced together with the new tent pitched upon it.

In each case the pragmatic 'it works' may be counter-effectuated as it entails considering what has been created in this case, the new 'it works' now resounding as an event which the convention, always this convention, effectuates. What may become a matter of consensus, the functions of the lived, settled interests and balance of power that will be used afterwards to explain or justify it, does not explain a convention. A convention is never explained by something else: it proceeds from a creation. As such, it fractalizes the ground following its own diverging demands, its own definition of what matters and how. And, yes, the scientific achievement of an experimental knot is a convention, but it is not 'only a convention'. As a 'marriage against nature', a tested knot between human argument and non-human, indifferent to the difference they make among humans, the 'convention' is a 'con-venire', a 'coming together' of heterogeneous modes of existence.

Deleuze and Guattari's celebration of the conventions at the heart of the empirico-pragmatic tradition does not give philosophers the task of being the defenders of silenced voices. Counter-effectuation has nothing to do with playing the righter of wrongs. But philosophers may create concepts that demand putting scientific achievements on the same plane of immanence as other diverging conventions, each demanding its own definition of what matters, and each related to the production of minorities with new actively diverging 'habits' that must be celebrated as something new entering the world and modifying it. When Deleuze and Guattari emphasized the need for a 'pedagogy of the concept', they were pragmatists because pedagogy means the creation of a habit, learning the 'taste' of concepts and the way one can be modified by the encounter with concepts.

An Acritical Turn

To escape the limits of scientific functions is probably what Félix Guattari had in mind when, in *Chaosmosis* (1995), he called for a new aesthetic paradigm. Such a paradigm does not give art, the third creative adventure characterized in *What is Philosophy?*, the status of an alternative paradigmatic model, but rather puts its emphasis on the power of feeling (*puissance du sentir*) and the pragmatics of assemblage and composition, which are indeed at the centre of what we call art, but concern the art of all deterritorializing processes, the capture and rendering perceptible, or thinkable, of forces that take the feeler or the thinker outside territories of settled habits. Guattari's paradigm is not art; rather it situates art as a practice of 'existential catalysis', bringing into existence what it cannot explain. As such, the 'aesthetic paradigm' is an ethical and a political one, as well as a processual one, as it concerns the coming into existence of new collective assemblages of enunciation and perception.

Guattari's aesthetic paradigm does not, however, conform to the tripartition of creation into philosophy, science and art that is proposed in *What is Philosophy?* Guattari's problem is rather to call for a rhizomatic population of creative, empowering processes, and he writes about a 'paradigm' because such a populated landscape would make impossible the 'objective' definition, by human and social sciences, of lived functions that look like scientific functions, and thus force new types of knots, which he associates with a nomad cartography, learning how to follow and not how to frame.

The contrast between Guattari's rhizomatic population and the triple definition of science, art and philosophy leads me back to the perplexity I experienced when reading *What is Philosophy?*. I could understand the way Deleuze and Guattari philosophically defined science and art. Those definitions did not express the philosopher's sovereign position, his or her ability to define human creation as such. Both art and science were defined from a double interrelated point of view. That is, on the one hand, from the point of view of the need to resist their own internal weakness, and, on the other hand, from the point of view of the way this weakness is threatening philosophy. It is vital for philosophy that both art and science learn to resist their own specific weaknesses and to affirm their creative divergence. But I could not defend myself against the feeling that this tripartition of creation was badly defended against a triumphant, progressive vision of our history, against an interpretation that would take them as the result of a purification process that would have finally led to the characterization of three pure lines of creation.

At the beginning of this chapter, I spoke about an 'acritical turn', which I associated with the Deleuzian figure of the 'idiot', the one for whom 'there is something more important', something that makes him or her resist and slow down. Here we must indeed slow down and remember that the very possibility of associating science, art and philosophy with creation, as done in *What is Philosophy?*, testifies to a depopulated world. These are practices that are now in danger of lacking resistance to the present, of being appropriated, but they are also the surviving ones, the ones that were tolerated and domesticated, 'encasted', while so many others were destroyed by what we call 'modernization'. As a result, these practices must also be considered from the point of view of the price they pay for their domesticated or tolerated survival. Deleuze considered the price philosophy paid when it was associated with the triple ideals, all of which imply judgement: contemplation, reflection and communication. I have considered the price science pays as presenting itself in the guise of the old fight of reason against mere opinion. I leave to artists the task to make explicit the price they pay, but will emphasize instead that Guattari's ethico-aesthetico-political paradigm may as well designate 'magic' as the neo-pagan and political activist witch Starhawk (1997) does: that is, not in supernatural terms but as an experiential and experimental art, daring to try to test what is required in order to produce ethico-aesthetico-political empowerment.

We no longer burn witches, but taking an interest in a processual paradigm may well mean facing accusations such as irrationality, superstition and regression. My conviction is that, as philosophers, we may, and indeed have to, 'counter-effectuate' this possible accusation. This is why I deliberately choose to take the risk of using the term 'magic', just as witches themselves take this risk. For them the very fact of naming magic as what they are doing is already an act of magic, producing the needed experience of discomfort that makes perceptible the power over us of the consensual functions of the lived. If contemporary witches take upon themselves such a shocking name, it is in order to produce the living, disturbing memory of the Time of Burning, the destruction of the Great Art, which happened in the very epoch when Man as the majority standard came to impose consensual functions of the lived, explaining away as illusions and superstitions every active divergence except the three surviving ones – philosophy, science and art.

Deleuze and Guattari, quoting Artaud, wrote that the writer writes 'for' the illiterate, 'for' an agonizing rat or a slaughtered calf (Deleuze and Guattari 1994: 109). Proposing that we have to think and feel 'for' those who did not survive, 'for' the burned witches, that is, to think and feel as survivors and as having paid the price for our survival, does not mean a nostalgic return to the past. It imposes the acritical decision to avoid ratifying both our survival and the brutal purification we inherit, according to which witches do indeed belong to the past: a ratification that precisely exhibits our lack of resistance to the great tale of progress that justified their destruction.

Naming witches and magic leads towards the question 'What is Philosophy?', with a doubly antagonistic move. On the one hand, it leads back to historical origins, which may certainly be associated with repudiation, with the choice of consensual arguments against dangerous powers. But, on the other hand, it leads towards the power of transformation that Deleuze associates with the philosophical concept when he writes, for instance, that thinking 'implies a sort of groping experimentation and its layout resorts to measures that are not very respectable, rational or reasonable … To think is always to follow the witch's flight' (Deleuze and Guattari 1994: 41). Indeed, it may be that if philosophy was able to survive its Greek origin, to resist many threatening moments, it was because it unwittingly and in a disguised manner captured something quite different from rational argumentation. It may be that the 'prephilosophical plane' it built on and secretly, unwittingly continued was inhabited not only by urban sophists, as is officially recorded, but also by those others whose not very respectable, rational or reasonable measures the sophist art of language had already urbanized. And it may well be that, if you separate philosophy from what it profits from and secretly, unwittingly continues, then you kill philosophy as surely as you kill science if you identify scientific functions with functions of the lived, and art, if you demystify it, that is, strip it of what recalls magic rituals.

Magic is one way to name what all creators know, that what empowers their creation is not theirs. The important point, whatever the name, is to affirm the debt, not an infinite one but one that must be honoured. This is the lesson I learned from the contemporary activist witches whose practices I discovered through Starhawk. They did not at all need to believe in a goddess as a supernatural, transcendent being. But they learned the pragmatic need for empowering rituals in order to honour a power that answers precisely to the process of empowering – the capacity to become able to resist the present – which the ritual is made to induce.

To honour is not to believe in a transcendent power but pragmatically to call forth what is required by any creation as a 'marriage against nature'. At the very end of his life, in *Some Problems of Philosophy*, William James wrote:

> We can and we may, as it were, jump with both feet off the ground into or towards a world of which we trust the other parts to meet our jump – and only so can the making of a perfected world of pluralistic pattern ever take place. Only through our precursive trust in it can it come into being. There is no inconstancy anywhere in this, and no 'vicious circle' unless a circle of poles holding themselves upright by leaning on one another, or a circle of dancers revolving by holding each other's hands, can be 'vicious'. The faith circle is so congruent with human nature that the only explanation of the veto that intellectualists pass upon it must be sought in the offensive character to them of the faiths of certain persons. (James 1996: 230)

Leaving the common, settled ground and the security of lived functions (including their critical deconstruction) is always a risk. This, however, is no argument against what William James calls the 'faith circle', which is not something we have 'discovered' but something non-modern traditions knew very well. They knew that parts of the world that come and meet some of our jumps may be devouring ones, and that some may become devouring ones if we do not know how to honour them when we have called them up. This may be why Félix Guattari called for an 'ecosophy', an active and creative care that is not limited to 'ecology' in the usual sense or to social processes. It includes those 'other parts of the world' that come and meet our jump, components we need to learn and call forth, but also to learn and care for.

Not wisdom but caution is what we need, wrote Deleuze and Guattari in *A Thousand Plateaus* (1992: 150). Even Antonin Artaud proceeded with caution. When writing, he was not a wild schizophrenic, he weighed and measured every word, and wrote about the danger of false sensations and perceptions. Not only did he experience such sensations and perceptions, but sometimes he believed in them:

> Staying stratified – organized, signified, subjected – is not the worst that can happen; the worst that can happen is if you throw the strata into demented or suicidal collapse, which brings them back down on us, heavier than ever. This is how it should be done: Lodge yourself on a stratum, experiment with the possibilities it offers, find an advantageous place on it, find potential lines of deterritorialization, possible lines of flight, experience them, produce flow conjunctions here and there. (Deleuze and Guattari 1992: 161)

This is the experience I tried to call forth, lodging myself on the stratum of royal science and its bearded dreams, as it is surprisingly celebrated in *What is Philosophy?*

Notes

This chapter is a substantially reworked version of the article 'Deleuze and Guattari's Last Enigmatic Message', published in *Angelaki: Journal of the Theoretical Humanities*, 2005, 10 (2), 151–67.

1. The *Abécédaire* was a series of interviews with Deleuze by Claire Parnet in 1988– 89. The interviews were filmed by Pierre-André Boutang, and were broadcast on the French Arte channel in 1994–95 (editors' note).
2. That is, in French, '*états de choses*', which the translators of *What is Philosophy?* have chosen to translate by 'states of affairs'. I would guess that this choice stems from the fact that the usual translation for '*états de choses*' is 'matter of fact,' and that 'matter of fact' belongs to an empiricist tradition that the translators could not endorse. 'States of affairs', in contrast, may refer to the lucid, social constructivist stance that, whatever scientists' claims, they can never escape states of affairs for some purified 'matter of fact'. Whatever their achievement, it will always refer to a 'state of affairs'. But then how to understand that 'things' or 'bodies' are not also 'states of affairs'? It may well be that here the translators stopped trying to understand (those pages are the most elliptical and obscure ones in the book, anyway).
3. Consensual does not mean dominant but may be connected with the majority/minority contrast proposed in *A Thousand Plateaus*. A position may well be affirmed by a quantitatively minority group and will be a majority position if it presents itself as what should be accepted by everyone, i.e. as a potentially consensual claim (even if such a consensus would mean defeating powerful illusions, ideologies, balances of power, and so on).

Bibliography

Deleuze, G. and F. Guattari. (1992). *A Thousand Plateaus*. London, Continuum.

———. (1994). *What is Philosophy?* London, Verso.

Guattari, F. (1995). *Chaosmosis: An Ethico-aesthetic Paradigm*, Bloomington, Indiana University Press.

Haraway, D. (1997). *Modest_Witness@Second_Millennium.FemaleMan_Meets_OncoMouse: Feminism and Technoscience.* New York and London, Routledge.

James, W. (1996). *Some Problems of Philosophy*. Lincoln and London, University of Nebraska Press.

Latour, B. (1987). *Science in Action: How to Follow Scientists and Engineers through Society*. Cambridge, MA, Harvard University Press.

———. (1999). *Pandora's Hope – Essays on the Reality of Science Studies*. Cambridge, MA, Harvard University Press.

Starhawk. (1997). *Dreaming the Dark*. Boston, Beacon.

Stengers, I. (2000). *The Invention of Modern Science*. Minneapolis, University of Minnesota Press.

Weinberg, S. (1993). *Dreams of a Final Theory*. London, Vintage.

Chapter 2
Facts, Ethics and Event

Mariam Fraser

> Of all the modern philosophers who tried to overcome matters of fact, Whitehead is the only one who, instead of taking the path of critique and directing his attention *away* from facts to what makes them possible as Kant did; or adding something to their bare bones as Husserl did; or avoiding the fate of their domination, their *Gestell*, as much as possible as Heidegger did; tried to get *closer* to them or, more exactly, to see through them the reality that requested a new respectful realist attitude. (Latour 2004a: 244)

This is the problem for Bruno Latour: that, in order to explore their conditions of possibility, science studies scholars in the end seem to have taken facts too much for granted and have assumed to know too well in advance what they are. In so doing, the facts that 'everyone else' could kick at, or bang on or sit down on seemed to disintegrate in their hands. This is the ironic conclusion, for '[t]he question was never to get *away* from facts but *closer* to them' (Latour 2004a: 231).

In this chapter I want to explore some of the sometimes different, sometimes overlapping ways in which the reality of facts is understood by Bruno Latour, Isabelle Stengers, Alfred North Whitehead and Gilles Deleuze. My intentions here are not at all to produce an exhaustive survey, or to come up with an ideal synthesis of these theorists' work in this area, or to 'compare and contrast' them. Instead, the argument in this chapter folds, unfolds and refolds around these authors with the aim of exploring what their different concepts or what the same concepts differently inflected can do. I want to ask where a few key terms – among them, relationality, exteriority, potentiality and virtuality – might lead, and how they might be made to matter. The discussion will be dominated by two attractors.[1] The first is event, the second is ethics.

Event has been used by many theorists – far more than I will refer to below – as a way of contesting the concept of bare fact that often dominates mechanistic (and common-sense) accounts of the world. An event in this context is not just something that happens. As a philosophical concept, it exists in relation to a specific set of problems, including the problem of how to conceive of modes of individuation that pertain not to being or to essences and representation but to becoming and effectivity. In this respect, event thinking can be understood to be part of an anti-reductionist project and, as such, is especially relevant with regard to the problematization of knowledge, and in particular to the philosophy of science.

Any discussion of the concept of event necessarily involves addressing far more than bare fact. In the first parts of this chapter, I explore how this concept aids Whitehead in his critique of the bifurcation of nature and, in particular, its role in Whitehead's, Latour's and Stengers' critique of the bifurcated relations between subjects and objects, primary and secondary qualities and facts and values, as these relations are often dramatized in the ideal of modern science. Having established, at some length, the use that these theorists make of event, I then push the analysis further by examining some of the implications of event thinking in relation not only to scientific facts, but also to value/s and ethics. Of course, this approach to ethics, via science, is not the only way to address the issue. Indeed, the limitations of the bifurcation of nature into facts and values that subtends much scientific thinking – in science as well as in other fields, such as economics (see, for example, Putnam 2004) – and the implications of that bifurcation for ethics are, and perhaps always have been, particularly noticeable. Ethical value is just as often (if not more often) identified, for instance, with the creativity of 'life *itself*' and not solely with values that are perceived to be imposed *upon* life. It is precisely because modern science claims a privileged relation to the facts of life, however (and, on this basis, its own privilege with regard to conceptions of the world), that I find it an especially fertile point of entry to value and, from there, to ethics. And indeed, the relation between ethics and science – or, more specifically, the relation of ethics to science – is a live and contested contemporary issue, as the burgeoning debates and critiques in and around bioethics suggest.[2]

Although my point of entry to ethics proceeds via science and turns, in large part, on Latour's and Stengers' different takes on Whitehead's notion of the bifurcation of nature, I should also add that there are themes in Whitehead's work that are rarely addressed by these two theorists, but which are more fully developed by Deleuze. I am thinking in particular of the points of resonance between Whitehead's concept of potentiality and Deleuze's concept of virtuality which, I shall argue, differ considerably from Latour's notion of exteriority. It is by way of Deleuze's conception of the relation between the virtual and the actual, a relation that also informs his understanding of the relation between problems and solutions, that I am able to further explore the implications of event thinking for ethics in ways that are relevant, but not exclusively so, to science. Indeed, in the final two sections of this chapter, I want to consider why it might be important for social scientists to attend not only to the actual domain but also to the virtual: first, because the concept of the virtual, in challenging the assumption that the social is the only valid level of explanation, extends critiques of social constructionist accounts of science; and, secondly, because this concept also provides a reason, a reason which is immanent to 'concrete fact', for asking about value. In this respect, as I shall argue, it further develops an ethics of social science research. My intentions in this chapter, then, are somewhat different from sociological critiques *of* ethics (and especially critiques of the style of ethical reasoning that is typical of a certain Anglo-American philosophy), of the 'ethicalisation process' (Barry 2004) and of bioethics (Evans 2002). Such studies are especially welcome in view of the increasingly important role that ethics is called upon to play in the contemporary scientific, technological and, especially, biomedical landscape. Nevertheless, in this chapter I want to make a positive argument *for* ethics, and to suggest that the concept of the event, augmented by the concept of the virtual, is useful in this task.

Prehension, Relationality, Reality

As Philip Rose explains, 'an absolute key to understanding Whitehead's work is the fallibility and revisability of his metaphysical scheme. Whitehead's attempt to develop a system of metaphysics should thus be seen not as a final statement concerning the nature of things, but rather as part of a larger ongoing historical project' (Rose 2002: 2). Necessarily so, for Whitehead was concerned not only with what he calls 'speculative metaphysics' – which addresses itself to the necessary conditions for the possibility of existence – but also with cosmology, with 'the *contingent* conditions of "things" as they happen to be' (Rose 2002: 3). One of the key contingencies installed by modern science, as far as Whitehead is concerned, is the 'bifurcation of nature' into subjects and objects and, relatedly, primary and secondary qualities. 'The sensationalist doctrine', as he calls it, rests on two problematic assumptions. In the first instance, it assumes that sense data do no more than signal (if they even manage that) to their existence. Passive and mute, they contribute nothing to meaning. The second dimension (the 'subjectivist principle') assumes that these inert facts are qualified and given meaning by a subject (a human mind, say), who organizes them according to a universal principle, such as rationality or morality.

Nevertheless, despite the 'genius' of the seventeenth century and the 'continued work of clearance' conducted in the eighteenth century, not everyone, Whitehead argues, accepts the opposition that underpins scientific realism (Whitehead 1985: 95). It is in English literature in particular that Whitehead finds representatives of 'the intuitive refusal seriously to accept the abstract materialism of science' (Whitehead 1985: 106) and, especially, the divorce of nature from value. His own aim therefore is to build a system of thought in which aesthetic value (for example) is as much a part of nature as is the mechanism of matter. For Whitehead, natural philosophy – and this well-known quote is cited in Latour's article in *Critical Inquiry* (2004a) – 'may not pick up and choose. For us the red glow of the sunset should be as much a part of nature as are the molecules and electric waves by which men of science would explain the phenomenon' (Latour 2004a: 244).

The complex historical genesis of the bifurcation of nature into primary and secondary qualities has been described, by Whitehead as well as by others, from a number of different angles (see, for example, Proctor 1991). In *The Concept of Nature*, however, Whitehead lays considerable emphasis on the part played by the systematic establishment of theories of light and sound in the seventeenth century, and in particular the connection that Newton made between light and colour. These transmission theories, as Whitehead calls them, put an end to 'the sweet simplicity' of 'the concept of matter as the substance whose attributes we perceive' (Whitehead 1920: 26) and dislodged the epistemological confidence that observation once guaranteed. For, while a colour may be perceived to be an attribute of matter, 'in fact' it is not. A gap thus opens up in Western philosophy and science between what *seems to be* (what is experienced by the subject) and what *is* (what is known as a fact), between the redness and the warmth of the fire on the one hand, and the conjectured system of agitated molecules of carbon and oxygen on the other. One of the principal aims of Whitehead's concept of nature is to address both the object of perception (which is the task that the philosophy of science sets itself), as well as the perceiver and the process, and the histories of their relations. For, if these entities are *not* understood to be related to each other, then, as Whitehead (taking the scientific neglect of aesthetic value to its logical conclusion) wryly notes:

> nature gets credit for what should in truth be reserved for ourselves: the rose for its scent: the nightingale for his song: and the sun for his radiance. The poets are entirely mistaken. They should address their lyrics to themselves, and should turn them into odes of self-congratulation on the excellency of the human mind. Nature is a dull affair, soundless, scentless, colourless; merely the hurrying of material, endlessly, meaninglessly. (Whitehead 1985: 68–69)

The concept of nature, therefore, must refer to *everything* that 'we observe in perception through the senses' (Whitehead 1920: 3). It must refer, as Isabelle Stengers says, not just to 'what we perceive and can identify, but [to] the whole indefinite complexity of what we are aware of, even if we have no words to name it' (Stengers 1999: 197). I shall be returning to this point in due course.

In an attempt to avoid the bifurcation of nature into subjects and objects and to illustrate instead their connectedness, Whitehead defines all things, or what he calls 'actual entities' or 'actual occasions', in terms of their relatedness. This is what an actual entity *is* in Whitehead's metaphysical system: a coalition into something concrete, a novel concrescence (or becoming) of relatedness or prehensions. Whitehead often calls prehensions 'feelings', although they are not emotions in any conventional sense and are not psychological, nor are they necessarily even associated with human subjects. Instead, prehension might be better understood as *a process of unifying*. It is by way of prehension, by way of processes of unification, that all actual entities and societies of actual entities come into existence. 'Feelings are variously specialized operations', Whitehead writes, 'effecting a transition into subjectivity ... An actual entity is a process, and is not describable in terms of the morphology of a "stuff"' (Whitehead 1978: 40–41). Nature is a complex not of 'things' per se, but of prehensive unifications. Importantly, the unity to which Whitehead refers is not given in a subject, a human mind, in consciousness or in cognition, but is rather 'placed in the unity of an event'

(Whitehead 1985: 114). Whitehead's prehensive unities 'precede' the bifurcation of nature not only into subject and object, but also into primary and secondary qualities. For this reason, the concept of subjective value also undergoes a radical transformation: '"Value"', Whitehead writes, 'is the word I use for the intrinsic reality of an event ... Realisation is in itself the attainment of value' (Whitehead 1985: 116).

Rather than pursue the implications of Whitehead's concepts of the event and of value now, I want to pause momentarily to consider some of the points of resonance between his and Isabelle Stengers' and Bruno Latour's understandings of reality, particularly in so far as they too privilege relationality (to a more or less radical degree). For instance, in her discussion of the notion of discovery in *The Invention of Modern Science,* Stengers describes the reality of America in terms that bear a striking similarity to the interwoven prehensions that are grasped and grasp themselves together in a unity: 'What other definition can we give to the reality of America, than that of having the power to hold together a *disparate* multiplicity of practices, each and every one of which bears witness, in a different mode, to the existence of what they group together?' (Stengers 2000: 98). Although Stengers' use of the notion of practices might be likened to Whitehead's emphasis on different modes of becoming (modes of becoming that will shape an entity's mode of achievement in its specificity), in fact, using a rather more Latourian vocabulary, she suggests that it is not the sheer number of witnesses that contributes to the reality of an entity such as America, but rather their heterogeneity: 'If the allies belong to a homogenous class, the stability of the reference only holds for a single type of test. America affirms its existence prior to the discovery of Columbus by the multiplicity of tests to which those who define their practice in reference to it have subjected it' (Stengers 2000: 98).

This is an 'answer' then – an answer that I shall be explicating in more detail below in relation to Stengers' understanding of the concept of an event – to the question as to whether America existed prior to its 'discovery', or whether 'the ferments (of the microbes)', in one of Latour's examples, 'exist[ed] before Pasteur' (Latour 1999: 147). It is the kind of question that haunts critiques, and especially constructionist critiques, of science and of the status of scientific objects (Are they real? Are they representations?) precisely because science aims 'at things that the passing of time cannot "make equal"' (Stengers 2000: 40). How can historians, Stengers ask, 'not think, *like the rest of us,* that the Earth revolves around the Sun'? (Stengers 2000: 42). And yet, she continues, the conception of reality in terms of bearing witness demands that the earth and the sun and the revolutions be understood to be absolutely specific to – and therefore contingent upon – the relations that constitute them. '[W]hoever doubts the existence of the Sun would have stacked against him or her not only the witness of astronomers and our everyday experience, but also the witness of our retinas, invented to detect light, and the chlorophyll of plants, invented to capture its energy' (Stengers 2000: 98). In so far as an entity is dependent upon relationality, upon its interconnectedness with other entities, its permanence – or endurance, as Whitehead puts it – cannot be guaranteed.

Latour ties, helpfully, I think, the problem of historicity to the bifurcation of nature into subjects and objects. The problem with the subject/object dichotomy, he writes, is that subjects and objects 'cannot share history equally' (Latour 1999: 149, emphasis omitted): 'Pasteur's statement may have a history – it appears in 1858 and not before – but the ferment cannot have such a history since it either has always been there or has never been there' (Latour 1999: 149). Herein, for him, lies the usefulness of the concept of an event:

> EVENT: A term borrowed from Whitehead to replace the notion of discovery and its very implausible philosophy of history (in which the object remains immobile while the human historicity of the discoverers receives all the attention). Defining the experiment as an event has consequences for the historicity of all the ingredients, including nonhumans, that are the circumstances of that experiment (see concrescence). (Latour 1999: 306)

Rather than conceding to the idea of bare and mute facts that lie waiting to be discovered by the active human agent and in order, instead, to grant activity to both actors *and* actants, Latour explores the associations and substitutions – that is, the connections and replacements – that occur between them as they come into existence. Reality is extracted, in Latour's terms, 'not from a one-to-one correspondence between an isolated statement and a state of affairs, but from the unique signature drawn by associations and substitutions through the conceptual space' (Latour 1999: 161–62). An entity does not secure a fixed ontological position by passing into an extrahistorical dimension. Rather, Latour is 'able to talk calmly about *relative existence'* (Latour 1999: 156), 'to define existence not as an all-or-nothing concept but as a gradient' (Latour 1999: 310). Both subjects and objects, or more accurately propositions, are characterized by a dynamic historicity, where historicity refers not simply to the moment of representation ('our contemporary "representation" of microorganisms dates from the mid-nineteenth century') or to evolution ('the ferments "evolve over time"'), but to '*the whole series* of transformations that make up the reference' (Latour 1999: 145, 146, 150). Each transformation defines an entity in its singularity: just as Whitehead claims that 'an electron within a living body is different from the electron outside it' (Stengers 1999: 202), so Latour suggests that '"air" will be different when associated with "Rouen" and "spontaneous generation" than when associated with "rue d'Ulm," "swan-neck experiment," and "germs"' (Latour 1999: 161).

Another, perhaps more technical, way of putting this would be to argue, as Whitehead does, that '[t]here is a becoming of continuity, but no continuity of becoming' (Whitehead 1978: 35). One of the implications of this claim is that it disputes the finality of those explanations of the world that 'privilege the continuity of the functions or patterns on which they depend' (Stengers 2002: 252) – a point that extends not only to tangible entities in the world, but also to space and to time. For Whitehead, once again, the recourse to time and space as a means of unifying nature

– for example the claim that the redness of the fire and the agitation of the molecules occur at the same time and in the same space – cannot suffice as an explanation for it demands that time and space be apprehended *independently* of the happenings that occur in time or of the objects that occupy space. Whitehead argues instead that, along with subjects and objects, space and time are also reified entities that are to be explained by the contingent and changing events from which they are abstracted.[3] An enduring entity – such as a molecule – does not move through time and space, nor do changes occur in space and time. Instead, motion and change are attributable to the differences between successive events, each with their own durations.

In so far as it is extensiveness that becomes (and not becoming that is extensive), there is, as Stengers points out, a strong contrast between the values of experimental science and of speculative philosophy: one is 'for' being and the other is 'for' becoming (Stengers 2002: 252–53). Indeed, Stengers argues that the atomicity of time was precisely the price, 'the speculative price', that had to be paid 'in order for philosophy to define itself "for" becoming' (Stengers 2002: 252). This does not mean, however, that the purpose of speculative philosophy is to act as a corrective, nor is it to *devalue* what scientists value (continuity, for instance). When Whitehead criticizes scientific method on the basis of the experiences that it fails to include, when he asks what it is that Wordsworth finds in nature that 'failed to receive expression in science', he does so, he underscores, 'in the interest of science itself; for one main position in these lectures is a protest against the idea that the abstractions of science are irreformable and unalterable' (Whitehead 1985: 103).

Although Whitehead might certainly have wanted to reform and alter scientific abstractions, it is arguable that his own level of abstraction and technicality makes his work difficult to translate into anything other than a most general programme. On the other hand, it may be precisely this 'difficulty' that enables Whitehead's work to be such a rich and influential resource for other critics.[4] There is no question, for example, as I shall be discussing below, of the impact of Whitehead's approach (or perhaps, more specifically, of the impact of a Stengers–Whitehead approach) on the expressly political – and, indeed, ethical – project that Latour outlines in *Politics of Nature*. Before beginning to address this project, I want to consider the way that Stengers deploys the concept of an event in order to re-conceive of the very relation between science and politics.

Politics, Science

The concept of the event is crucial in the context of Stengers' critique of scientists and of critics of science, for it offers a route out of 'the black hole' in which both parties, Stengers argues, often find themselves. The reason that they do so, she writes, is because scientists:

> if asked to explain, would describe the 'different from all other practices' in terms of privilege, and would distinguish science from other collective practices said to be stamped with subjectivity, instruments for the pursuit of different interests, guided by values that pose an obstacle to truth. Objectivity, neutrality, truth – all these terms, when used to characterize the singularity of the sciences, transform this singularity into a privilege. And this privilege, which confers on the sciences a position of judgement in relation to other collective practices, is also what the critics gathered together in the black hole transform, in their own way, into an instrument of judgement against the sciences. (Stengers 1997: 134–35)

It is not enough, for Stengers, for critics of science to draw attention to the ways in which this 'ideally' value-free discipline is 'in fact' riddled with various political, economic and other investments. Similarly, the claim that science is a social undertaking like any other (and here Stengers is undoubtedly referring to some of Latour's and other science studies scholars' early work) is problematic not only because it flattens science out and renders it equivalent to all other knowledges and practices – not 'different from all other practices' after all – but also because it establishes sociology as 'a superscience, the science that explains all others' (Stengers 2000: 4). For Stengers, the challenge is to respect the singularity of the sciences, without at the same time conceding to the perceived opposition between rationality, on the one hand, and 'illusion, ideology, and opinion', on the other (Stengers 1997: 135): 'Political engagement', she writes, 'is a choice, and not the result of a disappointment linked to the discovery of the political dimension of the practices that reason was supposed to regulate' (Stengers 2000: 60).

Rather than define science in opposition to politics, Stengers redefines politics. Or more accurately, she offers a definition of politics in terms of *cosmopolitics*. Her debt to Whitehead, in the following explanation of this term, is clear:

> The prefix 'cosmo' takes into account that the word *common* should not be restricted to our fellow humans, as politics since Plato has implied, but should entertain the problematic togetherness of the many concrete, heterogeneous, enduring shapes of value that compose actuality, thus including beings as disparate as 'neutrinos' (a part of the physicist's reality) and ancestors (a part of the reality for those whose traditions have taught them to communicate with the dead). (Stengers 2002: 248)

According to this definition, modern science is political through and through, not on account of its 'extra-rational' investments but because it has invented a new mode of 'togetherness', one which, specifically, problematizes the relation between fact and fiction. Rather than understanding this new 'use of reason' in terms of scientific discovery or progress, Stengers puts it under the sign of the event. The Gallilean event, she writes, was:

> capable of doing what it was no longer believed possible to do, celebrating the statements that lightheartedly cross the distance between 'nature' and polished balls rushing down a smooth, inclined plane. What is presented as having been reconquered in principle, if not (still) in fact, is precisely *something one believed to have been lost: the power to make nature speak*, that is, the power of assessing the difference between 'its' reasons and those of the fictions so easily created about it. (Stengers 2000: 81)[5]

At its most minimal an event, for Stengers, is the creator of a difference between a before and an after. Crucially, however, it is not the event itself that is the bearer of signification. Instead, all those who are touched by an event define and are defined by it, whether they are aligned with or opposed to it. In her words (and note how in keeping this description is with her conception of reality):

> [An event] has neither a privileged representative nor legitimate scope. The scope of the event is part of its effects, of the problem posed in the future it creates. Its measure is the object of multiple interpretations, but it can also be measured by the very multiplicity of these interpretations: all those who, in one way or another, refer to it or invent a way of using it to construct their own position, become part of the event's effects ... Only indifference 'proves' the limits of the scope of the event. (Stengers 2000: 67–68)

Indifference: feeling's own contrast. And yet the notion of an event is a provocation to feeling precisely in so far as it signals that something matters – that something has produced a variation or made a difference – *without* specifying what that something is or to whom or to what it will matter. It is impossible to draw up a list of the entities that enter an event in advance because identities and relations acquire definition through it.[6] Not only does the event not have a privileged representative therefore (science is not the domain of scientists alone), it is also impossible for any participant in an event, by definition, to stand outside it and pass judgement on it or to explain it *away* with reference to a history, culture or geographical area. As Stengers puts it, 'No account can have the status *of* explanation, conferring a logically deducible character to the event, without falling into the classic trap of giving to the reasons that one discovers a posteriori the power of making it occur, when, in other circumstances, they would have had no such power' (Stengers 1997: 217). Latour explicates this point about causality further: 'Not only should science studies abstain from using society to account for nature or vice versa, it should also abstain from using causality to explain anything. Causality *follows* from events and does not precede them' (Latour 1999: 152).

The above discussion has begun to address the ways in which the concept of event, as is used by Whitehead and, to a greater or lesser degree, by Stengers and Latour, can be mobilized as part of an anti-reductionist project that seeks to challenge the notion of bare and ahistorical facts, the distinction between primary and secondary qualities and the opposition between subjects and objects. If subjects and objects cannot be assumed to exist *prior* to the event and thus cannot claim any general validity, then the question of their existence and the nature of their identity and of their relations (their relation of opposition, for example) is no longer a philosophical one but, rather, a matter for practical investigation (Stengers 2000: 133–34). To argue thus is not to undermine or disrespect the achievements of science – be they methodological, epistemological or ontological – but rather to recognize the specificity of those achievements and the practices, risks and responsibility that enable them. This in itself serves to displace their privilege. As Stengers puts it, '"Science is different from all other practices!" For many scientists, this is a heartfelt cry, a cry that needs to be heard, *even if we remain free not to understand it exactly in the way that those who utter it would like*' (Stengers 1997: 134, emphasis added).

For Whitehead, as I have already noted, the singularity of an entity is derived from a multiplicity of diverse elements that are inextricably conjoined not by relations of cause and effect in space and time but by way of prehensive relations grasped in the unity of an event. One of the important implications of this point is that it displaces the need for any additional – or perhaps Whitehead would say any arbitrary – term to be introduced in order to explain the relations between things. Consider, for example, his critique of Newton's laws of motion:

> the notion of stresses, as essential connections between bodies, was a fundamental factor in the Newtonian concept of nature ... But [Newton] left no hint why, in the nature of things, there should be any stresses at all. The arbitrary motions of the bodies were thus explained by the arbitrary stresses between material bodies ... By introducing stresses ... [Newton] greatly increased the systematic aspect of Nature. But he left all the factors of the system ... in the position of detached facts devoid of any reason for their compresence. He thus illustrated a great philosophic truth, that a dead Nature can give no reasons. (Whitehead 1938: 184)

This critique has much contemporary relevance. One might consider its implications, for example, with respect to debates that address the usefulness, or not, of analytical terms such as 'the social', 'the natural' or 'discourse'. Some of the frustration that often surrounds the use of these terms can be put down to the recognition that they are abstractions, which, as such, cannot do the work of explanation: it is they themselves that instead *require* explanation. Hence Latour's suggestion, cited above, that 'science studies abstain from using society to account for nature or vice versa'. It is not surprising that Latour should advise his reader thus, for he is part of a sociological sub-discipline that has problematized grand narratives and reifying concepts. This is a valuable project, especially in so far as it draws attention to the singularity of each and every situation, a singularity that is not reducible to the individual components that can be identified within it, but is rather to be found in the unique combination of those components in a specific context.

Having said that, there are arguably two key problems with this focus on the singular and the specific, especially when it is pursued in isolation from any other conceptual construction. In the first instance, to concentrate on the specificity of the context does not in itself address the issue of how such components are connected beyond their circumstantial togetherness. Nature bifurcates: there is the togetherness, and there are the circumstances that led to it. Secondly, while it may be the case that 'concrete fact', as Whitehead would put it, 'is the only reason' and cannot therefore be explained with reference to another term (stresses, in Whitehead's example, society, in Latour's), it is also important to account for what Mackenzie calls 'the overflow', that is, the 'feeling or affect [that] overflows particular localities' (Mackenzie 2005: 4). I want to pursue these issues – in essence, *becoming* and *virtuality* – in the following section of this chapter, specifically in relation to the problem of ethics, and then go on in conclusion to suggest that the concept of virtuality is also useful in further extending Latour's critique of 'additional' – or 'arbitrary', as Whitehead might put it – analytical terms in the social sciences. For if a dead nature can give no reasons, as Whitehead claims, then it seems important, with regard to ethics, to explore what vital reasons there might be for focusing on value, as well as concrete fact, in social science research.

Exteriority, Potentiality

Conventionally, ethics concerns the application of moral principle to concrete social facts. To simplify in the crudest fashion, this understanding of ethics often rests, more or less explicitly, on the bifurcation of nature into subjects (who are active, moral and able to conceive of and establish value) and objects (which are passive, mute and indifferent, and which usually have no bearing on value at all). Clearly, this conception is a problem in the context of Whitehead's speculative metaphysics. In the first instance, as I have already noted, this is because Whitehead understands all entities to be constituted by way of their bonds or relations with the world. Thus the distinction between subjects and objects which subtends ethics, as it is usually understood, is impossible to uphold; indeed, it is impossible to conceive of any entity *in* the world being independent or autonomous *from* the world. As I mentioned briefly earlier, Whitehead further argues that all relations are *value relations*. This is how all real or actual relations (entities) are to be defined: by the value of their relations. Values, in other words, do not exist outside or beyond relations/things; they are not brought to them, nor can they be separated from them. Instead, an entity is the source of values for other entities and is the centre of values felt. Valuative relations, *being affected*, is a necessary condition of existence. Values are 'part of the very "matter" of fact – part of the very fabric of "things" in and of themselves' (Rose 2002: 2). This redefinition of the relation between facts and value is a particularly challenging one with regard to the question of ethics, for it suggests that all entities (regardless of their definition as subjects or objects) 'have' – or, strictly speaking, *are* – value. How, then, is it possible to adopt a normative position with regard to such entities and their relations?

These are issues that trouble Latour, often very explicitly, in his book *Politics of Nature* (2004b). Latour's agenda here is to rehabilitate political ecology through a detailed analysis (and rejection) of the concept of nature, where nature is understood not in terms of a domain of reality but as a particular function of politics (Latour 2004b: 133). For Latour, nature in the 'old regime' serves to make political assembly and the convening of the collective (associations of humans and non-humans) impossible. One of the ways in which it does this, he suggests, is by distributing the capacities of speech and representation along the lines of facts and values. Interestingly, Latour does not seek to critique this situation by dislodging the fact/value distinction, or even by conflating facts and values. Instead, he attempts to replace the vocabulary that describes facts and values and to recoordinate the axes on which they turn.

I do not wish to rehearse the details of Latour's position, which is comprehensively laid out in his chapter on this subject (2004b, see especially chapter 3). It is important to note at the outset, however, that Latour is, for the most part, concerned with propositions – literally, pro-positions, the movement and process prior to the point at which an entity becomes 'natural' (i.e. a 'position'), that is, a full-fledged member of the collective. With regard to this process of 'naturalization', Latour begins by drawing up a list of requirements that any replacement of the terms facts and values must meet, and reorganizes these requirements under two headings (or houses, as he calls them): the 'power to take in account: how many are we?' (which is the task of the upper house) and the 'power to arrange in rank order: can we live together?' (which is the task of the lower house). The key point about this reorganization of public life is that, by laying out the stages by which a candidate for existence becomes natural, Latour seeks to extend 'due process', to extend and enrol in other words, as much of the collective as possible in the fabrication of the common world.

Unlike in the old constitution, then, where the definition of nature required that facts be established before values are introduced, we all (and this 'we' includes non-humans as well as humans) participate in the tasks of the two houses, where some of these tasks refer to questions of fact and some to questions of value. So far, so unsurprising. If Latour's life work can be characterized as an exploration of the lengthy and complex ways in which facts are made, created, fabricated and invented, of the ways in which they are *not* given in the common world, then the idea that ethical questions are to be raised only *after* the facts have been established is bound to be a matter for critique. For Latour, it cannot be possible to build the best of possible worlds when the question of values (the common good) is separated from the question of facts (the common world). He argues instead that these questions must be conjoined – as the term 'the good common world', which Latour claims is synonymous with Stengers' 'cosmos', indicates (Latour 2004b: 93). The shift that Latour proposes, from the 'the normative requirement from foundations to the details of the deployment of matters of concern' (Latour 2004b: 118), is arguably not a pushing aside of ethics but rather an extension of it to all who/that are involved in world-making. In his words, 'All our requirements have the form of an imperative. In other words, they *all* involve the question of what *ought* to be done ... The question of what ought to be, as we can see now, is not a moment in the process; rather, it is coextensive with the entire process' (Latour 2004b: 125).

While Latour's position is not identical with Whitehead's (as I shall be discussing below), his claim that 'what ought to be' is coextensive with all world-making has something of the same effect as Whitehead's rather more blunt assertion – which I cited earlier in this chapter – that 'realization is in itself the attainment of value'. Both serve to extend the question of value to every aspect of the world/'worlding' (directly, in Whitehead's case, and more indirectly, via an extension of ethics, in Latour's). Whitehead's position is undoubtedly somewhat problematic, however, in so far as endurance itself – the sheer existence of a thing – is not an especially desirable basis for ethics. In Whitehead's schema, an actual entity will never fail to fulfil its obligation to produce itself and its own values, even though these values are not necessarily to be valued. It is for this reason, Stengers argues, that:

> specialists of human sciences who take advantage of the endurance of what they describe in order to claim resemblance with the lawful objects of natural sciences are doing a bad job. Each time they use their knowledge in order to claim that they know what humans and human societies may or may not achieve, they contribute to give to what exists the power over what could be. (Stengers 1999: 204)

Stengers is drawing attention here to a distinction between what can be known in and of the world and what the world could potentially be, a 'could' that can only, or at best, be imagined. I want to suggest that Latour's and Whitehead's 'answers' to the problems they raise by way of their extension of value to process lies here, in domains that pertain to the issue of 'could' – but also that their different conceptions of such a domain are suggestive of rather different ethical projects. I begin with Latour, and with the specific role that he ascribes to moralists[7] in the task of world-making.

The role of the moralist in Latour's new constitution is a particularly interesting one: it is 'to recognize that the collective is always a dangerous artifice' (Latour 2004b: 157), to recognize, that is, that the realization of things that hold an *essential* place, the work of what Latour calls 'internalization', is also always a work of 'externalization'. The notion of exteriority – of what is excluded or externalized – is an important one with regard to Latour's challenge to the concept of nature as 'stupid matters of fact' that surround society (Latour 2004b: 124). In place of the nature/society bifurcation, Latour suggests there is 'a collective producing a distinction between what it has internalized and what it has externalized' (Latour 2004b: 124). The entities that have been externalized, Latour reminds his reader, 'can be humans, but also animal species, research programs, concepts' – indeed they can be any rejected proposition at all

(Latour 2004b: 124). These rejected propositions represent something of a 'danger' since they might at any moment knock at the door of the good common world and, in demanding to be taken into account, not only modify the 'inside' but also, necessarily, invoke a new definition of the outside. The point here is that '[t]he outside is no longer fixed, no longer inert; it is no longer either a reserve or a court of appeal or a dumping ground, but it is what has constituted the object of an explicit procedure of externalization' (Latour 2004b: 125). It is the task of the moralist to '*go looking for [these entities]* outside the collective, in order to facilitate their reentry and accelerate their insertion' (Latour 2004b: 157).

It is tempting at this point to fold Latour into Whitehead and to suggest that the task of the moralist is to oblige others to be obliged to remember that 'every realization of value is the outcome of limitation' (Whitehead 1985: 116–17). For limitation, in Whitehead's metaphysics, is the price of becoming; specifically, becoming is enabled by the exclusion – and here a new conceptual construction must be introduced – 'of the boundless wealth of alternative potentiality' (Whitehead 1938: 207–8). Potentiality, for Whitehead, is an important concept, the correlative of what is 'given': '[t]he meaning of "givenness"', he writes, 'is that what *is* "given" might not have been "given"; and that what *is not* "given" *might have been* "given"' (Whitehead 1978: 44). Thus while concrete facts are for Whitehead the only reasons – which means that there can be nothing that is external to them that could possibly account for them (such as 'society') – they are not *wholly* given. Some parallels might be drawn here then, between Latour's concept of exteriority and Whitehead's potentiality. Both refer to an excluded exterior. Indeed, Adrian Mackenzie notes that in some of Latour's earlier work the concept of the collective is understood to be 'the outcome of an event in which some element of the pre-individual reserve associated with individuated beings in a domain is singularly structured. In this event, both the individuated beings (subjects, objects, assemblages) and the collective itself become something different' (Mackenzie 2005: 14).

Despite these similarities (which Mackenzie also goes on to question), for me the crucial distinction between Latour's and Whitehead's work on this point concerns their relation to what it is or is not possible to know of that excluded dimension. I cited Stengers earlier, who suggests that for Whitehead nature refers not just to 'what we perceive and can identify, but [to] the whole indefinite complexity of what we are aware of, *even if we have no words to name it*' (Stengers 1999: 197, emphasis added). This is a crucial point, and I would want to underline its relation to potentiality, the defining characteristic of which is that it cannot, by definition, be grasped in thought: 'by the nature of the case', Whitehead writes, 'you have abstracted from the remainder of things. In so far as the excluded things are important in your experience, your modes of thought are not fitted to deal with them' (Whitehead 1985: 73). It is precisely Latour's suggestion that moralists should *go looking* for excluded entities (which implies that something 'exists' that could be 'found'), indeed, his willingness to offer examples of the entities that are located in the exterior (such as the 8000 lives lost per year in France to speeding cars), which indicates, I think, its difference from the concept of potentiality.

Latour's concrete examples make it hard not to conclude that the outside to which he refers is not so much an exterior as a neglected interior.[8] Mackenzie's critical point, that science studies scholars have historically laid too much emphasis on 'social relations that could be rationally understood, and explicated' (Mackenzie 2005: 3), is relevant here also. 'Social structure', Mackenzie writes, 'does not exhaust the potentials of collective life' (Mackenzie 2005: 13). Not entirely dissimilarly, I would argue that Latour's examples point to a curious emphasis on what is already present in the world, on what can be known and what can be found, and on what is already able to be imagined.

Perhaps this should come as no surprise, since it is, ultimately, a politics of *reality* to which Latour is referring:

> Thanks to the moralists, every set has its complementary counterpart that comes to haunt it, every collective has its worry, every interior has a reminder of the artifice by means of which it was designed. There exists a *Realpolitik*, perhaps, but there is also a *politics of reality*: while the former is said to exclude moral preoccupations, the latter is nourished by them (Latour 2004b: 160).

Although I welcome the way that Latour seeks to revisit the question of value (and in doing so, to rehabilitate moralists), I want to propose that it is worth extending his politics of reality to a politics of *virtual* reality in order to attend to more than the processes – of exclusion and inclusion, externalization and internalization – by which things come into existence. Latour's point is that matters of concern, or Things, exist and maintain the sturdiness of their existence by way of the gathering together of participants, ingredients, humans and non-humans that are not necessarily physically or conceptually present in a specific spatio-temporal situation. By recognizing this point, and by launching 'a multifarious inquiry … with the tools of anthropology, philosophy, metaphysics, history, sociology to detect *how many participants* are gathered in a *thing*', critique, Latour argues, will no longer be confined to 'a flight into the conditions of possibility of a given matter of fact' (Latour 2004a: 245–46). Latour considers such matters to be 'simply a gathering that has failed – a fact that has not been assembled according to due process' (Latour 2004a: 246). One might ask, however, where critique might be led if due process referred not only to actual but also to virtual processes, if the critic was obliged to attend not only to those entities that are physically or conceptually present *somewhere* (just not here), but to virtual multiciplicities or singularities that have no corporeal presence at all. In other words, rather than focusing solely, as Manuel de Landa puts it, 'on the final product, or at best on the process of actualization but always in the direction of the final product', one might also (or, de Landa argues, one might *instead*) 'move in the opposite direction: from qualities and extensities to the intensive processes which produce them, and from there to the virtual' (de Landa 2002: 67–68).

My intentions in what follows are not to 'correct' Latour, and certainly not to do so by using the technical details of Whitehead's work as a primer in this task. Nevertheless, there is a richness in Whitehead's concepts and writing, as I suggested earlier, which acts as an invitation – or, as Stengers might put it, which functions as a lure (Stengers 2004) – to take up some his concepts and to explore where they might lead. It is notable that the concept of potentiality is one that Latour neglects, and yet it is here in particular that I find a provocation to ethics – specifically, to an ethics that is wedded to the virtual. This dimension of Whitehead's work was of special interest to Deleuze, and so it is with his reading – or more accurately, his inhabitation – of Whitehead that I begin.

Value, Ethics

In his chapter 'What is an Event?' in *The Fold*, Deleuze describes Whitehead's eternal objects – arguably his most developed concept of potentiality – as the 'last component of Whitehead's definition of the event' (Deleuze 2003: 79). A prehension, Deleuze writes, 'does not grasp other prehensions without apprehending eternal objects' (Deleuze 2003: 79). Eternal objects are 'pure virtualities that are actualized in prehensions' (Deleuze 2003: 79), or, as Whitehead puts it, 'the pure potentials of the universe' (Whitehead 1978: 149). The concept of eternal objects has a significant role to play in Whitehead's project, which is in part to return to nature the value (aesthetic value, for example) that he considers modern science to have misplaced. Eternal objects do some of this work inasmuch as they enable Whitehead to account for qualities and intensities without casting these as 'secondary'. This is because, with their physical ingression into an actual occasion, eternal objects become an actual and un-detachable property of a thing, defining it in its particularity. An eternal object, Deleuze writes, 'can thus cease becoming incarnate, just as new things – a new shade of colour, or a new figure – can finally find their conditions' (Deleuze 2003: 80). As the name suggests, eternal objects come close to being universals – 'though not quite', Whitehead adds (Whitehead 1978: 48). Not quite, because it is precisely through the 'realization' of eternal objects that actual entities differ from each another. Deleuze develops this point in *The Logic of Sense*[9] in relation to the infinitive verb, which he identifies as having two dimensions: on the one hand, it is virtual and incorporeal, it is a potentiality or becoming, while, on the other hand, it indicates a substantive relation to a 'state of affairs' that takes place in a physical time characterized by succession. Thus Deleuze writes of 'the verb "to green," distinct from the tree and its greenness, the verb "to eat" ... distinct from food and its consumable qualities, or the verb "to mate" distinct from bodies and their sexes' (Deleuze 2004: 221).

Whitehead is of particular interest to Deleuze because he rejected substance as the basic metaphysical category, choosing instead to privilege continuity. As the discussion of the becoming of continuity earlier indicated however, this is not the continuity of rectilinear tracks or of lines that could dissolve into independent points but of an infinite series of actual entities or coalitions of prehensions. Contra the oppositions between the figure of the sovereign subject and the inert object, between organic and inorganic matter, Deleuze also emphasizes continuous movement and activity, the constant enfolding, unfolding and refolding of matter, time and space. 'The unit of matter, the smallest element of the labyrinth, is the fold, not the point' (Deleuze 2003: 6). In arguing thus, Deleuze poses a challenge to any philosophy that rests on a distinction between the knowing subject and the object for knowledge. In Deleuze's 'objectless knowledge' (Badiou 1994: 67), the object refers not to a spatialized relation of form-matter, but to a *temporal* modulation, a variation, in a continuum. Correlatively, the subject, which also represents variation, is a 'point of view'. This does not mean that the subject 'has' a point of view (which would imply a pre-given subject) or that the truth varies from subject to subject (which would imply that the truth is relative), but rather that the point of view is 'the condition in which the truth of a variation appears to the subject' (Deleuze 2003: 20).

The concept of the event is especially important in the context of Deleuze's emphasis on continuity. As Alain Badiou explains, the event is what enables Deleuze to account for singularity, it is '*what singularizes continuity in each of its local folds*' (Badiou 1994: 56).[10] In this respect an event is always 'present' in a situation, at least in its virtual dimension. This is not to suggest that it is the cause of that situation, however, or that it precedes it *as such*, or that it should be thought of in terms of an original or model. On the contrary, the infinite number of contingencies that are introduced in processes of becoming ensure that a concrete fact does not amount to a realization of 'something that already existed in a nascent state' (Ansell Pearson 1999: 38). In so far as the world maintains the power of virtuality, it therefore also maintains the capacity to become differently. Able to be actualized in multiple ways (which is another way of saying that an event is not bound to a particular space and time, but may be experienced whenever and wherever it is actualised anew), an event retains an openness to reinventions. It is the inexhaustible reserve or excess that produces novelty. As Deleuze notes, the eternality of eternal objects 'is not opposed to creativity' (Deleuze 2003: 79).

Deleuze's concept of the event is especially useful in so far as it displaces the habituated notion that everything that is 'realized' (in Whitehead's terms) in a particular situation must be explained solely with reference to the participants in that situation. Against the mechanistic notions of cause and effect that underpin many scientific conceptions of the world, this is a notion of 'causation' in which what happens in a particular context cannot be explained or accounted for solely by it, or by the physical entities that compose it, nor can it be reduced to it. If one were to liken, for example, the qualities of subjectivity and objectivity to (a complex of) eternal objects, then these qualities would be understood to be both inside and outside the experiment, both 'universal' and particular, abstract and concrete.[11] Indeed, bearing in mind the relation between universality and particularity that the concept of eternal objects raises, or that is raised by the relation between the virtual and the actual, one might argue, as Whitehead does, that '[w]e are in the world, and the world is in us' (Whitehead 1938: 227). What is important here, however, is that the world to which

Whitehead refers includes a 'virtual' dimension. As such, the notion that 'we are in the world' (and vice versa) must be distinguished from Latour's ostensibly similar claim that, were we to give him 'one matter of concern', he would be able to show us 'the whole earth and heavens that have to be gathered to hold it firmly in place' (Latour 2004a: 246).

Although Whitehead's claim that 'realisation is in itself the attainment of value' and Latour's claim that world-marking is coextensive with 'what ought to be done' seem, at first glance, to have something of the same *effect* – both extend questions of value, directly or indirectly, to process – in fact, they give rise to rather important differences with regard to ethics. While Latour's argument in *Politics of Nature* undoubtedly addresses itself to key ethical issues (such as the relations between facts and values, and the task of moralists), in the end, ethics can hardly be distinguished from due process. If all praxis, all fabrication, is ethical, then it becomes difficult to understand what it might mean to think and act ethically, as opposed to what it might be to think and act at all. This is why the ascription of a specific role to moralists is one of the most confusing aspects of Latour's work in this area. Why is this necessary, if *every* question posed to the world, by whoever or whatever poses it, is always already ethical in character? Latour's answer – that moralists, in contrast to scientists, politicians and economists, do not have an investment in bringing closure to the discussion as to what should be taken into account – is hardly inherent to the profession. Indeed, in view of the many controversies that surround those who work in this field, and the complex networks of power that are invested in the institutionalization of ethics (and bioethics in particular), one might argue that there are others – artists, for example – who are far better qualified for the role as its requirements are defined by Latour.

It is difficult, in other words, to understand what the alignment of ethics and actualization offers to the critic in practice. Without wishing to collapse Deleuze into Whitehead or vice versa, it is notable that, for both, value – and this is the crucial point – does not pertain solely to processes of actualization or to actual(ized) entities. This is the lesson of potentiality in Whitehead: that it is not abstractions in themselves, whether they are internalized or externalized, which are relevant to ethics, but rather *the relation of those abstractions to unrealized potentialities*, to 'the remainder of things', which abstractions necessarily exclude but whose significance cannot be denied. While it may be the case, therefore, that for Whitehead endurance is itself the attainment of value, value is not identical to that which endures. It is notable, for example, that Whitehead defines life not in terms of an enduring entity, or as the property of an enduring entity (an entity that could, say, judge and be judged) but rather as 'a bid for freedom' from the 'shackle' of inherited ancestry to which an entity binds its occasions (Whitehead 1978: 104). For Whitehead, life 'lurks in the interstices' (Whitehead 1978: 105), it is 'a novelty of definiteness' (Whitehead 1978: 104), an *alteration* in value. This point is important because it provides a reason (a reason that is immanent to concrete fact) to develop a relation to the virtual, even if that relation is necessarily irreducible to it.

Not entirely dissimilarly, Deleuze argues that the properly ethical task is to try to 'ascend' to the virtual; 'to carry life to the state of non-personal power'; to 'carr[y] out the conjunction, the transmutation of fluxes, through which life escapes from the resentment of persons, societies and reigns' (Deleuze and Parnet 1987: 50). Evaluation, here, is not a question of judgement (defined in terms of pre-existing criteria) but rather is immanent to the mode of existence in question (Deleuze 1998: 134–35). Or, to put that differently, evaluation is evaluated by the extent to which it is 'creative of life' (Deleuze and Parnet 1987: 50). This is the difference between understanding the singularity of an event in terms of the coming together of relations in unique configurations, and understanding it in terms of a *becoming* together, that is, in terms of eliciting into being 'factors in the universe which antecedently to that process exist only in the mode of unrealized potentialities' (Whitehead 1938: 206–7). It is perhaps in this respect above all, then, that Deleuze is distinguished from Whitehead, Stengers and Latour. For Deleuze, event is not solely a conceptual tool by which to critique mechanistic and reductionist understandings of the world (for instance). More than this, being equal to an event – willing an event in a way that involves neither resignation nor resentment, that is affirmative and that transforms the quality of the will itself – is in itself an ethical task.[12]

It would be reasonable to point out here that Deleuze's project is also – and perhaps more importantly – distinguished from Latour's in so far as the former is philosophical while the latter is largely, and in keeping with the social sciences more generally, empirical. While there is certainly some truth in this, I nevertheless want to suggest in the final and concluding section of this chapter that the use of the virtual as a concept does not necessarily represent a radical departure from core social science concerns and that, as a 'methodological orientation device', it might even contribute to the continued 'life' of empirical social research. Deleuze's analysis of the relations between problems and solutions will be important in this context, especially in so far as problems and solutions are understood, as Manual de Landa (2002) argues they might be, as the epistemological counterpart of the ontological relation between the actual and the virtual.

Problems, Solutions

As I noted earlier, one of the most important aspects of Isabelle Stengers' contribution to the philosophy of science has been her analysis of the grounds on which science is critiqued. In this context, Stengers has been especially sceptical of the sociological approach to science. To quote her again on this, in full: 'In saying that science is a social undertaking, doesn't one subordinate it to the categories of sociology? Now, sociology is a science, and in this case it is a science that is trying to become a superscience, the science that explains all others. But how could it escape the very disqualification it brings on the other sciences?' (Stengers 2000: 3–4). Latour, as I have illustrated, has taken this claim seriously and shown how analytical terms such as 'the social' and 'the natural' may be used not to explain so much as to *explain away* the very facts that researchers have sought to get closer to. Although science studies scholars have generally been slow to apply the implications of their analyses to social theory, it is arguable that the 'mistaking [of] the analytical tool for the reality' (Haraway 1991: 143) often characterizes social science more broadly.

Consider, for example, a classic sociological text, *The Sociological Imagination* ([1959] 2000). Although C. Wright Mills claims in this book that 'no one is "outside society"', he nevertheless suggests that the sociologist is distinguished from 'the ordinary man' (Mills 2000: 184) in so far as he or she is uniquely positioned to make visible – that is, to make relevant – the relations between the individual's experience, which is here and now, and structures and forces (capitalism, power, patriarchy) that are not necessarily visible in themselves. For Mills, the sociological problem is the bridge between these two domains, between history and biography, and it is in formulating the problem that the sociological imagination realizes its full potential. This bridging, or making the connection, is, for Mills, transformative: it transforms the ordinary man's experience of his own experience. One of the problems with 'the sociological problem', however, as conceived by Mills, is that it takes historical social structures, on the one hand, and some variation of the subject, on the other, as *given*. These givens are abstractions, as Whitehead would put it; they are the fruit of sociology not only as a discipline, but also as a profession. One might speculate, therefore, that 'making the connection' between them is important not solely because it illustrates the relevance of history to biography, but because the activity of connecting makes sociology relevant to itself (to its own abstractions). Understood in this way, the sociological problem is its own solution: it transforms ordinary experience into sociological experience.

This is perhaps not surprising: as I discussed earlier, an enduring entity will never fail to produce its own values, whether they are of value or not. And it is also not necessarily problematic: if sociology is indeed a science, as Stengers pointedly implies, then one might confer on the discipline the respect that the singularity of any scientific endeavour deserves, that is, for inventing scientific objects under the strictest conditions. The 'post-constructivist' claim that the concepts and methods deployed by social scientists are productive of the very object they seek to investigate comes close – albeit with very different intentions – to confirming this view (see, for example, Law and Urry 2003). I say 'with very different intentions' because, although the aim of many of these arguments is to draw attention to the limits of social science and to demand, as is often demanded of the natural sciences, that researchers recognize the specificity of their objects (results, products, outcomes), it is also, simultaneously and perhaps in contradiction, to extend its ambitions. For unlike the 'ideal' of the natural sciences, the social sciences often come with an explicit aspiration to be relevant, even to make a difference, to something other than itself. The tensions that I am describing here are witnessed in John Law and John Urry's article, 'Enacting the Social' (a title that neatly captures their post-constructivist point): 'In a world where everything is performative, everything has consequences, there is, as Donna Haraway indicates, no innocence. And if this is right then two questions arise: what realities do the current methods of social science help to enact or erode? And what realities might they help to bring into being or strengthen?' (Law and Urry 2003: 5, footnotes omitted).

It is in this context – in the context of the delicate balancing act between, on the one hand, recognizing the role that social scientists play in creating the worlds they seek to investigate and, on the other, wishing to change worlds that include more than social scientists and their objects alone – that the concept of the virtual is of value. Specifically, it is of value as a tool or a technique that might orient the social researcher towards, as Whitehead puts it and as I cited earlier, that which is *not* given and that which *might have been* given; towards that which is *not* already known or even imagined; towards 'the whole indefinite complexity', to quote Stengers once again on this point, 'of what we are aware of, even if we have no words to name it'. In order to explicate this point, I want to consider what a social research project might look like if its basic commitments were not to historical social structures and the subject but to the virtual and the actual. And what Mills' sociological problem might look like, if it were refracted through the virtual problem.

Minimally defined as a dimension of the actual that is neither observable nor accessible in itself, the virtual offers a 'beyond' actual states of affairs for the social scientist to look to. This is important, I think, because the explication of what is not immediately, or indeed ever, accessible is how much of the 'magic' of sociology is generated, as Mills passionately (if somewhat polemically) illustrates. Unlike Mills' social structures, however, virtual structures or patterns cannot do 'explanatory work' because they are not determining in the way that social forces, or the material sedimentation of such forces over time, are often understood to be in sociology. The virtual does not govern the actual by rules or laws. Instead, as Deleuze puts it, 'The reality of the virtual is structure. We must avoid giving the elements and relations that form a structure an actuality which they do not have, and withdrawing them from a reality which they have' (Deleuze in de Landa 2006: 246). The structures, patterns or regularities of the virtual can be understood in terms of the distribution of singularities. These singularities are not determining not because the virtual has no relation to the actual (it is not an unintelligible outside), but because processes of actualization introduce many contingent divergences. Importantly, the social researcher's embodied participation, along with the concepts and methods that she or he deploys, are among the contingent divergences that will shape the actualization of the virtual as the research process unfolds. The virtual, in other words, is not a blueprint. What this means in practice is that the question as to whether something is (going to be) important or relevant in a piece of social research cannot be decided in advance. Indeed, the incommensurability of the relation between the virtual and the actual actively mitigates against this and arguably *institutes* an openness with regard to the question of what is and is not of value.

Deleuze's analysis of the relations between problems and solutions is informed by many of the conceptual themes that I have already introduced. As Philip Goodchild explains, for Deleuze, '[i]n the same way that events are different from the states of affairs in which they are actualized, problems are different from the solutions which they produce within thought' (Goodchild 1996: 54). '[T]he problem of "light"', Claire Colebrook writes, 'is posed, creatively, by different forms of life in different ways: photosynthesis for plants, the eye for animal organisms, colour for the artist' (Colebrook 2002: 21). As this example illustrates, there is no 'true' solution to a problem (although there are true problems). Photosynthesis, the eye and colour might have the problem of light in common, but their ancestry or, rather, the distribution of the singularities that determine them as solutions is clearly different. The best – and this is indeed the best, in value terms – that a solution can do is to develop a problem. 'It seems', Deleuze writes, 'that a problem always finds the solution it merits, according to the conditions which define it as a problem' (Deleuze 2004: 65).

Clearly, the ethical obligations that accompany the virtual problem and the sociological problem, as formulated by Mills, for example, are somewhat different. For Mills, making connections between *identifiable* domains is the foremost task of the sociologist, a task that should be exercised, he argues, in work, in educating and in life (Mills 2000: 187). For Deleuze, event thinking is ultimately judged by the extent to which it is able to invent new (i.e. previously *unidentifiable*) concepts that affirm and extend events. Although this gives way to a dramatic philosophy of living (as I noted towards the end of the last section), one that probably demands more than it is reasonable to expect from the average social researcher, it might also, less ambitiously, offer a practical orientation for the way that ethics itself, including Deleuze's own ethics, could be judged: that is, not in terms of inventive problem-solving, but rather in terms of inventive problem-making. The (ethical) obligation here, in other words, is not to solve a problem, or to explain it away but rather to try to enable it to 'speak' or to pose it in terms that enable it to play itself out in productively creative ways. The empirical evidence of a sociological 'ascent' to the virtual can be witnessed in the extent to which sociology is open to an engagement that is transformative not only of the object but also of itself. For, when *the problem* (rather than the social scientist, and rather than the 'ordinary man') is enabled to make things that cannot be identified in advance relevant to each other, *both* the social scientist and the 'ordinary man' are likely to be transformed.

There is a tendency with event thinkers to focus on remarkable points and on the creative aspects of an event, to argue, for instance, that regardless of the rhetoric of reductionism that may take hold of an experimental event, an event will always imply 'something excessive in relation to its actualization, something that overthrows worlds, individuals and persons' (Deleuze in Halewood 2003: 241). As I have suggested throughout, an event is irreducible to the concrete facts that are actualized in process and in this respect there will inevitably be in any actualization a dimension of creativity and novelty. Nevertheless, while it is possible to discern ancestries that 'differenciate'[13] the virtual in inventive and creative ways – and in these instances a problem 'is a way of creating a future' (Colebrook 2002: 1) – Whitehead also reminds his reader, as I noted briefly earlier, of the shackle or burden of inherited ancestries, ancestries for which 'the uniformity along the historic route increases the degree of conformity which that route exacts from the future. In particular each historic route of like occasions tends to prolong itself, by reason of the weight of uniform inheritance derivable from its members' (Whitehead 1978: 56).[14]

One might apply this point to disciplinary abstractions and to the inheritances that serve to limit what a discipline can or cannot become. As I noted earlier in this chapter, this issue was of considerable concern to Whitehead. His intentions in developing a speculative metaphysics were, precisely, to produce both 'a restraint upon specialists, and also ... an enlargement of their imaginations' (Whitehead 1978: 17). There are many ways to do this. One of them might be to pursue the 'minor history' of sociology, which, as Wolf Lepenies (1992) illustrates, is literary rather than scientific.[15] In this chapter, I have chosen to foreground the concept of the virtual, which is also commonly linked to the vitality of creativity. I do not think, however, that 'pursuing' the virtual – making a real difference, producing a variation in value – is an easy task. There is a difference, as Stengers puts it, between sophisticated observation and an event. Or, as Deleuze writes, 'what is ... frequently found – and worse – are nonsensical sentences, remarks without interest or importance, banalities mistaken for profundities, ordinary "points" confused with singular points, badly posed or distorted problems – all heavy with dangers, yet the fate of us all' (Deleuze 2004: 191). At the very least, then, the virtual serves as a reminder that not all experiments, not all assemblages, not all gatherings develop a problem that is worth trying to extract from actuality. Not all actualizations are, in themselves, ethical. And articulating a sociological problem is not in itself necessarily the agent of transformation or the mark of novelty. To redistribute the singularities that determine a solution is to truly transform an event. While such transformations are undoubtedly rare, the aspiration towards them provides a reason for continuing to ask questions about value, including the value of social research.

Notes

I am very grateful to the editors of this volume, to the anonymous reviewer and to Noortje Marres for their insightful comments on and constructive criticism of earlier drafts of this chapter. All errors etc. are my own.

1. 'What this means is that a large number of different trajectories ... may end up in exactly the same final state (the attractor), as long as all of them begin somewhere within the "sphere of influence"' (de Landa 2002: 15).
2. These debates cut across a range of disciplines and sub-disciplines, as exemplified by the 2004 4S and EASST meeting in Paris (on Public Proofs: Science, Technology and Democracy), where pleas were made for an engagement with the role and place of ethics in social scientific studies of science to begin in earnest (for example, Mol 2004).
3. In this respect, event thinking is a protest against the notion that time is an ordered succession of instants without duration and that space is a system of points without extension.
4. This point is indebted to the numerous conversations I have had with Andrew Barry on this subject.
5. This is relevant to my earlier discussion of primary and secondary qualities because it accounts for what are perceived to be some of the most significant 'fictions' to have been created about nature: the fictions

that stem from the subjectivity of the human senses. Like other modern natural philosophers, Galileo distinguishes between 'qualities absolute and fixed, which form the object of mathematical analysis, and qualities subjective and in flux, which derive from the constitution of the observer' (Proctor 1991: 54). While the former alone are real, necessary and essential to knowledge of an object, the latter are spurious distortions. And herein, in the recognition of the subjectivity of the observer, lies the significance of experimentation, for the experimental method is cast as an important – if not the most important – technique for eliminating bias and appearance, and for gaining access to the essence of things.

6. Consider in this context Latour's claim that 'the stock drawn upon *before* the experimental event is not the same as the stock drawn upon *after* it' (Latour 1999: 126).
7. This is a somewhat controversial term, as is the notion of 'ethicist' in contemporary science and especially biomedicine. It would be interesting to address the question as to what name might be given to the group of people who undertake the 'moralists' task' (see below) as Latour understands it, but this is beyond the scope of this chapter.
8. My thanks to Michael Parker for encapsulating this point so elegantly and, in so doing, helping me to better understand the implications of it.
9. Which Paul Patton suggests 'might equally have been entitled "The Logic of the Event"' (Patton 1996: 13).
10. Which is precisely Badiou's problem. The event, understood by Deleuze as that which emerges out of an ontological univocity, is too much *of* the world, is so much a part of the world, in fact, that Badiou feels obliged to call its singularity into question: how is it possible to distinguish an event from a fact if 'everything is event'? (Badiou 1994: 56). Deleuze's concept of the fold is so profoundly anti-extensional, Badiou argues, so labyrinthine and directly qualitative, that he is unable to account for the singularity of an event or rupture at all.
11. For a detailed illustration of this point, in the context of an analysis of early experiments on serotonin, see Fraser (2003).
12. In arguing thus, Deleuze owes as much to Nietzsche as he does to the Stoics. Indeed, Philip Goodchild suggests that the eternal return should be understood 'not [as] a theory of time, but [as] a technique for living the event' (Goodchild 1996: 53).
13. In Deleuze's work, differenciate, with a 'c', refers to processes that relate to the virtual, while differentiate refers to processes that relate to the actual.
14. Technically speaking, Whitehead makes this claim on the basis of his distinction between pure potentiality and real potentiality. Philip Rose explains the difference thus: 'Pure potentiality is an aspect of the "mere" continuum while real potentiality is an aspect of the "real" continuum ... Where the mere continuum includes the entire spectrum of potentiality, the real or extensive continuum represents the general field of *real* potentiality, that is, the field of objectified or Past Actual Occasions (and their relations)' (Rose 2002: 50–51). This is essentially how all actual entities must be understood following the cessation of immediacy or concrescence. Having 'passed away' or 'perished' as Whitehead puts it, the actual entity functions as a resource or, more accurately, as the real potential for the becomings of subsequent actualities: 'The pragmatic use of the actual entity, constituting its static life, lies in the future. The creature perishes *and* is immortal' (Whitehead 1978: 82).
15. Note also that, like Whitehead, Deleuze finds in literature – specifically, in Anglo-American literature – not the 'terrible mania for judging and being judged [that] runs through [French] literature' but an ability 'to be a flux that combines with other fluxes' (Deleuze and Parnet 1987: 50) (what Deleuze, elsewhere, calls 'becoming-imperceptible'). The aim of writing, for Deleuze, is thus not to establish one's authorship/authority, nor is it to flee from life into the imagination or into art.

Bibliography

Ansell Pearson, K. (1999). *Germinal Life: The Difference and Repetition of Deleuze.* London and New York, Routledge.

Badiou, A. (1994). 'Gilles Deleuze, The Fold: Leibniz and the Baroque', in C.V. Boundas and D. Olkowski (eds) *Gilles Deleuze and the Theater of Philosophy*. New York and London, Routledge, pp. 51–69.

Barry, A. (2004). 'Ethical Capitalism', in W. Larner and W. Walters (eds) *Global Governmentality.* London, Routledge, pp. 195–211.

Colebrook, C. (2002). *Gilles Deleuze*. London and New York, Routledge.

de Landa, M. (2002). *Intensive Science and Virtual Philosophy*. London and New York, Continuum.

———. (2006). 'Deleuze in Phase Space', in S. Duffy and P. Patton (eds) *Virtual Mathematics: The Logic of Difference*. Manchester, Clinamen Press, pp. 235–48.

Deleuze, G. (1998). *Essays Critical and Clinical.* London and New York, Verso.

———. (2003). *The Fold: Leibniz and the Baroque*. London and New York, Continuum.

———. (2004). *The Logic of Sense*. London and New York, Continuum.

Deleuze, G. and C. Parnet. (1987). *Dialogues*. London, Athlone Press.

Evans, J. (2002). *Playing God!: Human Genetic Engineering and the Rationalization of Public Bioethical Debate 1959–1995*. Chicago, University of Chicago Press.

Fraser, M. (2003). 'Material Theory: Duration and the Serotonin Hypothesis of Depression', *Theory, Culture and Society* 20 (5), 1–26.

Goodchild, P. (1996). *Gilles Deleuze and the Question of Philosophy*. London, Associated University Presses.

Halewood, M. (2003). 'Subjectivity and Matter in the Work of A.N. Whitehead and Gilles Deleuze: Developing a Non-essentialist Ontology for Social Theory'. Unpublished PhD thesis, Goldsmiths College, University of London.

Haraway, D. 1991. *Simians, Cyborgs, and Women: The Reinvention of Nature*. London: Free Association Books.

Latour, B. (1999). *Pandora's Hope: Essays on the Reality of Science Studies*, Cambridge, MA, and London, Harvard University Press.

———. (2004a). 'Why Has Critique Run Out of Steam?: From Matters of Fact to Matters of Concern', *Critical Enquiry* 30, 225–48.

Latour, B. (2004b). *Politics of Nature: How to Bring the Sciences into Democracy*, Cambridge, MA, and London, Harvard University Press.

Law, J. and J. Urry. 2003. 'Enacting the Social', published by the Department of Sociology and the Centre for Science Studies, Lancaster University, Lancaster LA1 4YN, UK. Available at http://www.comp.lancs.ac.uk/sociology/papers/Law-Urry-Enacting-the-Social.pdf

Lepenies, W. 1992. *Between Literature and Science: The Rise of Sociology*. Cambridge: Cambridge University Press.

Mackenzie, A. (2005). 'Problematising the Technological: The Object as Event?', *Social Epistemology* 19 (2–3), 1–19.

Mills, C.W. 2000. *The Sociological Imagination*. Oxford: Oxford University Press.

Mol, A. (2004). 'Good and Bad Realities: On Appreciation', paper presented at 4S and EASST, Public Proofs: Science, Technology, and Democracy, Paris, 25–28 August.

Patton, P. (ed.). (1996). 'Introduction', *The Deleuze Reader*. Oxford, Blackwell, 2–17.

Proctor, R. (1991). *Value-Free Science?: Purity and Power in Modern Knowledge*. Cambridge, MA, and London, Harvard University Press.

Putnam, H. (2004). *The Collapse of the Fact/Value Dichotomy and Other Essays*. Cambridge, MA, and London, Harvard University Press.

Rose, P. (2002). *On Whitehead*. Belmont, CA, Wadsworth.

Stengers, I. (1997). *Power and Invention: Situating Science*. Minneapolis and London, University of Minnesota Press.

———. (1999). 'Whitehead and the Laws of Nature', *SaThZ* 3, 193–206.

———. (2000). *The Invention of Modern Science*. Minneapolis and London, University of Minnesota Press.

———. (2002). 'Beyond Conversation: the Risks of Peace', in C. Keller and A. Daniell (eds) *Process and Difference: Between Cosmological and Poststructuralist Postmodernisms*. New York, State University of New York Press, pp. 235–56.

———. (2004). 'A Constructivist Reading of Process and Reality'. Paper presented at 'Whitehead, Invention and Social Process', Centre for the Study of Invention and Social Process, Goldsmiths College, London June.

———. (2005). 'Deleuze and Guattari's Last Enigmatic Message', *Angelaki* 10 (2), 151–67.

Whitehead, A.N. (1920). *The Concept of Nature*, Cambridge, Cambridge University Press.

———. (1938). *Modes of Thought*. Cambridge, Cambridge University Press.

———. (1978). *Process and Reality*. New York, Free Press.

———. (1985). *Science and the Modern World*. London, Free Association Books.

Chapter 3

Irony and Humour, Toward a Deleuzian Science Studies

Katie Vann

> The mask, the costume, the covered is everywhere the truth of the uncovered.
> The mask is the true subject of repetition.
> Because repetition differs in kind from representation,
> the repeated cannot be represented;
> rather, it must always be signified,
> masked by what signifies it,
> itself masking what it signifies.
> (Deleuze 2004a [1968]: 20)

Irony and Humour – an Introduction

In an enigmatic passage of *The Invention of Modern Science,* Isabelle Stengers differentiates irony and humour as two different modes of engagement, two political logics available to students of the socialilities of science. She defines humour as 'the capacity to recognize oneself as a product of the history whose construction one is trying to follow – and this in a sense in which humour is first of all distinguished from irony' (Stengers 2000: 66). The passage goes on to flesh out the comparison between irony and humour, which in turn is positioned as an invitation for science studies to orient to the politics of scientific writing in a particular (humorous) way.

'As Steve Woolgar has shown,' she writes, 'the sociological reading of the sciences of the relativist type puts its specialists in the position of being "ironists". They are those who will not let themselves count, who will bring to light the claims of the sciences. They know they will always encounter the same difference in point of view between themselves and scientists, which guarantees that they have conquered, once and for all, the means for listening to scientists without letting themselves be impressed by them. Some authors can advocate an "ironic" reading of their own texts because the latter are equally scientific (dynamic irony). The fact remains that the position in principle requires a reference by the author to transcendence (stable or dynamic), to a more lucid and more universal power to judge that assures his or her difference from those being studied' (Stengers 2000: 66).

In the text that Stengers alludes to here (Woolgar 1983), Steve Woolgar orients to an epistemological problematic of 1980s constructivist ethnographic laboratory studies of 'science as it actually happens'. Tarja Knuuttila has likened the problematic addressed by Woolgar to arguments from self-refutation in philosophy, and characterizes the particular 'irony' targeted by Woolgar as follows: 'Woolgar criticized this urge of laboratory studies to describe science as it happens of an instrumental conception of ethnography, which applies relativist epistemology only selectively – to other scientists' accounts – *whereas* one's own accounts are presented realistically' (Knuuttila 2002, emphasis added).

The connection that Stengers makes to her point about irony, is Woolgar's recognition that laboratory studies presuppose that they operate from the standpoint of 'a more lucid and more universal power to judge that assures [their] difference from those being studied'. And when Stengers hearkens to Woolgar in the preceding passage, she seems to be warning those who study the sociality of scientific practices against instigating the situational irony that the 'whereas' inaugurates. It would be ironic to presuppose the constructedness of scientific knowledge, and then to proceed with the purported certainty of ethnographic research to disclose it in the case of laboratory science; the implicit appeal to a privileged access to truth from the standpoint of ethnographic method betrays the supposition of the constructedness of scientific knowledge as such. To claim a privileged access to the processes that attend anti-foundational scientific truths would inaugurate an irony when coupled with the foundationalist premise that justifies ethnographic privilege per se.

Stengers goes on to foil the logic of irony for the purposes of advocating the logic of humour: 'Humour, by contrast,' she continues, 'is an art of immanence. The difference between science and nonscience cannot be judged in the name of a transcendence, in relation to which we would designate ourselves as free, and where only those who remain indifferent to it are free. For our dependence on this transcendence in no way reduces our degrees of liberty, our choice as to the way we will attend to the problems created by the constitution of this difference. The situation is the same as that of politologists, who know that their problem would have no meaning had not the Greeks invented an "art of politics". They are themselves a product of this invention, which they thus cannot reduce to nothingness. But they remain free to put this invention in history. In this sense, irony and humour constitute two distinct political projects, two ways of discussing the sciences and of producing debate with scientists. Irony opposes power to power. Humour produces (to the degree it itself manages to be produced) the possibility of shared perplexity, which effectively turns those it brings together into equals. To these two projects, there correspond two distinct versions of the principle of symmetry: an instrument of reduction or a vector of uncertainty' (Stengers 2000: 66–67).

Stengers' gesture to the distinction between irony and humour works to point out two different modes of social practice on the part of those who study scientific practices. And, in discussing irony and humour in this manner, the two categories have begun to be applied in an analytical or classificatory manner in such a way as to characterize modes of social practice: irony and humour are distinctive ways in which she who studies scientific practices orients to the historicity of her own practices and their potential contiguity with that of those she studies.

While Stengers' explicit citation of Woolgar does the important work of mapping her understanding of the distinction between irony and humour to the practices of science studies scholars,[1] it also has the limiting consequence of characterizing ironic practices in the somewhat general way in which irony can be understood – a dynamic in which an opposition emerges between what is said or claimed and what is undertaken. Irony, in that general sense, is likened to practices of hypocrisy. While this vernacular lends preliminary sense and attraction to Stengers' call, it is also limiting in that it manifests an opacity around the historical specificity of Stengers' own discourse, which is much more technical.

Stengers suggests the rather fine-grained reading of humour as a mode of practice that entails not so much the overcoming of irony through the creation of an adherence between what is claimed and what is undertaken, as the realization that the practice of those who undertake an enquiry into the socialities of scientific practice is itself part of the history of the sciences, a contiguity that cannot be reduced. In its humorous mode, science studies remains free to put this invention in history.

Yet the difference between irony and humour remains enigmatic in this passage in so far as it leaves us wondering what to do with the uncomfortable recognition that an appeal to a more universal power to judge may be one of the constitutive dimensions of the history of the sciences to which science studies itself belongs. If humour is the putting of such an appeal in history as an aspect of the invention of science that one shares with her interlocutors, how does that differ from an ironic appeal as such?

Stengers writes of irony as an appeal to transcendence, and of humour as an art of immanence. And it is perhaps worth our appreciation that Stengers' uses of these tropes of transcendence and immanence – at once spatial and cosmological – echo a powerful Deleuzian trace that animates her work. That trace takes her quite beyond general uses of the terms irony and humour, where they manifest a fruitful space in which Deleuze and science and technology studies intersect.

To be sure, by the time *The Logic of Sense* arrives (1969), Deleuze will have begun to draw on the comparative figures of irony and humour in such a way as to take on the position of humorist. Resonant with Stengers' later invitation to engage the arts of immanence, he writes, 'There is a difficult relation, which rejects the false Platonic duality of the essence and the example. This exercise, which consists in substituting designations, monstrations, consumptions, and pure destructions for significations, requires an odd inspiration – that one know how to "descend". What is required is humour, as opposed to the Socratic irony or to the technique of ascent' (Deleuze 2004b [1969]: 154).

In both *The Logic of Sense* and *The Invention of Modern Science*, both Deleuze and Stengers invite readers to engage in the process of humour. For Deleuze, irony is associated with 'ascent' while humour is associated with 'descent'. The latter would reject the false Platonic duality of the essence and the example. For Stengers too, the distinction between irony and humour is set in an analogous relation with a distinction between transcendence and immanence. To ascend to the transcendent, to the more universal, to oppose power to power, would somehow be an instrument of reduction rather than an instrument of uncertainty; to appeal to a transcendent would be to choose a political project that lacks the capacity to recognize oneself as a product of the history whose construction one is trying to follow, and that is ill-fitted to turn mutual interlocutors into equals.

Deleuzian Passages

Students of Deleuzian science and technology studies are free to read Stengers' enigmatic passage as an invitation to be humorous that we had better choose to accept (not least in so far as humour is a properly Deleuzian, if not Woolgarian, trait). Yet, in doing so, are we not invited to become more intimate with irony and humour and the histories through which they too have come into being? For these invitations to humour – technical as they themselves are – actually render questions that tease us into the genealogies of the categories of irony and humour. With what situation, indeed, are students of the sociality of the sciences thus confronted, when an aspect of the invention of the sciences is precisely an appeal to a transcendence that would justify one's power to judge? How will a humorous relation with such an appeal on the part of the scientist be brought into being as shared? In short, what is the difference between the ironic ascent (appealing to a more universal power to judge) and the humorous descent (putting that very invention into motion as shared history)? Such conundrums call up opacities in the genealogies of the categories of irony and humour as they are deployed in Stengers' text, opacities that we crafters of Deleuzian science studies might pursue.

Relevant to my suggestion here is a peculiar moment in the histories of irony and humour as devices built by Deleuze. The moment is a comparative reading of the literary works of the Marquis de Sade and Leopold von Sacher-Masoch, which Deleuze published in 1967, just before the publication of *Difference and Repetition* in 1968. The work, entitled 'Coldness and Cruelty', was framed as an effort to overcome the prevalence of what he called the presumption of a 'sadomasochistic entity' in psychoanalytic studies, notably in the work of Richard Freiherr von Krafft-Ebing.[2] For Krafft-Ebing, sadomasochism is a process in which two characters engage in a reciprocal relationship from which each derives an equal pleasure: the sadist takes pleasure in inflicting pain, the masochist in receiving it.[3] In such a logic there can be no better reciprocity than that brought forth in the union of opposites, as the pleasure of each, though inversely, is brought forth in direct proportion to that of its mate. Do not the top and the bottom make perfect bedfellows?

'Coldness and Cruelty', perhaps minor in Deleuze's repertoire, renders the category 'sadomasochism' a 'misbegotten name' that had presupposed 'the principle of the unity of opposites' (Deleuze 1991 [1967]: 13). Deleuze lamented the analytical status that had been attained by Krafft-Ebing's clinical category, which had supposed the dialectical unity of sadism and masochism. Deleuze saw that Krafft-Ebing had drawn on the works of Sade as a kind of template for the diagnosis of a pathology, which Krafft-Ebing then read as dialectically expressed by two characters – the sadist and the masochist – in relation. Deleuze's urgent concern was that Krafft-Ebing's dialectical reasoning had been contingent upon an eclipse of the works of Masoch and the associated specificity of masochism as realized in them: Krafft-Ebing assumed that the 'masochist' in Sade's work and the 'masochist' in Masoch's work were identical. And, in a precisely formulated corrective gesture, Deleuze asked whether sadistic and masochistic complexes might be better understood as two profoundly original clinical pictures and scenic logics rendered, respectively, by Sade and Masoch.

Rather than accepting a 'sadomasochism' that would ground and envelop the dialectical union of sadist and masochist, Deleuze attempted to call attention to sadism and masochism as two radically different kinds of processes, each with its own proper 'bottom' and 'top'. And, by enacting a strategy of anti-dialectical reasoning that would come to mark subsequent texts as Deleuzian, Deleuze not only questioned the supposition of difference (you seek to beat, I seek to be beaten) as grounded in higher unity (the sadomasochistic pathology of which we are the dialectical expression), but also had begun to hone a mode of literary enquiry in which the novel is poised as a philosophical and diagnostic venue through which writers create 'a counterpart of the world capable of containing its violence and excesses' (Deleuze 1991 [1967]: 37).

One of the conceptual achievements of 'Coldness and Cruelty' is Deleuze's development of the categories 'irony' and 'humour' as comparative literary devices that could be used to mark the respective singularity of Sade and Masoch as two writers in whose stories Deleuze reads peculiar theatres or scenographies in which distinctive relations of power take shape.[4] 'Irony' is set in a chain of other 'symptoms', of which sadistic scenography is both diagnostic and characteristic, whereas 'humour' is set in a chain respectively with masochistic scenography. And, while 'Coldness and Cruelty' has tended not yet to manifest as a marked textual resource for social research at the intersection of Deleuze and science and technology studies, the categories of *irony* and *humour* have had some interest. As mentioned earlier, Stengers invokes and distinguishes irony and humour as two ways of engaging scientific practices and invites readers to take up humour as an orientation and distinctive political project.

But what is so fascinating about 'Coldness and Cruelty'as a moment in the genealogies of the categories of irony and humour is that, while Deleuze develops these categories for the purposes of comparative diagnosis, he refrains from advocating either as a political strategy. Although it is clear that he is already getting a thing for Masoch in 'Coldness and Cruelty' (he laments the eclipse of Masoch by Sade, which is supported by the 'etiological fallacy' of the 'sadomasochistic entity'), he offers himself to us as a writer who, rather than preferring humour as a desired mode of politics, engages in a reconstruction of the singularities of sadistic irony and masochistic humour as markers of peculiar *political* theatres in their own right. This descriptive orientation enables Deleuze to create a fine-grained schematic of the categories of irony and humour, which paves the way for a subtle transition to a normative orientation in subsequent texts, for example, as was intimated in the passage shown above from *Logic of Sense* (2004b [1969]).

And what is important here in its relation to the eventual normative orientation that Deleuze (and Stengers) seems to pursue is that in 'Coldness and Cruelty', both sadistic irony and masochistic humour are characterized as particularly *modern* political theatres, which are contrasted with the irony and humour proper to the political theatres achieved in the dialogues of Plato, which Deleuze reads as proper to a pre-modern, *classical* milieu. As we shall see below, this has the interesting effect of constructing a grid of possibilities – not just two general political strategies named irony and humour, but rather two political strategies that are proper to two distinctive political milieux which are marked respectively by the classical and modern conceptions of law. In effect, by differentiating two conceptions of law, the otherwise dichotomous 'irony' versus 'humour' is transformed into a fourfold schematic (classical irony; classical humour; modern irony; modern humour).

It is my interest in this chapter to open a conversation that considers whether and in what ways such differences will matter for how we understand, for example, Deleuze's eventual reference to humour as an alternative to Socratic irony or the technique of ascent, which we find in the *Logic of Sense* (2004b [1969]), as noted above, as well as Stenger's call to humour in *Invention*. I believe that it is worthwhile, in other words, to visit Deleuze's descriptive treatment of Sade and Masoch, because it entails attention to details that are relevant to the genealogy of irony and humour as Deleuzian devices that could be invoked in subsequent treatments such as Stengers'.

In the next section I turn to Deleuze's comparative reading of Sade and Masoch in 'Coldness and Cruelty' in an effort to specify irony and humour as he develops them there. I then turn to a particular instance in *Difference and Repetition* (2004a [1968]), a subsequent work that mediates the passage from 'Coldness and Cruelty' to *The Logic of Sense* and in which Deleuze raises the distinction between irony and humour. My aim here will be to consider how Deleuze renders the distinction in this textual setting, and how that rendering is contextualized by his concerns in the broader passage in which it occurs.[5] In light of these considerations, I ask readers to return to the enigma of Stengers' call and to think about its implications at the intersection of Deleuze and science studies.

The Ethical Hollow of Modern Law

As figures of Deleuzian enquiry, irony and humour are manifested foremost in 'Coldness and Cruelty' as markers of the respective scenographic logics created by Sade and Masoch. The work presents many features of their differences. My focus here, because it specifies the political logic peculiar to each writer, will be Deleuze's reading of the approaches that Sade and Masoch use to create a distinctive literary counterpart to what Deleuze calls 'the modern conception of the law' as described in Kant's *Critique of Practical Reason* (Kant 2004 [1788]). For Deleuze, Kant's tract marks a distinctive milieu for the works of Sade and Masoch; as such, it provides a standpoint from which to ascertain the differences between the works of Sade and Masoch, and the differences between their works and those of Plato. Deleuze specifies the modern conception of law by contrasting it with what he calls the classical conception.

'The classical concept of the law', he writes, 'found its perfect expression in Plato and in that form gained universal acceptance throughout the Christian world. According to this conception, the law may be viewed either in the light of its underlying principles or in the light of its consequences. From the first point of view, the law itself is not a primary but only a secondary or delegated power dependent on a supreme principle, which is the Good … from the standpoint of its consequences, obedience to the law is "best," the best being in the image of the Good … This conception, which is seemingly so conventional, nevertheless conceals elements of irony and humour which made political philosophy possible, for it allows the free play of thought at the upper and lower limits of the scale of law. The death of Socrates is exemplary … the laws place their fate in the hands of the condemned man, and ask that he should sanction their authority by submitting to them as a rational man. There is indeed a great deal of irony in the operation that seeks to trace the laws back to an absolute Good as the necessary principle of their foundation. Equally, there is considerable humour in the attempt to reduce the laws to a relative Best in order to persuade us that we should obey them. Thus it appears that the notion of law is not self-sufficient unless backed by force; ideally it needs to rest on a higher principle as well as on a consideration of its remote consequences … Irony and humour are the essential forms through which we apprehend the law. It is in this essential relation to the law that they acquire their function and their significance. Irony is the process of thought whereby the law is made to depend on an infinitely superior Good, just as humour is the attempt to sanction the law by recourse to an infinitely more righteous best' (Deleuze 1991 [1967]: 81–82).

We can see elements of both Deleuze's (2004b [1969]) and Stengers' (2000) appeals to the distinction between irony and humour in this passage. The spatial metaphors of ascent and descent are here, and so is the centrality of 'the supreme principle' that grounds law and to which ironic strategy entails an ascent. Such a supreme and grounding principle is unique to the classical conception of law as characterized by Deleuze in 'Coldness and Cruelty'. With respect to the classical conception, Deleuze identifies the literary logics of irony and humour in the story of the death of Socrates. And what seems to be key here is that the classical conception affords movement at both the 'upper and lower limits of the scale of law'. This expansive possibility stems from the reliance of both ascent and descent, both irony and humour, on the existence of the grounding principle of the Good, of which the law is a reflection; the law is an image of and reflects the higher principle of the Good, which in turn grounds the law and provides the foundation of its validity.

Although differentiated in terms of the metaphors of ascent and descent, humour also comes into being in relation to the same, grounding principle of the Good; that is, even in the case of humour, which entails descent to consequences, the 'Best' so arrived at is itself in the image of the underlying principle of the Good upon which the validity of law is secured; humour is a descent to the best of the righteous, itself a reflection of the grounding principle of the Good. Deleuze thus speaks of irony and humour in the same breath when he characterizes the scenographic logic of the death of Socrates; the principle of the Good allows even for the condemned man to sanction the authority of the law in light of the best that is established in its image.

With the distinctiveness of this classical conception in place, Deleuze goes on to specify what he calls the *modern* conception of law, a conception that is both presupposed by and contained in the texts that bring forth the peculiar scenographic logics of *sadistic* irony and *masochistic* humour. Both sadistic irony and masochistic humour will be differentiated from the Socratic irony/humour that is proper to the classical milieu.

'Kant gave a rigorous formulation of a radically new conception,' writes Deleuze, 'in which the law is no longer regarded as dependent on the Good, but on the contrary, the Good itself is made to depend on the law. This means that the law no longer has its foundation in some higher principle from which it would derive its authority, but that it is self-grounded and valid solely by virtue of its own form. For the first time we can now speak of THE LAW, regarded as an absolute, without further specification or reference to an object. Whereas the classical conception only dealt with the laws according to the various spheres of the Good or the various circumstances attending the Best, Kant can speak of the moral law, and of its application to what otherwise remains totally undetermined. The moral law is the representation of a pure form and is independent of content or object, spheres of activity or circumstances ...

'But there is yet a further dimension ... which is correlated with and complementary to the first. The law can no longer be grounded on the superior principle of the Good, but neither can it be sanctioned any more by recourse to the idea of the Best as representing the good will of the righteous. Clearly THE LAW, as defined by its pure form, without substance or object or any determination whatsoever, is such that no one knows nor can know what it is. It operates without making itself known. It defines a realm of transgression where one is already guilty, and where one oversteps the bounds without knowing what they are, as in the case of Oedipus. Even guilt and punishment do not tell us what the law is, but leave it in a state of indeterminacy equalled only by the extreme specificity of the punishment' (Deleuze 1991 [1967]: 82–83).

Deleuze differentiates between classical law and modern law; and he will pursue this difference as a standpoint from which to ascertain the unique qualities of the political theatres brought to life through Sade and Masoch's texts. And the key to understanding the modernity of Sade and Masoch from this standpoint is to appreciate the ways in which their stories differently create counterparts of a world that is shaped by the groundlessness of modern LAW, counterparts capable of containing that world's violence and excesses. Each writer takes up a relation with the absence of any principle of the Good upon which the authority of law could be grounded. I will call this groundlessness *the ethical hollow of modern law*, which will have set parameters by which Deleuze differentiates the scenography of Socrates' death, on the one hand, from the scenographies of Sade and Masoch, on the other.

'The classical irony and humour of Plato that had for so long dominated all thinking on the subject of the law', writes Deleuze, 'are thus turned upside down. The upper and lower limits of the law, that is to say the superior principle of the Good and the sanction of the righteous in light of the Best are reduced to nothingness. All that remains is the indeterminate character of the law on the one hand and the specificity of the punishment on the other. Irony and humour immediately take on a different, modern aspect. They still represent a way of conceiving the law, but the law is now seen in terms of the indeterminacy of its content and of the guilt of the person who submits to it' (Deleuze 1991 [1967]: 85).

Deleuze alludes to a kind of predicament of the subject in its relation to the ethical hollow of modern law, and he will work through its implications for how the logics of specifically modern irony and humour manifest themselves. And what he reasons is that Sade and Masoch put into motion scenographic logics that orient differently to the predicament of a necessary but indeterminate *guilt* wrought by the ethical hollow of modern law. The delicacy of thinking here is vintage Deleuze, because with it he will go on to establish a conceptual passageway, from the distinctiveness of ascent and descent that are proper to the modern milieu, to the specificity of the administration of pain in Sade and Masoch's respective scenographies: 'We now note a new attempt to transcend the law, this time no longer in the direction of the Good as a superior principle and ground of the law, but in the direction of its opposite, the Idea of Evil, the supreme principle of wickedness, which subverts the law and turns Platonism upside down' (Deleuze 1991 [1967]: 87).

Counter-intuitively, Deleuze characterizes Sade as ironic in virtue of his ascent to a transcendent principle – Evil; and what is so strange about this ascent is that it is actually not available as such, given the specificity of the modern conception. Without a principle of the Good, a principle of Evil is clearly a ruse. Yet Sade creates a scene in which a principle of Evil can flourish, and as they engage it Sade's characters inhabit a scene that actualizes both the realm of a grounding principle (which is actually absent as such in the modern conception) and the guilt that is proper to law's ethical hollow. They 'turn' the (absent) realm of Platonic principles 'upside down' by ascending to an (absent) Evil that is the (implied) opposite of the Platonic Good as an (absent) grounding principle of law.

How do Sade's heroes ascend to an absent principle, of Evil no less? They manifest Evil in their arbitrarily dirty deeds. Effectively, both the top and the bottom of sadistic play come to embody a principle of Evil through acts of indifferent obscenity.[6] The administration of pain works in these scenes not as an instrument of punishment, but as a medium of a transgressive play that could ground the necessary but indeterminate guilt of the subject of law's ethical hollow: *if I am always already guilty, for LAW has no grounding principle, I shall bring forth and revel in an Evil, similarly indeterminate, which can ground it.* The administration of pain is thus a theatrical medium for an Evil that could ground or justify the guilt that is proper to the ethical hollow of modern law. It is from this standpoint that the administration of pain realizes the peculiar ironic logics of Sade's scenes as a theatre of ascent, grasped as a response to the predicament of the subject in its relation to the ethical hollow and self-grounding of modern law. The irony consists in the palpable gesture to an Evil that is itself ungrounded, arbitrary, hollow. And in this ironical gesture, evil speaks back as the fictitious grounding principle of the ethical hollow of modern law.

The administration of pain will be manifested in Masoch's scenographies as well, but through a completely different logic. For Masoch, writes Deleuze, 'attacks the law on another flank. What we call humour – in contradistinction to the upward movement of irony toward a transcendent higher principle [whether of Good or of Evil] – is a downward movement from the law to its consequences. ... The law is no longer subverted by the upward movement of irony to a principle that overrides it, but by the downward movement of humour, which seeks to reduce the law to its furthest consequences' (Deleuze 1991 [1967]: 87–89).

A descent through the consequences of the ethical hollow of modern law will look very different from the standpoint of the administration of pain. Like Sade, Masoch also orients to the realm of indeterminate guilt that is proper to the predicament of the subject's relation to the ethical hollow of modern law. But instead of ascending to a theatrically manifest Evil that could ground that guilt, his characters manifest and revel in a *punishment* that such guilt would demand. Abstaining from an appeal or ascent to principles (whether of the Good or of Evil), his characters instead actualize a scenography of presupposed guilt and associated punishment, which are the necessary correlate or consequence of the ethical hollow. The assertive bottom, Masoch's hero, pursues the consent of another subject, who will in turn become the agent of her own punishment and humiliation.

And what a reader will notice in the stories of Masoch is a subject garnering her own punishment for evidently *absent* crimes. The administration of pain is a theatrical medium for the punishment of an indeterminate, but necessary, guilt. Having pain inflicted on one's self is an act of punishment proper to the guilt secured by the indeterminacy of self-grounding law: *if I am always already guilty, I shall seek to be punished for it (though we do not know what it is).* It is from this standpoint, suggests Deleuze, that the administration of pain in the stories of Masoch is humorous – a descent to the consequences of the subject's relation to the ethical hollow of modern law.

For Deleuze, then, Sade and Masoch are not just the authors of particular stories, but names given to the movement of particular social relations that are 'contained' in the scenes that their texts create. With sadistic irony and masochistic humour we should read two distinct scenographic logics, each with its own proper top and bottom, in which relations between subjects are configured – the infliction of pain in each is in an important sense a medium for the realization of relational movements that emerge between the respective heroes.

The specifically masochistic bottom actively seeks out a potential but at first unwilling top, educates and persuades her to enter into a contract and engages her in a social relation of the bottom's own design. From this standpoint, the masochistic bottom is 'in control' of the process of securing her punishment, not because she exploits the existing desires of the top to inflict pain, but rather in so far as she is the active agent who sets into motion the contractual relation that could secure the submission of the top to her position as its administrator. A sadistic bottom, in contrast, enacts no such agency or design. Rather, she is the instrument of a willing other, for whom the infliction of pain manifests as a site of obscenity; the sadistic bottom isn't quite 'into it', like, or for the same reasons as, the masochistic bottom.

Whereas the masochistic pair enter into a kind of suspenseful script to which both subjects have, albeit begrudgingly, given their mutual consent, the sadistic pair enter rather directly into a process of clinical execution of just about all that emerges as a (de)neg(r)ation of propriety. Put bluntly, sadistic activity is engaged in a spirit of obscenity, through bodily acts undertaken with such precision that the possibilities for play are enumerated in exact proportion to those in relation to which they could be defined as a transgression. Sadistic eroticism is characteristically devoid of the kind of romance that emerges in the masochistic bottom who feels love in the coldness of her top. Crystallizing the difference, as Deleuze put it, 'a genuine sadist could never tolerate a masochistic victim' (Deleuze 1991 [1967]: 40).

Theatrical Philosophy?

In light of Deleuze's treatment of the distinction between the classical and modern conceptions of law, we actually have something like a fourfold typology of the irony – humour distinction. On the one hand, we have Socratic irony and humour as ascent and descent conjoined through the upper and lower limits of the scale of law, which is grounded by a transcendent principle of the Good, apropos the classical conception. On the other hand, we have a split that is predicated upon the absence of a grounding principle of the Good and the indeterminacy of the guilt that is the correlate of the ethical hollow of modern law. Sadean irony responds through an ascent to an absent principle of Evil, which it manifests theatrically, and which could properly ground modern, indeterminate guilt. Masochistic humour responds through a descent through a punishment that is necessary to and follows from the ethical hollow, a punishment that is manifest theatrically as enjoyment, in spite of the absence of any apparent crime.

In 'Coldness and Cruelty', the difference between the classical and modern conceptions matters to Deleuze, in that the latter conception calls up particular theatrical practices through which a relation to the ethical hollow is brought about.

Plato's story of the death of Socrates can presuppose a principle of the Good, which grounds both ascent and descent; but neither Sade nor Masoch can presuppose such a principle. Indeed, for Deleuze, the exercise in which they are engaged is very precisely to respond to the predicament of that absence. In this sense, for example, Sade's heroes must enact a realm of grounding principles through their deeds. That the principle that they enact is Evil is, from the standpoint of the difference between the classical and modern conception, perhaps less crucial than that the principle must be enacted, posited theatrically as such. Similarly, Masoch's heroes must enact a guilt and punishment, which is a descent through the consequences of the ethical hollow of modern law. The theatrical nature of this guilt and punishment consists significantly in that they are manifest in the face of no apparent crime.

Resonant with Stengers' eventual uses of the categories of irony and humour in *Invention*, Deleuze's subsequent references to the literary distinction suggest that the categories may undergo a recrafting towards the possibility of serving, in turn, as more general philosophical categories that can be used to characterize phenomena beyond those of the literary text.

For example, in early pages of *Difference and Repetition* (2004a [1968]) Deleuze writes, 'If repetition is possible, it is as much opposed to moral law as it is to natural law. There are two known ways to overturn moral law. One is by ascending towards the principles: challenging the law as secondary, derived, borrowed or "general"; denouncing it as involving a second-hand principle which diverts an original force or usurps an original power. The other way, by contrast, is to overturn the law by descending towards the consequences, to which one submits with a too-perfect attention to detail. By adopting the law, a falsely submissive soul manages to evade it and to taste pleasures it was supposed to forbid. We can see this in demonstration by absurdity and working to rule, but also in some forms of masochistic behaviour which mock by submission. The first way of overturning the law is ironic, where irony appears as an art of principles, of ascent towards the principles and of overturning principles. The second is humour, which is an art of consequences and descents, of suspensions and falls' (Deleuze 2004a [1968]: 5–6).

In this passage, Deleuze continues to differentiate irony and humour – as an ascent to principles, on the one hand, and as a descent through consequences, on the other – as he had done in 'Coldness and Cruelty', and as he would continue to do in *The Logic of Sense*. He also extends the category of humour in such a way as to characterize an empirical phenomenon – the work-to-rule strike. But, if we read this passage through the conceptual achievements of 'Coldness and Cruelty', there appears to be a possible reduction of the fourfold possibility to two,[7] and a possible, associated slippage within the comparison across the classical and modern conceptions of law as milieu that set the frame for the practices of irony and humour. The slippage across these two milieux is buttressed by the practice of highlighting only the distinction between ascent and descent, which understates the differences between the classical and modern conceptions of law and how they shape and give meaning to the practices of irony and humour.

In the passage above, subtle cues are given which suggest that different conceptions are at work in the comparison. Irony entails an ascent towards a grounding principle, which exposes the law as second-hand and as a diversion of an original force. The effect of the ascent to such a principle is to expose the derivative nature of law. This suggests a framing by the classical conception: there is a transcendent principle of the Good to which one might appeal through ascent, so as to overturn the legitimacy of an existing moral law.

Humour, in contrast, is described as a descent through consequences and entails the adoption of the law. Here, there is no allusion to a relation between the subject and a transcendent principle in its relation to law, but rather to a relation that is reduced to the subject and law alone. Also significant here is that the relation between the subject and law is realized through the practices of what Deleuze calls a *falsely* submissive soul. This suggests a modern conception: rather than the law and a realm of principles in which it might be grounded, we have a law alone, to which the subject pretends to submit. And Deleuze makes the point that she submits falsely, suggesting that the difference between false and true submission matters.

The falsely submissive soul engaged in descent is given literary texture, moreover, by reference to the vernacular 'masochist behaviour'. No symmetrical literary texture by reference to sadistic behaviour is lent to the ironist. And if we recall that in 'Coldness and Cruelty' sadism and masochism are reserved to refer to the scenographic logics proper to the ethical hollow of modern law, then the possibility arises that Deleuze is implicitly contrasting Socratic irony (classical conception) and Masochistic humour (modern conception) in the passage above. Which is fine, of course, though it does suggest that the specificity of sadistic or theatrical irony (as contrasted both to Socratic irony and to masochistic humour) is somehow being subjected to a kind of erasure.

If we read a bit further moreover, it becomes possible that such an erasure is buttressed by Deleuze's extended reading of both irony and humour under the sign of repetition.[8] He immediately continues: 'Must we understand that repetition appears in both this suspense and this ascent, as though existence recommenced and "reiterated" itself once it is no longer constrained by laws? Repetition belongs to humour and irony; it is by nature transgression or exception, always revealing a singularity opposed to the particulars subsumed under laws, a universal opposed to the generalities which give rise to laws' (Deleuze 2004a [1968]: 6).

In the passages that follow, Deleuze develops and contextualizes the prior discussion of irony and humour through an extended commentary on the philosophical strategies of Nietzsche and Kierkegaard as philosophers of repetition who 'bring philosophy a new means of expression' (Deleuze 2004a [1968]: 9). And what we want to notice here is that each of them does philosophy as a kind of theatrical exercise, as a scenography. This resonates with the practices that Deleuze earlier recognized in the works of Sade and Masoch. To be sure, notes Deleuze, there are differences between Kierkegaard and Nietzsche, primarily in terms of the tactics of their heroes (Kierkegaard's Job and Abraham, Nietzsche's Zarathustra). But importantly, what they share is a relation to Kant, and, more precisely, a relation to Kant that is in opposition to the philosophical strategies of Hegel in his relation to Kant. Their opposition to Hegel in his relation to Kant is achieved through the quality of inventing 'an incredible equivalent of theatre within philosophy, thereby founding simultaneously this theatre of the future and a new philosophy' (Deleuze 2004a [1968]: 9).

Whereas Hegel 'remains in the reflected element of "representation", within simple generality ... represents concepts instead of dramatizing ideas', Nietzsche and Kierkegaard set in movement theatres of repetition: 'When Kierkegaard explains that the knight of faith so resembles a bourgeoisie in his Sunday best as to be capable of being mistaken for one, this philosophical instruction must be taken as the remark of a director showing how the knight of faith should be *played*' (Deleuze 2004a [1968]: 10); 'When Nietzsche says that the Overman resembles Borgia rather than Parsifal, or when he suggests that the Overman belongs at once to both the Jesuit Order and the Prussian officer corps, we can understand these texts only by taking them for what they are: the remarks of a director indicating how the Overman should be "played"' (Deleuze 2004a [1968]: 11).

The crucial distinction between the Hegelian response to Kant and those of Nietzsche and Kierkegaard, then, will be the preponderance of the philosophy of representation in the former, and the preponderance of a theatrical philosophy in the latter. Put bluntly, Hegel responds like a Platonist to Kant. And in a radical transformation, philosophers of repetition work through a theatrical space, by which Deleuze meant 'the emptiness of that space, and the manner in which it is filled and determined by the signs and masks through which the actor plays a role which plays other roles' (Deleuze 2004a [1968]: 11). Significantly, both Nietzsche and Kierkegaard are capable of constructing theatres through which even the role of the Platonist can itself be played by other roles.

This positioning of Nietzsche and Kierkegaard as philosophers of repetition in opposition to Hegel's Platonic representationalist frame strikes me as reminiscent of Deleuze's differentiation of Sade and Masoch's theatres from those of Plato in 'Coldness and Cruelty', in the very precise sense that there he reads them as a contrast to Socrates' death from the standpoint of the specificity of the classical conception of law. In Deleuze's hands, Plato does not need the theatrical realization of a principle to which Socrates might ascend, for example, because the classical conception itself implies that the principle to which he ascends (and from which he descends, for that matter) is the principle of the Good, which motivates Socrates' death as such. In short, Socrates' death itself occurs within and is motivated by a world in which law is conceived representationally, as reflection, and which is as such indistinguishable from and therefore without need for any scenographic counterpart. Sade and Masoch, in contrast, work precisely in relation to the absence of such a grounding principle; they work in relation to a milieu for which there is no such Platonic commitment.

If we set these two texts next to each other – 'Coldness and Cruelty' (1991 [1967]) and *Difference and Repetition* (2004a [1968]) – then, we find both a resonance and an augmentation. The resonance is that each book marks repetition as a theatrical practice and as a response to Kant, and pits that response against its Platonic representationalist alternative. The augmentation is that, in *Difference and Repetition*, rather than accentuating the distinction between classical law as set out in the works of Plato and modern law as set out in Sade and Masoch, there is an amplification of a distinction between the (Platonic) representational philosophical strategy of Hegel in his response to Kant's moral law and the theatrical philosophical strategies of Nietzsche and Kierkegaard in their responses to Hegel's strategy. Deleuze differentiates Nietzsche and Kierkegaard's theatrical strategies from the philosophical frame of representation, which is traced to Plato and found at work in Hegel's critique of Kant. Although the figures of Sade and Masoch do not manifest explicitly in these passages of *Difference and Repetition*, we find a treatment of Nietzsche and Kierkegaard that resonates with the treatments of Sade and Masoch that are offered in 'Coldness and Cruelty'.

Thinking of the two together, then, it is possible that Deleuze will have created two sets of oppositions: sadistic irony and masochistic humour opposing Socratic irony/humour; Nietzschean and Kierkegaardian theatrical philosophy opposing Hegelian philosophical representationalism. And our question must be whether sadistic irony and masochistic humour are subsequently merged under the sign of the theatrical philosophy as a strategy that arises in opposition to philosophical representationalism as such. Our question must also be whether such theatrical philosophy subsequently manifests as the site and medium of Deleuzian 'humour', as contrasted to 'irony', which by now works as a blunt code for Platonism, or the classical conception of law.

Returning to Deleuze's subsequent characterization of humour in *Logic* (2004b [1969]), the suggestion now seems plausible. He writes, 'This exercise, which consists in substituting designations, monstrations, consumptions, and pure destructions for significations, requires an odd inspiration – that one know how to "descend". What is required is humour, as opposed to the Socratic irony or to the technique of ascent' (Deleuze 2004b [1969]: 154). Might not Stengers' subsequent differentiation of irony and humour further such a classificatory practice?

Such a reading is at least worth speculation and debate: that through the passages of irony and humour from 'Coldness and Cruelty' to *The Logic of Sense* Deleuze becomes an advocate of humour who has situated himself within a more overarching philosophical frame of repetition – under the sign of which *both modern irony and modern humour are framed as humorous strategies*. Are we thus offered in his later advocation of humour the possibility of sadistic irony and masochistic humour as strategies proper to a philosophy of repetition, each of which differs from the Platonic politico-philosophical frame, now read exclusively under the sign of 'irony', in terms of which the ascent and descent of Socratic irony and humour had been specified? If so, even appeals to a transcendence, appeals to a principle that grounds one's power to judge – so long as they are the expression of a role that plays other roles – might be recast as arts of immanence proper to theatrical practice.

Conclusion

In this chapter I have visited Deleuze's characterization of sadistic irony and masochistic humour in 'Coldness and Cruelty' (1991 [1967]) as markers of distinctly *modern* political scenographies. I have also considered that distinctness in light of his subsequent discussion of Nietzsche and Kierkegaard in *Difference and Repetition* (2004a [1968]) as theatrical philosophers of repetition, who put metaphysics into motion. My suggestion has been that the Deleuze who advocates humour as a political strategy in *The Logic of Sense* (2004b [1969]), the Deleuze whose echo we hear in Stengers' invitation to humour, has situated himself in the philosophical frame of repetition, under the sign of which *both sadistic irony and masochistic humour are possible strategies.*

But is there a merging of sadistic irony and masochistic humour under the sign of theatrical philosophy as specifically *Deleuzian* humour? I remain open to what that might mean for Deleuze's subsequent works and for Stengers' Deleuzian call that students of the socialities of science engage in humour rather than irony. But, if history matters, then it is worth our consideration that Deleuze's attention to the specificities of the classical and modern conceptions does some work for him in 'Coldness and Cruelty'. That difference suggests that the metaphysical commitments of a milieu have consequences for the ways in which its participants enact the practices of irony and humour, ascent and descent, that occur in relation to them. Irony is achieved in particular ways when undertaken in a milieu in which a metaphysical commitment to the principled groundedness of law obtains; irony is achieved in different ways when undertaken in a milieu in which such a metaphysical commitment does not obtain.

But perhaps the difference between the two conceptions of law and their respective metaphysics begins to break down, no longer matters. Is there any meaningful difference between appealing to a transcendent principle, and theatrically enacting such an appeal? Is it perhaps the mark of a particular milieu when the difference between these cannot itself be established? Is this our milieu?

Notes

I thank Sisse Finken, Casper Bruun Jensen, Kjetil Rödje, Geof Bowker and an anonymous reviewer for inspiring feedback on earlier drafts. Writing was supported by a generous grant from the Netherlands Organization for Scientific Research, Shifts in Governance Programme Award 450–04–135.

1. It is worth mentioning here that Woolgar's piece works strictly with 'irony' and not the distinction between irony and humour.
2. The book was originally published in French under the title 'Le Froid et le Cruel' in *Presentation de Sacher-Masoch*, Editions de Minuit, Paris 1967. I have been working with the English translation (Deleuze 1991).
3. This view of sadomasochism remains a conception of many practising sadomasochists and their commentators today; we might notice the reproduction of Krafft-Ebing's own clinical construction through their practices.

4. Sources for Deleuze here were commentaries on Sade and Masoch that had been contributed by Georges Bataille and Maurice Blanchot. Citing Blanchot, Deleuze writes: 'in spite of the similarity of descriptions, it seems fair to grant the paternity of masochism to Sacher-Masoch and that of sadism to Sade. Pleasure in humiliation never detracts from the mastery of Sade's heroes; debasement exalts them; emotions such as shame, remorse or the desire for punishment are quite unknown to them' (Deleuze 1991 [1967]: 39).
5. My strategy here turns on the idea that it is more fruitful to understand how Deleuze uses technical terms in particular textual contexts, rather than trying to ascertain either what those terms must really mean for him or whether his uses of the terms are consistent across texts.
6. That they are obscene is obviously arbitrary, since there is no actual Good that provides the principle on the basis of which transgression is established.
7. That is, Deleuze alludes to a distinction between irony and humour, without specifying them with respect to the distinction between the classical and modern conceptions of law.
8. The trajectory of the extended reading is actually indicated in the opening sentence of the passage, as quoted above.

Bibilography

Deleuze, G. (1991) [1967]. *Masochism: Coldness and Cruelty and Venus in Furs.* New York, Zone Books.

———. (2004a) [1968]. *Difference and Repetition.* London, Continuum.

———. (2004b) [1969]. *The Logic of Sense.* London, Continuum.

Kant. I. (2004 [1788]). *The Critique of Practical Reason.* Available at http://www.gutenberg.org/etext/5683

Knuuttila, T. (2002). 'Signing for Reflexivity: Constructionist Rhetorics and Its Reflexive Critique in Science and Technology Studies' [52 paragraphs]. Forum: Qualitative Research. [Online Journal], 3 (3). Available at http://www.qualitative-research.net/fqs/fqs-eng.htm

Stengers, I. (2000). *The Invention of Modern Science.* Minneapolis, University of Minnesota Press.

Woolgar, S. (1983). 'Irony in the Social Study of Science', in K. Knorr-Cetina and M. Mulkay (eds) *Science Observed.* London, Sage Publications, pp. 239–66.

Chapter 4
Between the Planes: Deleuze and Social Science

Steven D. Brown

> Books against structuralism ... are strictly without importance; they cannot prevent structuralism from exerting a productivity which is that of our era. No book *against* anything ever has any importance; all that counts are books for something, and that know how to produce it. (Deleuze 2004: 192)

The 'Social' and the 'Scientific' in the Social-Scientific Procedure of 'Social Science'

Electric glare of a strip-lit lecture room without windows. Too many people in the audience to make the seating comfortable. The speaker dwarfed by the gigantic power-point screen. She wanted a laser pointer, but finding none is now resorting to expansive hand-waving gestures for emphasis. The next slide. A brief animation, she announces. Two small triangles, one slightly bigger than the other are contained in a larger square with one side removed. 'Who would say that these shapes have intentions?' she asks, with obvious rhetorical flourish. The audience obliges by remaining silent. Click. The animation begins. The triangles now move around one another. They follow a pattern of movement that mimics human or animal behaviour. The small triangle appears to 'push past' the larger, which 'blocks' its 'attempts' to 'leave' the square. It takes little effort to attribute a script of parent–child relations. Eventually the two triangles leave the square 'house' and 'dance' together, spinning around one another. 'And now?' the speaker asks again. The audience is pleased to go along with her point.

This is social science, twentyfirst century style. The topic of the talk is the cognitive-neuroscientific basis of autism. The animations are not central to the main thrust of the talk. They are there to exemplify an approach to autism known as 'theory of mind' (see Baron-Cohen et al. 1985; Baron-Cohen 1997). According to this approach, the 'problem' of the autist is that they the lack to ability to ascribe the fundamental properties of conscious mental life, which they themselves experience, to other people. The autist would never see mummy and baby triangle fighting and dancing. They would merely see a very poorly executed animation of geometric shapes. Having thus introduced theory of mind the speaker can move to the main business – discussing regions of the brain where the various cognitive operations that she thinks are involved in the process of ascribing mental attributes to others must reside (and which must therefore be impaired or damaged in autism). For example, one site must, she claims, be involved or 'active' during the perception of movement, which curiously is related to the simultaneous activation of other sites that govern the execution of those same movements within the perceiver. In other words, it seems to be the case that there is a hard-wired circuit in the human brain for recognizing and mimicking the behaviour of others.

How is the speaker able to make these weighty claims? By showing a series of beautifully coloured pictures of brains that are produced by functional magnetic resonance imaging (fMRI). The 'owners' of these brains have been invited to lie in the fMRI scanning equipment whilst observing images of various kinds of movement (e.g. tennis playing). The equipment is then able to discern higher and lower patterns of neural activity, which can be represented on coloured scans of the brain. These scans can then be compared with current cartographic knowledge of neural pathways and regions. If two parts of the brain are simultaneously activated, this indicates the existence of some kind of circuit, which can then be fitted onto a theoretically derived account of the cognitive processes implicated in perception and action. The speaker can then seemingly move from biology to psychology in the single gesture of showing the coloured fMRI scan. From recognition/representation/action to light patterns/neural firing/muscular activation. Or, more simply, from philosophy to matter.

In the course of an essay on Henri Bergson's *Matter and Memory,* Maria Cariou (1999) remarks that medical imaging and neuroscience are effectively performing metaphysical speculation in their claims about brain functions. She then goes on to assert that the demolition of neural-based theories of representation that Bergson accomplishes in his work applies equally well to neuroscientific arguments that authorize themselves on the basis of fMRI scanning. To paraphrase Bergson, the problem is not necessarily with the accuracy of the scans themselves but rather with the metaphysical complement that is bolted on to the images. As is well known, Bergson's mistrust of an inadequate conceptualization of time leads him to suspect that the metaphysics commonly used in relation to neurology incorrectly attempts to spatialize thought and project it directly on to the cartography of the brain itself. The brain is then misunderstood to be the container of experience. Against this view, Bergson argues for an understanding of the brain as the mediational means through which time, as duration, is inserted into action, which unfolds in relation to spatial coordinates. The brain is the 'and' in 'matter and memory', as Worms (1999) beautifully observes.

The case above amply demonstrates Cariou's and Worms' points. 'Theory of mind' is a peculiarly truncated species of argument that belongs in the general orbit of the 'other minds problem' within the philosophy of mind. Classically this problem is concerned with reasoning how it is that we might support the claims we might want to make about the mental states of others, and, more fundamentally, the means through which it might be possible to ascribe a mental state to the other (fashioned on our own) in the first place (see Ryle 1946; Strawson 1959, Wisdom 1968). The particular difficulty with the application of theory of mind to autism arrives when the small change of its philosophical argument is inspected. It is all very well to argue that mental qualities are inferred from behaviour and subsequently ascribed to the 'target' object by the subject (assuming we can manage the immediate twitchiness that this problematic dualism sets off). But it is quite another thing to argue that the kinds of qualities we might want to momentarily attribute to a moving triangle are seriously akin to those we would want to apply to our interactions with other humans. Moreover, it is quite a stretch to say that we would want to claim that the triangle has 'intentionality' in the full-blown philosophical sense. Or, put slightly differently, we are being asked here, on the basis of the loose coupling of a very simple, albeit ingenious, illusion and a very complex, albeit simplified, image of the human cortex, to take a position on a long-standing metaphysical problem that would have professional philosophers running for cover.

We cannot then take our speakers claims seriously in formal philosophical terms. But what about in relation to the standards of exact science? Here again the argument is somewhat problematic. The coloured regions of the fMRI image that the speaker points to may indeed be reasonably glossed as representing 'neural activity' (although we might be tempted to historicize the formulation of this notion – see Star 1989). But the characterization of such activity as 'involved in the perception of movement' and going on to claim that this hard-wired neural capacity is involved in a cognitive process of attributing intentionality can only hold if we are first convinced of a 'dual-aspect' approach to the neuropsychological, where the empirical claims of brain imaging are conjugated with the empirical claims arising from experimental psychology (in this case evidenced by questionnaire responses made by participants to an inventory designed to test attributions of intentionality). This leap between the brain and theoretical claims about the kind of emergent properties of the cognitive architecture that constitutes 'mind' is submerged in the overall argument.

What we are confronted with in this particular application of 'theory of mind' is an argument that mobilizes both philosophy and exact science in the service of constructing an account of autism, but simultaneously fails to be entirely plausible in either formal philosophical or neuroscientific terms. However, this is for a good reason, because 'theory of mind' is in actuality solving neither a philosophical nor a neuroscientific problem. What it instead addresses, I would venture, is the social problem that the autist, and in particular the autistic child, creates for others: 'Why won't she eat properly?', 'Why doesn't he appear to love me?', 'Why won't they get dressed?', 'Who will stop all this screaming and crying?' This discomfort is itself shorthand for the extremely complex dynamics that are structured by the cultural demands of adult–child interactions, the ways these interactions become enveloped by the institutional axioms of education and the manner in which social welfare policies shape the possibilities for living that are made available for autistic children and their parents or carers. Theory of mind articulates all this very real discomfort by offering an account of autism that safely locates the causes of 'autism' within the child themselves, through the dual epistemic technologies of a special version of the 'other minds' problem and highly particular interpretations of fMRI images. In so far as parents, educators, welfare providers and state agencies can all see in 'theory of mind' a settlement (if not a resolution or a proper explanation) to their particular concerns, then the account may usefully serve as a way of coding the various flows of love, learning and finance that envelop the autistic child.

What I have been seeking to illustrate in this extended example is a general dilemma that social science confronts. The objects of social science are in every sense 'overdetermined'. They are complex admixtures of cultural, political and moral ferment, bereft of any clear ontological surety, and unmoored from any definite epistemological framework. In this sense 'the autistic child' is every bit as ambiguous and contested an object as 'the authoritarian personality', 'division of labour', 'the market' or 'the end of history'. Moreover, as Ian Hacking, for one, has described it, these are objects that are not stable kinds of things but are rather 'interactive kinds', whose very nature is shaped by the unfolding epistemic debates or 'looping effects' in which they are perpetually submerged (Hacking 1999, 2004). In order to stabilize such recalcitrant objects, social scientists have typically found themselves dependent upon the tools and technologies of philosophy and the exact sciences. Philosophy provides the basic vocabulary in which to speak of social scientific objects – e.g. motives, intentions, goals – and underpins this with a formalized common sense embedded in the philosophical grammar of 'the mind–body problem', 'the other minds problem', 'the correspondence theory of truth' and so on. For its part, science supplies the technologies in which objects thus described become amenable to measurement, calculation, division and distribution.

Social science then finds itself tasked with providing 'political settlements' – quasi-explanations that can recruit stakeholders (Latour 1999) – regarding complex objects over which it is able to exercise neither adequate control (i.e. as the chemist does at the laboratory bench or the particle physicist at the generator) nor convincing authority (i.e. as with the philosopher or mathematician who delivers the generally accepted proof or solution). Small wonder then that social scientists have typically entered into denial over the political nature of the settlements they make, and seen themselves as involved in providing adequate conceptual accounts of social phenomena (i.e. philosophy manqué) or imputing clear laws or regularities based on the calculation of supposedly clear chains of causal relations (i.e. faux exact science), or, in the classic instance, as performing a deft sleight of hand that combines both gestures.

To characterize social science in this way is to arrive at a thoroughly negative definition of an epistemic practice caught uncomfortably between two better formalized practices. Clearly this is insufficient. A positive definition of social science is required, not least as a means of authorizing the subject positions of individual social scientists. But how might this be accomplished? Classical epistemology insists on treating social science within the overarching rubric of 'scientificity'. From this perspective social science is perpetually doomed to be regarded as lacking formalization and/or an appropriate paradigm (Kuhn 1962; Popper 1968). Alternatively, social science may be characterized as departing from Husserlian roots as a project aimed at clarifying and describing human experience: in which case it is inevitable that social science be seen as the applied wing of philosophical movements (e.g. Wittgenstein, Heidegger, Foucault) that are inevitably superior in terms of their rigour and precision (this I take to be the message of Winch 1958). A positive definition is not easily accomplished because it can seemingly only be gained through recourse to those practices through which the negative definition is entrenched.

What is seemingly required is a philosophical authority who demands a complete inversion of logical and metaphysical common sense – an 'anti-philosopher' – whilst simultaneously levelling the epistemological differences, the bifurcations in knowledge that erect science and philosophy as the twin sources of legitimation for social science. The groundswell of interest amongst social scientists in the work of Gilles Deleuze (and correspondingly, although to a lesser extent, Felix Guattari) is then understandable. In the rest of this chapter I want briefly discuss what I take to be the major features of this appeal before describing some of the immediate obstacles to a social scientific engagement with Deleuze (and Guattari). I shall then go on to explore how Deleuze's 'assemblage theory' has been directly mobilized in the work of Manuel de Landa, whom I shall treat as an exemplary case of Deleuzian interpretation in and for social/cultural theory. I shall argue that, despite the often dazzling quality of his analyses, the thrust of his contribution is to efface rather than clarify the place of social science. By way of avoiding a dispiriting conclusion, I offer a few possibilities for reconsidering what it is that social science makes of Deleuze.

The Plane of ... Social Science?

As the contributors to this and several other recent edited collections (e.g. Buchanan and Colebrook 2000; Massumi 2001; Buchanan and Lambert 2005; Fuglsang and Sørensen 2006) make clear, there is a great deal in Deleuze's work that is of interest to social science. At the risk of extreme simplification let me note four general areas.

First, the appeal of Deleuze's 'flat ontology' as promised in the enigmatic demand to 'arrive at the magic formula we all seek – PLURALISM = MONISM – via all of the dualisms that are the enemy, an entirely necessary enemy, the furniture we are forever rearranging' (Deleuze and Guattari 1988: 20–21, see also the introduction). Social scientists have spent a great deal of time rearranging their ontic furniture – see Archer (1995) and Edwards, Ashmore and Potter (1995) for two very different design tastes. What is at stake here is managing the idealism-realism tension in such a way that social science can claim to be both analysing objects 'out there' in the world, whilst simultaneously claiming that such analysis 'adds something' to the object that transforms its role in human affairs (see Woolgar 1988; Lynch 2004). Thus, in the case of autism, it is vital to maintain the essential 'reality' of the condition, but also emphasizing the way it is transformed as it is differentially configured in various social practices (hence talk of an 'autistic spectrum'). The notion that social objects might be considered as intrinsically 'unfinished', 'processual' and subject to 'continuous variation', but at the same time subject to codings and repetitions, clearly promises much (see Deleuze's (1991a) discussion of the work of Sacher-Masoch as a paradigm case of how to treat a social object in terms of its capturing by formal knowledge and as simultaneously eluding and exceeding such knowledge).

Secondly, the critique of common sense developed at length in Deleuze's key work *Difference and Repetition*. Social science in a post-Husserlian mode only avoids grand claims to have imputed law-like regularities by inviting the risk of being seen as merely providing an alternative, and often extremely complex, language to redescribe common wisdom (a claim that has been made, for example, regarding Latour and Woolgar's *Laboratory Life* (1979)). As Deleuze shows at length, such a claim is premised on the notion that analysis amounts to the exercise of thought endowed with intrinsic 'good will' or 'common sense' toward a world that will deliver itself up to 'sound thinking'. Deleuze inverts this scheme by arguing that thinking only really begins when its faculty is confronted by a 'sign' that throws it into 'discord' or a 'violence which brings it face to face with its own element' (Deleuze 1994: 141). Out of such confrontations 'sense' emerges as para-cognitive perception of a 'problem'. In other words, problems are not delivered up to perception, but are continuously created and recreated by an exercise of thought that overturns common sense. With autism, approaches such as 'theory of mind' would be immediately viewed with suspicion since they merely affirm common sense rather than seeking to overcome it to genuinely start thinking about what defines the complex problem space of 'becoming autistic'.

Thirdly, the demolition of the 'subject' accomplished in *Anti-Oedipus* offers a comprehensive rebuttal to the idea, commonly found in social science, that it is necessary to constitute a theory of the subject as a means of accounting for how persons 'invest' in discursive fields (see for example the newly emerging field of psychosocial studies –Hollway and Jefferson 2000; Frosh 2003). In the royal place of the subject, Deleuze and Guattari (1984) offer instead three transpersonal 'syntheses of the unconscious' which are indexed to singular 'desiring machines'. Individuality is then dramatically reinstated as a 'machinic effect' of how these three syntheses act to connect, divide and turn around on material flows. It then becomes unnecessary to ask how, for instance, 'the autistic child' is forced to adopt a subjectposition in relation to medico-welfare discourse, since this narrows analysis to subjects and language. We ought instead to ask how the child is distributed across the 'factory-machines' of school and home, and how they are plugged into the various flows of care, learning, control, etc. that link each site.

Fourthly, Deleuze (and Guattari) offer an account of power that transcends many of the limitations of both classical neo-Marxist and more recent Foucauldian notions.

With regard to the former, *A Thousand Plateaus* introduces a set of terms such as molar/molecular, arboreal/rhizomatic and Body without Organs/Strata that facilitate the characterization of a given assemblage in terms of its relative degrees of territorialization (i.e. boundedness, consistency, regimes of coding). This enables the structural oppositions of classical Marxism (e.g. base/superstructure) to be understood as emergent effects with varying degrees of scale rather than causes. With respect to Foucault, Deleuze emphasizes that 'desire' is primary in relation to power – 'for me, power is an affection of desire (granted that desire is never a neutral reality)' (Deleuze 2006a: 125). It is desire – the productive onwards drive to connect – that animates an assemblage (hence the use of the term 'desiring machines' in *Anti-Oedipus*). Power arises secondarily, as Spinoza once described it, as the form taken by desire when it turns around on its own connections and elaborates and re-entrenches them.

Taken together, these four hastily sketched domains indicate many fruitful ways in which Deleuzian thought can animate social science. But there is an immediate problem. What kind of status can be given to 'Deleuzian social science'? Is it constituted merely by the borrowing of a set of terms that although they have precise philosophical value, function only in a purely arbitrary and descriptive sense when mobilized in relation to social scientific objects? Matters are not helped by the distinctions that are drawn by Deleuze and Guattari in *What is Philosophy?*. In this text an apparently retrograde move is made to characterize philosophy as a project of inventing concepts. However, the concept is not here treated in the usual sense of a concrete set of relations that exist in the form of 'a given knowledge or representation that can be explained by the faculties able to form it (abstraction or generalization)' (Deleuze and Guattari 1994: 11) and which correspond (more or less) to some state of affairs in the world. Instead Deleuze and Guattari insist that 'the concept is not given, it is created; it is to be created. It is not formed but posits itself in itself – it is a self-positing' (ibid.). The upshot of this self-positing is that the work of constructing a concept is at once the work of imposing a field of order – or laying out a 'plane of immanence' – in which the concept can be seen to have some purchase. Concept and plane are posited together, since the one directly implies the other. For example, the Cartesian *cogito* only posits itself as a concept once 'doubting', 'thinking' and 'being' are used as categories to order human activity (or 'intensive ordinates' for the concept). The *cogito* then 'throws bridges' between these ordered 'zones' on the plane.

In addition to philosophy, Deleuze and Guattari describe science as the process of laying out a 'plane of reference', a mode of organizing that describes the world as 'slowed down' into possible variables and relations of speed and slowness (i.e., 'states of affairs'), to which correspond 'functives' or 'systems of coordinates' that can be used to connect together such states of affairs. Finally, they offer an account of art in terms of the dual positing of a 'plane of composition', or a field of possible sensations, with 'affects' and 'percepts' – very roughly, structured ways of engaging with this field – being the other half of artistic production.

Nowhere in *What is Philosophy?* does one find any complementary 'plane of social science'. Indeed the sole references to social science occur in the opening chapter, where

Deleuze and Guattari discuss the historical 'rivals' with which philosophy has found itself confronted. These are currently 'computer science, marketing, design and advertising, all the disciplines of communication, seized hold of the word *concept* itself and said: "This is our concern, we are the creative ones, we are the *ideas men*! We are the friends of the concept, we put it in our computers"' (Deleuze and Guattari 1994: 10, emphasis in original).

The disdain that Deleuze and Guattari here express for the 'disciplines of communication' (which I take to include modern social science) comes from what they see as the corruption of the concept, which in terms of their threefold scheme of planes can be said to properly belong to philosophy alone (although the latter chapters of the book do hint that concepts might productively traverse other planes). The treatment of social science is reproduced elsewhere in Deleuze's work. For example, in his introduction to Jacques Donzelot's *The Policing of Families* (reproduced as Deleuze 2006b), Deleuze writes that:

> The question here [the rise of the social] has nothing to do with the adjective that used to qualify the group of phenomena encountered in sociology: THE social refers to a *particular sector* in which very diverse problems are categorized as needed, special cases, specific institutions, a category of qualified personnel ('social' assistants, 'social workers'). We speak of social plagues, like alcoholism and drugs; social programs, from repopulation to birth control; social adjustment or maladjustment (of the pre-delinquent, emotionally disturbed or handicapped, including different types of advancement). (Deleuze 2006b: 113, emphasis in original)

In this preface Deleuze positions 'the social' not as the overarching term in which human affairs are to be grasped, but rather as a 'particular sector' or practice that initially emerges alongside 'law' and 'economics'. This sector comes historically to subsume these other practices in order to constitute new modalities of ordering – singular 'social' assemblages – where heterogeneous sets of problems become instituted and reworked. The social is, then, not an ontic term for denoting human affairs (classically contrasted to *physis*, or non-human affairs), but rather a 'hybrid domain' of intersecting assemblages, whose emergence and proliferation can be subject to historical and philosophical analysis. The practice of 'social science' can then only be that specialized form of philosophical anthropology that takes the 'rise of the social' as its particular object (see the recent work of Nikolas Rose, e.g. Rose (2001) as a good illustration of what such a practice might resemble).

Now in one sense what Deleuze does here is broadly in line with his treatment of philosophy, science and art. *What is Philosophy?* insists that these three epistemic practices are necessarily indexed to the historically determined 'milieux' in which they operate (thus Greek philosophy formalized the concept of 'friendship' in order to articulate the particular problematic of governance of the citystate). Philosophy is, at base, geo-epistemic, it is a geo-philosophical project. However, the particular treatment of philosophy, science and art by Deleuze and Guattari implies that they have a genealogy that is in some sense exemplary – they are trans-geo-epistemic. For instance, in the case of art, Deleuze and Guattari devote the entirety of chapter 11 of *A Thousand Plateaus* to exploring the link between art and territoriality, making a direct comparison between the ethological study of animal 'signs' and artistic production – 'The artist: the first person to set out a boundary stone, or to make a mark' (Deleuze and Guattari 1988: 316). Art, then, is different from social science, not merely in being accorded its own 'plane' but also in being seen as a practice that exceeds the confines of those human aesthetic-economic activities we now call 'the art world'. Much the same can be said of philosophy. Although *What is Philosophy?* claims, in a way mockingly similar to Heidegger, that philosophy was invented by 'the Greeks', elsewhere Deleuze expounds at length on Bergson's notion that 'concepts' arise from the selective engagements that the living organism maintains with the world (see Deleuze 1991b: chapter 5) and on Spinoza's pivotal postulate of 'common notions' as the means by which encounters between bodies can reorder relations between ideas (see Deleuze 1992: chapter 17). Indeed, clarifying the status of the 'idea' is a thread that extends from Deleuze's earliest work on Hume (Deleuze 2001), through *Difference and Repetition* to the late work on Leibniz (Deleuze 1993). Although this thread does not culminate in the questioning of the philosophical concept as such, it is clear that Deleuze situates his 'concept of the concept', and thus the business of philosophy itself, in a trans-historical reading of the status of the ideational. No corresponding line of thought exists in his work for the 'business of social science'.

Here then we have a strange puzzle. At no point does Deleuze offer a positive determination for what 'social science' might mean as practice in the way he does for philosophy, science and art. Social science falls between the three planes Deleuze and Guattari define. As I shall go on to describe in the next section, this creates real difficulties for social scientists who wish to engage with his work. But this is not to say that Deleuze shuns social science. Quite the reverse! A cursory glance at the organization and, in particular, the footnotes of *A Thousand Plateaus* displays a rich engagement with a varied range of social scientists, including, but not limited to, Johann von Uexküll, Emile Benviniste and Claude Lévi-Strauss, who are clearly taken seriously as thinkers (see the introduction for a more comprehensive list). Elsewhere, in a lengthy essay 'How Do We Recognize Structuralism?' (Deleuze 2004), the extent to which developments in late 1960s/early 1970s social science fed into Deleuze's evolving thinking of ordering practice and difference is made clear. As the editors point out in the introduction, it is not that Deleuze refuses to engage with social science (and in particular anthropology) but rather the terms under which this engagement is conducted.

We might, then, modify our question slightly. If Deleuze does not offer a positive determination of social science, does he offer something that makes the very question of a 'positive determination' manageable by transforming the terms in which social science might be conducted? For those who dislike the style of the faux protracted argument, let me give away my answer in advance: yes and no.

The Limitations of Assemblage Theory

As we have seen, Deleuze refuses to accord 'the social' the status of a 'plane', and moreover disconnects social science, as a practice, from his trans-historical thinking of ideas, perceptions and affections. In so doing, Deleuze rejects the idea that 'the social' is a naturalized set of transcendent organizing principles that stabilize a clear ontic domain of things. He wishes instead to situate 'the social' as an effect of more general, transpersonal and material modes of organizing, which only comes to take on a transcendent character through a secondary, retroactive process.

In *Anti-Oedipus* this is described through use of the terms 'socius' and 'earth'. Deleuze and Guattari's engagement with Marx and with the notion of 'universal history' makes it necessary to articulate the ground upon which capitalism can apparently project itself as a necessary force (rather than a contingency, a series of 'great accidents'). 'The earth' fulfils this role:

> The earth is the primitive, savage unity of desire and production. For the earth is not merely the multiple and divided object of labour, it is also the unique, indivisible entity, the full body that falls back on the forces of production and appropriates them for its own as the natural or divine precondition ... It is the surface on which the whole process of production is inscribed, on which the forces and means of labor are recorded, and the agents and products distributed. (Deleuze and Guattari 1984: 140–41)

It is the earth as 'full body' that allows us to see 'desiring production' – the exercise of living powers to connect with other bodies – as commingled with social production, as part of a 'unique, indivisible entity'. But this vision is very rapidly subverted as the earth becomes transformed into a 'recording surface' upon which desiring and social production are separated and coded as distinct. 'Socius' denotes this state of the earth as reorganized into distinct segments and sectors, 'territorialised' as a social field of differences and relations: 'the *territorial machine* is therefore the first form of socius, the machine of primitive inscription, the "megamachine" that covers a social field' (ibid.:141). In their self-consciously ironic universal history, Deleuze and Guattari describe the *socius* as coming to bloom through the extension of a system of inscription that, whilst semiotic, is quite literally carved into the body of the earth. Whereas bodies (human, animal, non-organic) were previously organized through intensive relations that did not pick out gross differences, the *socius*, as a territorial machine, necessarily 'codes flows, invests organs and marks bodies' (ibid.: 144). It does this in order to secure a basis upon which exchange, substitution and accumulation may subsequently be possible (again, note that this is an 'ironic' universal history, since it recognizes its collusion in explaining the emergence of an exchangist model of society, which capitalism seeks to naturalize – see also Vann, this volume). At base the *socius* is a massive recording process: 'The essence of the recording, inscribing socius, insofar as it lays claim to the productive forces and distributes the agents of production, resides in these operations: tattooing, excising, incising, carving, scarifying, mutilating, encircling, and initiating' (ibid.: 144).

This explicit nod towards Nietzsche allows Deleuze and Guattari to approach universal history as a genealogy of cruelty. The *socius* 'traces its signs directly on the body, constitutes a system of cruelty, a terrible alphabet' (ibid.: 144–45). At the heart, then, of what we call 'culture' is this brutal system of inscription, which forces bodies to be both visible as distinct unto themselves and one another and emblazons them with signs that facilitate their distribution and circulation across a territorialized social field – 'it makes men or their organs into the parts and wheels of the social machine' (ibid.: 145).

The remainder of part 3 of *Anti-Oedipus* is spent describing how this original form of *socius* becomes developed and reorganized in a series of other emergent social machines. In each case, the grid of inscriptions laid down by what they call the 'primitive territorial machine' becomes 'recoded' or 'overcoded' such that transcendental values are now projected backwards as the supposed grounds of the social field itself. For example, Deleuze and Guattari reread Marx's account of 'asiatic production' in terms of the emergence of a particular body – 'the despot' – that splits apart previous alliances and forces all possible relations to proceed by way of its own presence. What is significant in Deleuze and Guattari's argument is the way they move between semiotics, economics and psychoanalysis. The significance of the despot is not merely in terms of capital accumulation (in the form of tributes), it is also with respect to how it recodes familial relations, such that 'the father' becomes modelled as a secondary form of despotic signification.

What is of particular interest in all of this is the mode in which Deleuze and Guattari approach Marx. They do not engage with Marx as a social scientist – that is, as someone who may or may not be offering more or less accurate accounts of historical transformations. To do so would – as Eugene Holland (2003) points out so vociferously – constitute a lapse into a form of representational thinking that is antithetical to their entire philosophical project (see introduction to this volume). But here is the rub. As social scientists who read Deleuze, we are then forced into the alternatives of either 1) treating them purely as philosophers who engage with social science solely as material for philosophical speculation, or 2) accepting that the construction of concepts of 'the social' is primarily a philosophical business that is only tangentially related to the business of social science. And, since Deleuze does not deign at any point to accord social science a positivity that he is prepared to accord to philosophy, science and art, we are again thrown back on to the agonising position of treating social science as pre-paradigmatic science or impoverished philosophy.

Matters both improve and decline when it comes to *A Thousand Plateaus*. Whilst *Anti-Oedipus* was primarily concerned with Freudo-Marxism and was thus forced to engage a version of 'the social', the second volume of *Capitalism and Schizophrenia* authorizes itself to veer wildly between the most varied sources and problematics (here nomadic metallurgy, there medieval courtly love). Being no longer obliged to take into account a coherent version of human science, in either a Freudian or a Marxist dialect, they roam freely across topics without having to either situate themselves in relation to a tradition of work or clarify their relationship to the empirical details they mobilize. We might say that in this work Deleuze and Guattari properly demonstrate that philosophy, through its relation to the concept, renders social science redundant save for its role as empirical under-labourer.

However, a completely different and entirely positive relationship to *A Thousand Plateaus* is mapped out by Deleuzians who have taken up the trans-geo-epistemic thrust of this work and seen in it the basis for a project of working across disciplinary and epistemic boundaries in order to develop new concepts for social and cultural theory. Brian Massumi (2002), for example, uses the notion of affect to re-situate social science as the study of 'movement'. That is, rather than seeing human action as analysable into discrete factors or systems of signification, Massumi wants to emphasize the unfinished, incomplete, 'virtual' aspects of action. In order to do so, Massumi draws on neuroscience and psychology to speculate on the way that bodies, through the medium of feeling and sensation, are opened out to potentialities that exceed the givens of actual experience. Similarly, Keith Ansell-Pearson (2002) takes Deleuze's 'memories of a Bergsonian' (Deleuze and Guattari 1988: 237–39) as the point of departure to explore how the complex question of the relationship between memory and temporality might be configured. Ansell-Pearson then works towards a revised conception of memory that is sensitive to a neovitalist tradition that has seen a recent resurgence across critical social science (see Fraser et al. 2005; Greco 2005).

For my current purposes, the exemplary demonstration of how to stage a social scientific encounter with Deleuze is performed by the philosopher Manuel de Landa (2006a,b). I shall focus on his work at some length because it illustrates particularly well a tendency that is shared across Deleuzian scholarship. De Landa claims to discern in *A Thousand Plateaus* the basis for a coherent approach to the complexity of the social that he calls 'assemblage theory'. The basis of this approach is to be found in the subtle shift made between the two volumes of *Capitalism and Schizophrenia*. In the second volume, the terms 'socius' and 'desiring machine' become replaced by 'strata' and 'assemblage'. In one sense this change is entirely arbitrary, since the terms are roughly equivalent – 'strata' refer to the formal organization of inscribed bodies, and 'assemblage' denotes a multiplicity that holds together singular bodies in relations of exteriority (i.e. it holds together bodies without summating them as a seamless whole). De Landa summarizes the notion of assemblage in the following way:

> In an assemblage components have a certain autonomy from the whole they compose, that is, they may be detached from it and plugged into another assemblage ... Deleuze and Guattari characterise assemblages along two dimensions: on the one axis or dimension, they distinguish the role which the different components of an assemblage may play, a role which can be either

> material or expressive; on the other axis, they distinguish processes which stabilise the emergent identity of the assemblage … from those which destabilise this identity, hence opening the assemblage to change. These are processes of territorialisation and detteritorialisation respectively. (De Landa, 2006a: 253)

This summary beautifully establishes two classes of 'components' (material/expressive) and two sets of related processes (territorialisation/deterritorialisation), which can be used to organize a given empirical field. In fact, Deleuze himself uses this kind of characterization of assemblage to similar ends in his commentary on Foucault's work around 'panopticism':

> The content has both a form and a substance: for example, the form is the prison, and the substance is those who are locked up (who? why? how?). The expression also has a form and a substance: for example, the form is penal law and the substance is 'delinquency' in so far as it is the object of statements. Just as the penal law as a form of expression defines a field of sayability (the statements of delinquency), so prison as a form of content defines a place of visibility ('panopticism', that is to say a place where at any moment one can see everything without being seen). (Deleuze 1988: 47)

De Landa promotes 'assemblage theory' as a way of relating to the empirical. Yet in doing so he reproduces the particular (and quite peculiar) way in which Deleuze uses concepts to organize empirical fields without seeking to clarify the status of this creative conceptual work in relation to the objects that are thereby grasped and the epistemic traditions that have already caught hold of them. Consider, for example, the oddly truncated discussion of the work of Erving Goffman. De Landa summarizes the key points of this work – the focus on face-to-face interaction, the presentation of self, the reproduction of social order and so on – but without seeking to critically engage either with the work or the broader problematic that it is seeking to solve. Take the following gloss:

> As an assemblage, a conversation possesses components performing both material and expressive roles. The main material component is *co-presence*: human bodies correctly assembled in space, close enough to hear each other and physically oriented towards one another. Another material component is the attention and involvement needed to keep the conversation going, as well as the labor spent repairing ritual disequilibrium. While in routine conversations this labor may consist of simple habits, other occasions may demand the exercise of skills, such as tact (the capacity to prevent causing embarrassment to others) and poise (the capacity to maintain one's composure under potentially embarrassing circumstances). These are the minimal components playing a

> material role. But technological inventions (such as telephones or computer networks) may make strict physical co-presence unnecessary, leading to the loss of some material components (spatial proximity), but adding others, the technological devices themselves as well as the infrastructure needed to link many such devices. (De Landa, 2006b: 53–54)

Despite his clear interest in Goffman, de Landa here shows a disregard for what Goffman is actually seeking to achieve. Copresence is not simply a question of the correct material organization of bodies in space, but is a means of addressing the question of intersubjectivity (how is it that we respond to the other?). Goffman's interest in 'ritual' is part of a wider attempt to develop a position on how social structure is maintained. The uniqueness of Goffman's notion of an 'interactional order' with respect to its sociological forebears is completely ignored. Similarly the 'labor' of routine conversation and, in particular, efforts around 'embarrassment' are not simply functional but are part of an orientation to a 'local moral order', which constitutes a complex matter of grasping the relationship between a multiplicity of ethno-moralities rather than a sweeping notion of the 'social good'. In this respect Goffman's empirical overcoming of the Kantian tradition of moral foundations in the social good is overlooked. Finally, the question of how technology does (or does not) transform all of this has been subject to around thirty years of careful empirical examination, which has arrived at some very nuanced observations of how interaction becomes extended beyond its immediate settings (see Gackenbach 2006, for instance).

De Landa thus simultaneously strips social science of both its actual philosophical value (i.e. its relationship to substantive conceptual problems) and the complex nature of its empirical observations (i.e. by ignoring the development of research programmes). There is, then, an extreme selectivity in the particular bits of social science that de Landa chooses to engage with. For example, de Landa promotes 'network theory' as 'the only part of theoretical sociology which has been successfully formalised' (2006a: 256). The point is moot, since much turns on what is here meant by 'formalisation'. It seems from the gloss de Landa gives to network theory that formalization is to be equated with offering a set of interlinked analytical categories, which maintain a kind of validity whilst remaining operant in relation to an (unspecified) empirical operation (namely, 'the links in a network may be characterised in a variety of ways: by their presence or absence … by their strength, that is by the frequency of interaction among the persons occupying the nodes, as well as by the emotional content of the relation, and by their reciprocity' [2006a: 256]). Yet a great many social science approaches safely meet and exceed formalization if it is constituted by this listing of polyvalent terms ('presence', 'absence', 'interaction', 'reciprocity') qualified by modifiers ('strength', 'frequency', 'emotional content') that seem defined so as to ensure maximum ambiguity and interpretative flexibility. In using the term formalization, de Landa is clearly situating social science in terms of the functives which constitute the scientific plane of consistency. Here formalisation has a clear meaning since functives are posited in relation to a clearly defined set of mathematical–empirical coordinates. But, in the absence of specifying a corresponding plane of social science, de Landa's appeal to formalization is difficult to grasp, since it is not worked out in relation to the concrete specificities of social scientific work.

Although much is quite rightly made in de Landa's work of the 'individuality' and 'singularity' of a given assemblage ('In other words, unlike taxonomic essentialism in which genus, species and individual are separate ontological categories, the ontology of assemblages is flat since it contains nothing but differently scaled *individual singularities* (or hecceities)' [de Landa 2006b: 28]), he chooses not to recognize social science as itself a singular assemblage, with its own processes of territorialization and deterritorialization shaping its particular history. Not recognizing this means that de Landa tends towards imposing his own problematics on social science in general. For example, he claims that the 'micro–macro' problem is the defining feature of social science. He presents this as classically understood in terms of the relation between a global entity (i.e. 'society') and a collection of small-scale component entities (i.e. 'persons'). Now such a formulation is so broad in its sweep that it is almost intrinsically insolvable, leading de Landa to reason that: 'This recalcitrant problem has resisted solution for decades because it has been consistently badly posed. Assemblage theory can help to frame the problem correctly, thus clearing the way to its eventual solution, a solution that will involve giving the details of every mechanism involved' (2006b: 32).

Despite my immediate sympathy for both the premise and some aspects of the conclusion, this proposition misses the mark. The 'micro–macro' link is rarely treated or even hailed as a problem in itself, but is part of a complex set of debates about 'scaling', which take on very different forms in particular social science settings (in social anthropology, for instance, Marilyn Strathern's work (1991; 1999) is exemplary in the systematic working through of questions of scale and analysis). The problem, then, is only badly posed when it is secondarily (and only occasionally) taken as a meta-concern of social science in general. Thus, assemblage theory does not frame a problem that exists in the form which it asserts, but rather imposes a problem space which unifies singularities which neither need nor require such unification. The solution which is deemed to follow as a consequence – 'giving details of every mechanism involved' – is misconceived, in this sense, since it amounts to saying that a specification of the particularities of each given object of social science will be required. But these particularities only appear to be lacking since they are first erased by the imposition of a problem that is both badly formed and, to some extent, externally derived.

Again, I want to argue that what underpins all these difficulties is that de Landa, following Deleuze, is unwilling to accord positivity to social science. Hence he fails to adopt the 'trans-geo-epistemic' approach that serves Deleuze so well when it comes to unfolding the historicity and materiality of philosophy. The costs here are clear – de Landa imposes a 'badly formed problem' that does not emerge from the plane of social science itself. His solutions are then, properly speaking, idealist rather than realist, since they are not describing the conditions of actual experience, but rather the conditions of a summative account of experience that is dislocated from the particularities described by social science. Moreover, de Landa clearly positions social science as an immature version of physics (erected in familiar fashion as science in its purest, most instructive form). Promoting the idea of 'phase space' as a way of clarifying Weber's idealtypical structure, for example, de Landa comments: 'In the biological and social sciences, on the other hand, we do not yet have the appropriate formal tools to investigate the structure of their much more complex possibility spaces' (2006b: 29). The overall benefit, it then appears, to the encounter with Deleuze is that social scientists can finally clarify their confused and 'badly formed' impressions of the social world by adopting the language of assemblage theory, even though this neither results in a clear research programme nor apparently opens an empirical field.

Building the Body-without-Organs of Social Science

Much of this chapter has adopted a tone of almost relentless criticality. Such a tone is not readily compatible with the approach that Deleuze himself suggests, as captured in my opening quotation. Deleuze demands that we overcome critique (a 'reactive power') in favour of continuous and relentless acts of creation ('active powers') (Deleuze 1983). A social science that confines itself to the business of critique without creation is one trapped in ressentiment, turning its powers back on itself reactively and hopelessly. Perhaps it is the intimation of such generalized reactivity that leads Deleuze to pass over the social sciences without comment, for the most part.

This is certainly a position adopted by Deleuzians such as de Landa. They promote a distinctive version of social science by inventing concepts that wilfully traverse and to some extent ignore disciplinary boundaries (see Massumi's 2002 work on affect in particular). Now, in one sense, I find this project to be absolutely necessary and thoroughly compelling. There is a widely shared sense that twenty-first-century social science is a confused and problematic project. As work around 'theory of mind' demonstrates, science is invoked as a means of doing philosophy, and, rather than creating concepts per se, philosophy is deployed as a way of arriving at political settlements, which necessarily obviate the full range or relevant empirical details (for example, by rendering autism as a problem of cognitive processes, rather than as a set of complex processes that occur between schools, families and social welfare agencies).

Seen in this context, the work of Deleuze (and Guattari) with its restless imagination for conceptual innovation and turning 'common sense' on its head can only be welcome. What de Landa and other Deleuzians then accomplish is to show how a very different encounter with philosophy is possible and the ways that such an encounter may lead to an apparently wholesale reconfiguration of the way social scientists segment and conceptualize their particular empirical fields. There is no question that this is a necessary gesture. But, as I have tried to spell out, the lack of interest in social science as a geo-epistemic project with its own forms of historicity, which stems from Deleuze and is reproduced by de Landa and others, creates serious difficulties since it negates the singularity of social scientific work. At its best, this offers useful support for debates already occurring in the field (e.g. how to get past the dominant concern with language/discourse). At its worst, such work can seem as though it harbours the ambition to relegate social science as merely an empirical underlabourer who mines gems to be properly polished and honed by philosophers.

Yet I want to finish by asserting that I do not believe this exhausts the possibilities for the engagement of social science with Deleuze. It seems to me that the central obstacle is with the failure to describe properly the 'plane' on which social scientific work is posited. How then might such a plane be constructed? Let me speculatively offer three tasks.

First, if we are prepared to allow that only philosophy creates concepts, we need to name the particular 'ideas' in which social science is implicated, and describe the process of their self-positing. Science may have its functives and art its percepts, so social science must have its own particular defining relationship to the objects it organizes. I would suggest that an interesting point of departure here would be in characterizing the kinds of 'political settlements' that social science is obliged to make. Theory of mind, for example, is a compound object whose apparent simplicity is designed to hold together an otherwise unmanageable complexity. Yet it does so by systematically negating the conditions of real experience, which must, when they inevitably re-emerge, be fixed through auxiliary modifications of the object. In this way, social science objects always veer towards a kind of arbitrariness that is experienced by the social scientist as a tension between realism and idealism, or alternatively as the difficulty of distinguishing 'giving voice' from 'just theorizing'. We might perhaps call such unstable compound objects 'implicatives'.

Secondly, the geo-epistemic move that Deleuze makes in relation to philosophy, science and art needs to be instated for social science. Consider, for example, the problem of 'suicide' as it emerges in the statistical data of late nineteenth-century social science (see Hacking 1990). This is an 'event' through which new relationship between the impersonal forces of a statistically actualized social field and the intensities of collective affect emerge. Suicide comes to be seen as a threshold crossing made by collectivities, which must necessarily be 'owned' by the individual who is 'picked out' or 'selected' by these impersonal forces. Ian Hacking goes some way towards articulating this 'Durkheim event'. This way of reflexively turning around on the geo-epistemic character of social science adopts Deleuzian vocabulary without figuring the social as reducible as a particular moment on the development of territorial–machinic arrangements (i.e. as 'socius').

Thirdly, if we take seriously the notion of distinct planes, which may nevertheless communicate concepts, functives, percepts, etc. then the question of how social science relates to science and philosophy becomes problematized in an interesting way. Typically we regard social science as tasked with integrating science with philosophy.

But, if social science has its own plane, then it is a question instead of how its particular 'ideas' relate to, that is, circulate, conjugate, become commingled, with the concepts, functives and percepts of other planes. A potentially helpful notion here is a 'principle of non-contradiction'. A social science theory need not be thought of as compatible with other planes (indeed, if it is to have any positivity, then it must remain singular). But at the same time it ought not to directly contradict the functioning of other planes. Isabelle Stengers (2000) characterizes this as the relationship a practice maintains with the 'established sentiments' of another practice. We should not, she argues, go against such sentiments. For example, a candidate social science theory of autism must certainly attend to what neurology knows of the brain and what philosophy claims for 'intentionality'. But it should not seek to integrate such knowledge. It ought instead to find modes of actualizing autism that do not directly contradict other concepts and functives since it is fundamentally not trying to do their work. The singularity of a social science theory, then, arises from separating itself from science and philosophy without seeking to either integrate or oppose.

Finally, following from the above, we may say that each 'event' of social science requires its own corresponding idea. The way is certainly indicated here by *A Thousand Plateaus*, with its unique structure of actualizing empirical fields in relation to singular concepts. To take this work as delivering to social science an 'assemblage theory' is, then, misguided. Social science must instead find ways of engaging in a 'one event, one idea' mode of practice, rather than seeking an overarching framework in which to situate its particular theories (examples of this might include Latour's (1988) notion of 'throwaway' concepts or Strathern's (1991) emphasis on fitting concepts to particular forms of scaling). What, then, is the 'idea' that corresponds to such a practice? It is nothing other than what is currently withheld from social science by the event of 'Deleuzianism': the plane of social science itself. Which we must make.

Bibliography

Ansell-Pearson, K. (2002). *Philosophy and the Adventure of the Virtual*. London, Routledge.

Archer, M. (1995). *Realist Social Theory: the Morphogenetic Approach*. Cambridge, Cambridge University Press.

Baron-Cohen, S., A.M. Leslie and U. Frith. (1985). 'Does the Autistic Child Have a "Theory of Mind"?' *Cognition* 21, 37–46.

Baron-Cohen, S. (1997). *Mindblindfulness: An Essay on Autism and Theory of Mind*. Cambridge, MA, MIT Press.

Buchanan, I and C. Colebrook (eds). (2000). *Deleuze and Feminist Theory*. Edinburgh, Edinburgh University Press.

Buchanan, I. and G. Lambert (eds). (2005). *Deleuze and Space*, Edinburgh. Edinburgh University Press.

Cariou, M. (1999). 'Bergson: the Keyboards of Forgetting', in J. Mullarkey (ed.) *The New Bergson*. Manchester, Manchester University Press, pp. 99–117.

De Landa, M. (2006a). 'Deleuzian Social Ontology and Assemblage Theory, in M. Fuglsang and B.M. Sørensen (eds) *Deleuze and the Social*. Edinburgh, Edinburgh University Press, pp. 250–66.

———. (2006b). *A New Philosophy of Society: Assemblage Theory and Social Complexity*. London, Continuum Press.

Deleuze, G. (1983). *Nietzsche and Philosophy*. London, Continuum.

———. (1988) *Foucault*. Minneapolis. University of Minnesota Press.
———. (1991a) *Masochism: Coldness and Cruelty and Venus in Furs*. New York, Zone Books.
———. (1991b). *Bergsonism*. New York, Zone.
———. (1992). *Expressionism in Philosophy: Spinoza*. New York, Zone.
———. (1993). *The Fold: Leibniz and the Baroque*. London, Continuum.
———. (1994). *Difference and Repetition*. London, Continuum.
———. (2001). *Empiricism and Subjectivity: An Essay on Hume's Theory of Human Nature*. New York, Columbia University Press.
———. (2004). 'How Do We Recognize Structuralism?', in G. Deleuze, *Desert islands and other texts 1953–1974*. New York, Semiotext(e), pp. 170–93.
———. (2006a). 'Desire and Pleasure', in D. Lapoujade (ed.) *Two Regimes of Madness: Texts and Interviews 1975–1995*. New York, Semiotext(e), pp. 122–35.
———. (2006b). 'The Rise of the Social', in D. Lapoujade (ed.) *Two Regimes of Madness: Texts and Interviews 1975–1995*. New York, Semiotext(e), pp. 113–22.
Deleuze, G. and F. Guattari. (1984). *Anti-Oedipus: Capitalism and Schizophrenia*. London, Continuum.
———. (1988). *A Thousand Plateaus: Capitalism and Schizophrenia*, London, Continuum.
———. (1994). *What is Philosophy?* London, Verso.
Edwards, D., M. Ashmore and J. Potter. (1995). 'Death and Furniture: the Rhetoric, Politics and Theology of Bottom Line Arguments', *History of the Human Sciences* 8, 25–49.
Fraser, M., S. Kember and C. Lury (2005). 'Inventive Life: Approaches to the New Vitalism', *Theory, Culture and Society* 22 (1), 1–14.
Frosh, S. (2003). 'Psychosocial Studies and Psychology: Is a Critical Approach Emerging?', *Human Relations* 56 (12), 1545–67.
Fuglsang, M. and B.M. Sørensen. (eds). (2006). *Deleuze and the Social*, Edinburgh, Edinburgh University Press.
Gackenbach, J. (ed.). (2006). *Psychology and the Internet: Intrapersonal, Interpersonal and Transpersonal Implications*. New York, Academic.
Greco, M. (2005). 'On the Vitality of Vitalism', *Theory, Culture and Society* 22 (1), 15–27.
Hacking, I. (1990). *The Taming of Chance*. Cambridge, Cambridge University Press.
———. (1999). *The Social Construction of What?* Cambridge, MA, Harvard University Press.
———. (2004). *Historical Ontology*. Cambridge, MA, Harvard University Press.
Holland, E.W. (2003). 'Representation and Misrepresentation in Postcolonial Literature and Theory', *Research in African Literatures* 34 (1), 159–73.
Hollway, W. and T. Jefferson. (2000). *Doing Qualitative Research Differently: Free Association, Narrative and the Interview Method*. London, Sage.
Kuhn, T. (1962). *The Structure of Scientific Revolutions*. Chicago, Chicago University Press.
Latour, B. (1988). 'The Politics of Explanation', in S. Woolgar and M. Ashmore (eds) *Knowledge and Reflexivity: New Frontiers in the Sociology of Scientific Knowledge*. London, Sage, pp. 155–77.
———. (1999). *Pandora's Hope – Essays on the Reality of Science Studies*. Cambridge, MA, Harvard University Press.
Latour, B. and S. Woolgar. (1979). *Laboratory Life: The Construction of Scientific Facts*. Princeton, Princeton University Press.
Lynch, M. (2004). 'Science as Vacation: Deficits, Surfeits, PUSS and Doing Your Own Job', conference presentation to 'Does STS Mean Business?', Said Business School, Oxford.
Massumi, B. (ed.). (2001). *A Shock to Thought: Expressions after Deleuze and Guattari*. London, Routledge.
.(2002). *Parables for the Virtual: Movement, Affect, Sensation*. Durham NC, Duke University Press.
Popper, K. (1968). *Conjectures and Refutations*. London, Routledge.
Rose, N. (2001). 'The Politics of Life Itself', *Theory, Culture and Society* 18, 1–30.
Ryle, G. (1946). *The Concept of Mind*. Harmondsworth, Penguin.
Star, S.L. (1989). *Regions of the Mind: Brain Research and the Quest for Scientific Certainty*. Stanford, Stanford University Press.
Stengers, I. (2000). *The Invention of Modern Science*. Minneapolis, University of Minnesota Press.

Strathern, M. (1991). *Partial Connections*. Lanham, MD, AltaMira.

———. (1999). *Property, Substance and Effect: Anthropological Essays on Persons and Things.* London, Continuum.

Strawson, P.F. (1959). *Individuals.* London, Methuen.

Winch, P. (1958). *The Idea of a Social Science and its Relation to Philosophy*. London, Routledge.

Wisdom, J. (1968). *Other Minds*, 2nd edn. Oxford, Blackwell.

Woolgar, S. (1988). *Science: The Very Idea*. London, Routledge.

Worms, F. (1999). 'Matter and Memory on Mind and Body: Final Statements and New Perspectives', in J. Mullarkey (ed.) *The New Bergson*. Manchester, Manchester University Press, pp. 88–98.

Part II
Sociotechnical Becomings

Chapter 5
A Plea for Pleats

Geoffrey C. Bowker

Introduction

There are many folds in Deleuze's *Le Pli* (Deleuze 1988). In this short book, he reads Leibniz as a quintessential baroque philosopher – exploring his work through its resonances in contemporary music, architecture, art, mathematics and sculpture. This is not a philosophy of Enlightenment, with bright lights transfixing a personified Nature – all order is accompanied by confusion, light by shade and bliss by melancholy in this best of all possible worlds. I offer here a reading of *Le Pli* as a model for a critical reading of scientific texts – centred on an exploration of the relationship between the individual and the collective. In particular, I draw attention to ways in which political, philosophical and scientific discourses are imbricated one in the other – and to how Deleuze's fold gives us ways to work analytically with this imbrication without getting lost in vapid assertions that it is all one.

This chapter is written in counterpoint to some tendencies in science studies to eschew both the philosophical and the political within the science being studied: as a discipline, STS is good at showing the political context within which science is produced but not at getting inside the science itself. Works in social constructivism and actor–network theory – my two reference traditions – have shown that the scientist or technologist is a political actor, and yet it often seems that it is enough to show the rational actor gaining funds for her lab while black-boxing the science. As a corollary, for both traditions there has been a tendency to avoid philosophizing and valuing: our task, it seems, as social students of science is to represent *their* categories, not to develop our own. We must, it is said, 'follow the actors', who are better social theorists than we are. This fundamental error means that we are trapped in the 'poverty of empiricism'. If we accept the actors' categories, then we cannot see what they cannot, and we cannot – with appropriate anthropological strangeness – read their complex mythological, sociological and political work. Deleuze's work suggests ways of reading scientific texts that honour this complexity.

Deleuzian philosophy enables this by bringing ontology into central focus, whereas much of science studies either avoids relating to the issues brought to life by the entities we people the world with or turns its attention – in a hangover from the 1970s and the science wars – towards epistemological questions. Actor-network theory once promised this kind of ontological examination (not surprisingly, given the influence of Leibniz on Michel Serres and Bruno Latour) but has generally failed to follow through.

Scientists gain their political power precisely through their claims to keep politics outside the door of the laboratory (Shapin and Schaffer 1985) – and yet they cannot theoretically and do not practically achieve this. You cannot do interesting politics about science without engaging it philosophically. Gone are the halcyon days when we could just refer to the *interest* of specific theories – though clearly modern-day eugenics and euphenics are amenable to interest theory. The task now is to explore the entities – scallops, electrons, dark matter – we people the world with in order to recognize the limits to our own ways of knowing (a political imperative if we are not to continue the current age of imperialism). In order to move in this direction, I believe that there is room in the field of science studies for new kinds of reading of science that are philosophically and politically rich – and that Deleuze offers great analytic purchase on the shape of that reading.

Much of Deleuze's work consists of readings – of Proust (Deleuze 1996), Foucault (Deleuze 1986) and Leibniz, for example – each being inflected by his own philosophical quests. His is an expository reading, which permits a movement between multiple registers, akin to those of Lévi-Strauss' mythologies (Lévi-Strauss 1971): it is this practice that I concentrate on. I do not therefore distinguish between Deleuze and Leibniz – apart from a critical sentiment, the very last in the book, in which Deleuze gives a counterpoint to baroque certainty: 'We are still Leibnizian, even though harmonies are not what express our world or our text. We are discovering new ways of folding and new envelopes, but we are still Leibnizian because we are always folding, unfolding, folding again' (Deleuze 1988).[1]

In *Le Pli*, Deleuze imagines a beautiful machine, which this final move somewhat disarticulates. This machine treats in part the question of the relationship between the individual and the collective; and it is this relationship that will ground my development of the fold in science studies.

This chapter is in three parts. The first is an examination of the nature of the fold in Deleuze, centred around three ontological themes (foregrounds and backgrounds, identity and multiplicity and insides and outsides). Secondly, I provide a reading of some recent literature in biogeography that instantiates these themes and in so doing provides an integrated political, social and mythological reading of the science. Finally, I discuss the value of such readings for the field of science studies.

What is a Fold?

To understand the fold, we start with the monad. Monads are infinitesimally small, but they are not easy things to write about. Yet we need a little monadology in order to begin to make sense of Deleuze's work. For a first approximation, we can say that monads are the stuff of life and are its smallest unit. They have a dual structure: an upper chamber and a lower chamber; and they are completely sealed from the world. The upper chamber is filled with light, the lower is dark, obscure. As a living entity, I have a monad, which is the expression of my being in the world. That being itself is constituted by the events/arenas in which I can see clearly, and those which for me are dark or in shade. Only God can comprehend the world at one moment. Now my body is made up of organs, such as a liver and a heart. These organs themselves have monads, enabling them to perform their livery or hearty functions as best they can. And each separate part of the liver has monads and so on all the way down. There is no bottom line here – each 'world' contains sub-worlds, which contain others. (One can see immediately here the connection with the infinitesimal calculus that Leibniz helped develop.) Now, strangely, these monads, which act in the world, have no direct communication with others – the appearance of communication is a reflection of a pre-existing divine harmony. God has chosen one world (and therefore the best of all possible), in which that harmony guarantees the complex phenomena we experience. Evil is possible in this best of worlds as an evil monad is one filled only with rage against the Lord. But this also constricts its range of clarity to a point, thereby freeing up energy and space for virtuous monads. All light entrains shadow. Each monad contains a reflection of the entire world within it (just as in David Bohm's holographic universe). The light part is that which it can see and act on clearly (not much for a liver monad, much more for me) and the dark part is its area of confusion, distortion, shadow.

A moment here on what is going on ontologically. In general, analytic philosophy does not trade in metaphor and allegory (Deleuze reads Leibniz as writing allegory) and so it can appear unusual to explore seriously a position that is fairly demonstrably not the case – or, at the very least, which is unprovable. The trick in making this interesting – as with imaginary numbers – is to take the impossible to be true and see what can be usefully done with it. What emerges – around issues of the collective and the individual, sameness and difference – is just such an outcome.

So on to the fold. The world has two levels or moments: 'the one by which it is enveloped by or folded into monads, and the other by which it is engaged in or unfolded into matter' (Deleuze 1988).

Rather than an absolute separation between the organic and the inorganic (ibid.: 139), the inside and the outside, this view entails a dynamic philosophy of folding and unfolding. And, crucially, it's turtles all the way down – there is no point at which you can say you have found pure matter or pure spirit: each is always, infinitely folded into the other: 'The soul and the body are always really distinct, but they are inseparable as a result of the coming and going between the two stages: my unique monad has a body; the parts of my body has masses of monads; each of these monads has a body' (Deleuze 1988).

There really are in the principle of the fold no kinds of things, species, since everything is radically singular, incorporating its own sets of monads with their own unfoldings into the world.

The best analogy for the fold in the world of science studies is the actor-network position whereby the act of intermediation is central – unsurprisingly, given the centrality of Leibniz to Michel Serres (Serres 1968) and Bruno Latour (Serres and Latour 1995) (and of Whitehead to Stengers [Stengers 2002]). There is, in this view, no nature apart from science – both are constructed simultaneously in this act of coming and going between the one and the other. They are outcomes of a dynamic process, not pre-existing realities. The fold is the guarantor of irreducibility; the act of folding is the dynamic we analysts need to understand.

Many binaries dissolve as soon as one uses the language of the fold – as with Latour's observation in *Science in Action*, entities like Nature and Science can only be seen as changeable outcomes of specific historical processes (Latour 1987). Three sets of binaries and their dissolution – foregrounds and backgrounds, insides and outsides and identity and multiplicity – will allow us to trace the historiographical force of Deleuze's vision for our understanding of the sciences.

Foregrounds and Backgrounds

When you ask an academic how much work they got done on a given day, they'll generally respond in terms of their 'real' work – time spent researching or writing. The rest falls away into the background. (The *reductio ad absurdum* is Dick Boland and Ulrike Schultze's imaginary executive jacket, tailored to catch the actual moment of knowledge production and completely blind to the distributed nature of that work [Boland and Schultze 1995]). Similarly, when academics talk about papers that they write, there is little recognition of the collective nature of writing: the role of peer reviewers and editors in honing, shaping, altering ones productions.

To give another illustration, when we talk about remembering an event, we tend to foreground our mental processes and let slip into the nebulous background the social and physical inscriptions of our highly distributed memory processes (Halbwachs 1968; Hutchins 1995). Yet remembering is inexorably material and social.

And, likewise, when we foreground the technical artefact – the impact – of computers on society then we lose the computer itself as an expression of the principle of division of labour (Yates 1989; Bowker 1994). The first object-oriented program (Simula) was designed to mimic organizations (Khoshafian 1993); now organizations attempt to mimic object orientation (Vann 2001).

From this vantage point, also, it might be said that there is no biodiversity problem today. At the microscopic level, viruses, parasites, bacteria proliferate even in the most poisonous sludges we have been able to concoct (Staley and Reysenbach 2002), and these facilitate new sources of life. This paradoxical situation indexes a problem in contemporary visions of nature, where the big, the beautiful and the economically salient are invariably foregrounded. If we let the microscopic background into our reckoning of nature, we would have a radically different vision of natural processes –

parasitism of all sorts would be a dominant mode of relationship, not an annoying backdrop to the pageant of life (Serres 1980; Margulis 1992; Zimmer 2000). This foregrounding articulates the tree of life as a complex rhizome, in which genes may cross genera and species. Indeed, Deleuze and Guattari's use of the rhizome is as accurate for the natural world (with lateral gene transfer being more the rule than the exception) as it is for the worlds they described.

In each of these brief illustrations the act of foregrounding thus creates a divide between an ideal, self-contained unit and a set of background activities from a different realm – in turn the thinking academic, the remembering brain, the ideal machine, the separate species (of which we humans constitute the *ne plus ultra*). In a world of folds, foregrounds and backgrounds are not particularly useful meso-constructs – they propose a form of separation that is precisely belied by the dance of the fold.

Identity and Multiplicity

The concept of 'identity' is often presented as timeless but this poses ontological problems. Objects in general are not identical with themselves over time – indeed, the practice of mereology (the relationship between parts and wholes) is partly predicated on this non-equivalence (Martin 1988). I am other at different moments. I am physically multiple through the constant shedding and accretion of cells, flora, fauna, sugars, toxins and salts in my body. I am also psychically multiple. There is no single thread of consciousness that travels from cradle to grave. At different times I have access to radically different selves: on my bicycle rides a self without much memory (I think ...) except perhaps of the day before, along the same route; or in a lecture course when I am enveloped only by previous lectures, on the one hand, and how to string together some variegated riffs, on the other. This is a staccato form of discontinuity. There is a spatial form when I go to another place. England clenches my stomach with its damp, mildly depressed culture – I tap into that dimension of my being while I am there and not a whit of it when I am not. Rather, I know about it when I am not there, but from the outside – I see it through a glass darkly.

The self is also discontinuous over more extended periods of time (Linde 1993). I remember now (from the outside) when my trajectory was towards living in a permaculture commune. The points along the way (stations of the cross) for that self included a tram in blue-hazed mountains in Victoria, uncomfortable politics and a hot lead press in the outback and so forth: points that have no particular significance in the most recent trajectories I have conjured for myself. I am processually multiple. As I accrete more possible presents, futures and pasts – in Marcel Proust's terms, as time tends to the infinitely slow as we age – I am more multiply other in the present than I was a decade ago.

People are multiple rather than self-identical, then. Historical objects are no less so. For example, as Paul Veyne so beautifully demonstrates, no trans-historical object such as 'democracy', the 'liberal subject' or the 'state' exists (Veyne 1971). There are two specific historiographical problems with such formulas as 'the rise of the state', 'the formation of the liberal subject', for our purposes. First, they operate willy-nilly on the assumption that at the present time these entities have achieved their fullest realization. History is too often written as if the historian were poised at the end of history – this tendency is recognizable from the early historicism of the eighteenth century culminating in Marx's proclaimed end of history to the more recent, more empty exercise by Fukuyama (Fukuyama 1992). This Whiggish tendency is doubly true of many scientists' conceptions of their own objects – 'They used to think but now we know', in Howie Becker's phrase. The first effect of this analytical move is to simultaneously remove the present from history (and multiplicity) and to tie the past down to a single thread. Yet recent work on the Middle Ages in the annalist tradition (Goff and Schmitt 1999) has shown how we can only understand ways of thinking and being in those times with a radical historicization and multiplication of terms like God, purgatory, the state and empire. The second effect is that, in assuming gradually emerging identities such studies deny real discontinuity. Discovering actual discontinuities is a really difficult and yet important exercise, which Foucault carried out with great aplomb.

Nature is similarly multiple. The 'ecosystem' concept, for example, is a useful approximation but it is one that assumes the identity of flora and fauna and place over relatively stable periods – especially in 'mature' systems. Recent calls to bury the concept point to the fact that flora and fauna are persistently moving in and out of the presumed boundaries of the ecosystem. Thus narrowing the concept can have deleterious consequences. As Western points out for the African elephant, the creation of national parks to preserve their natural habitat has meant the enclosure of wetlands that the elephants need very occasionally – when there has been a particularly extended period of drought (David Western, personal communication; see also Gunderson and Holling 2002). Rick Jonasse makes this point beautifully for environmental planning in British Columbia, where caribou were tracked over one season and these tracks were used to set their boundaries for the subsequent two hundred (Jonasse 2001). Even when abstracting climate change and other forms of variability, this is absurd. But it does follow naturally from a science and a social science that attempt to remove the present from the flow of history and create an eternal present (Bowker 2006). Through the analytic of the fold, Deleuze provides tools for allowing STS to eschew entities falsely posited as trans-historical.

Insides and Outsides

I was somewhat shaken when I read the observation that the intestinal tract is an outside within – material passes through it, to be selectively accreted by the body, with most never coming into what is considered real contact with it.[2] Our bodies become different places if we start to look at the outside within. As Arthur Bentley wrote about this in his beautifully entitled 'The Human Skin: Philosophy's Last Line of Defense': '"Inner" and "outer" are ever present distinctions, however camouflaged, in philosophical procedure as well as in conventional speech-forms' (Bentley 1941). His argument was that, if we take skin as the separator between myself and my environment physiologically or my self and other souls psychologically, then we are making a strong ontological commitment that cannot be justified either scientifically or philosophically. Michel Serres' *Cinq Sens* resonates with this position: he places the soul not in a Cartesian temple in the body but distributed throughout our senses, where the outside gets folded into our bodies (Serres 1985). As Deleuze adumbrates in *Le Pli*, this is also the nature of the Leibnizian soul: it contains the world within it; and its individuality is comprised in what it can perceive clearly outside that world within.

Insides and outsides extend to our artefacts, our history, as well as external nature. In *The Great Wall of China: From History to Myth*, Arthur Waldron explores the myth of the fence as a barrier that keeps the barbarian hordes from ravaging an unprotected countryside (Waldron 1989). First, he notes, the barrier itself acted also as an entrance. Several times central powers lost wars to the nomads beyond, who then became the central powers, who then ... At any given temporal scale, movements of lesser or greater magnitude can be seen. Secondly, the wall was never a continuous stretch of patrollable boundary; it was shored up wherever the current threat was and elsewhere fell into desuetude. The apprehension of a single fence stretched across the landscape has always been a political statement in the present rather than a historical representation. Thirdly, the site of exclusion – the wall itself – was a rich and stable trading zone between the nomad barbarians beyond and the civilized within. It was a site of interpenetration of people, artefacts, food, disease.

As with the skin and the border, so with our understanding of nature. Most notably, when we put ourselves in one category and nature as that which is other, we are missing the ways in which we incorporate multiple forms of life – ourselves as part of ecosystems rather than external forces acting on them and ecosystems as temporally and spatially highly variable.

Across these three sets of insides and outsides – self, society, nature – my historiographical interest is in alternatives to the creation of the inside as a place removed from time, flow, the world. We murder to dissect; we don't do a lot of good when we transect. When the body gets trapped behind the skin, the soul within the body or the state within the wall, then they become trans-historical entities about which one can speak lasting truth. The scientific laboratory is archetypal (rather than sole exemplar) of this act of enclosure, whereby the inside is sequestered (to use Susan Leigh Star's phrase) from the vagaries of the outside. The nature of the fold is to perpetually bring the outside within (folding) and the inside without (unfolding) – the false dichotomy between the two is both politically and philosophically charged: politically because we have a tendency to smuggle our 'inside without' back into our world as transcendent truth about the nature of reality, as discussed in Latour (1993); philosophically, because it breaks up a world that by its very nature should be understood as one.

The Virtues of Categories and Problems with Origins

There is a whole set of similar binaries that structure our categorical systems through systematic exclusions. For example, we might mention the relationship between visible and invisible work (Star 1991a,b) or between ourselves and the Other. In these two cases, as in each of the binaries above, there is an analytical point to be made of the value of recognizing the multiple, the outside, the Other and the invisible. The underdog of the binary relationship should, it is said, be cherished on its own grounds. Deleuze goes further, by underscoring the value of a theoretical language to match these several ontological commitments.

It is hard to get away from categories, to imagine a language not predicated on some set of classification systems. Just like standards in our built infrastructure, classifications proliferate at every level of our interaction with the world: which is what makes it interesting to explore the sets of commonalities across the classifications we use, and which makes it important to consider whether, in Watson-Verran's terms, there are techniques for avoiding hardening of the categories (Watson-Verran and Turnbull 1995). A historical view within Western society takes us from classifications by essence to classifications by origin – genetic classifications (Tort 1989). Today we classify species, soils, people, languages by where they come from – what they were originally. Tort's careful historical analysis demonstrates how principles of genetic classification travelled from discipline to discipline through a series of happenstance events – a mineralogist taking a course in biology, a linguist knowing someone interested in evolution and so forth.

However, this technical feature of the way we cut up the world is synchronized with our profound historicism. I could say our historicism stemming from great thinkers from two centuries past such as Hegel (Yerushalmi 1996) – but this assertion itself would but instantiate the problem. Thought only stems from thought if we accept a separable realm whose internal logic resists the outside world; it assumes that 'our' historicism is a consequence of theirs.

The historian's search for and obsession with origins has been challenged over the past forty years and for good reasons related to the themes of identity and multiplicity and insides and outsides. First, for identity and multiplicity. The first 'x' – be it the first member of the bourgeoisie, the first modern state, the first trade union – assumes that things are identical with themselves over time. It is really not the same to be a member of the bourgeoisie in twelfth century France (the earliest sighting I know of) as to be one today. If there are no trans-historical concepts, then the pursuit of origins is moot. Secondly, if at any one moment we recognize that historical objects interpenetrate (are folded into each other), then again it is only at a first level of approximation that we can talk of stand-alone origins.

This can sometimes be a useful level, though. How else can we talk about the world? There have been numerous calls to change our own language to bring process rather than fixity into central place (Bohm 2002) or to cast categories adrift on a sea of local distinctions.[3] The language of the fold gives us a way of understanding the world as inherently multiple and porous, without losing the value of categories.

Reading Biogeography

Ontological recommendations such as this often seem to make little practical difference: if I can internalize the Leibnizian monad and use the concept to build up a world pretty much like the one I knew beforehand in, say, mechanistic terms, then arguably the new ontology does not have much traction. If all the fold does is to put scare quotes around insides, outsides, identity and categories without changing practice, then it is not so interesting. Through their very proliferation, scare quotes slowly fade into the background, leaving old habits of thought unchanged.

I shall turn now to some recent work in biogeography to see what a folded reading of science looks like. As a preface, I note that it has in general been difficult to get people in the field of science studies to actually read science and interpret it. The baleful calls to use only actors' categories, to preserve anthropological strangeness and to worship at the Temple of Symmetry have rendered it very difficult for the science studier to actually approach the text. 'Actors' categories' are one but only one way of parsing the world. However, we fall into the 'poverty of empiricism' trap (Jones 1972) if we assume that only what they see and say is going on. Indeed, actors do not in general recognize their own work and the entities therein as being folded into each other, into the socio-political realm and into a dynamic ontology. The 'principle of symmetry', in any of its many forms, is an invocation for the science studies practitioner to remove themselves from direct engagement with the flow of history and from the insides of sociotechnical exchange. In general, any attempt to create a neutral observer (the God trick, in Haraway's (1997) term is one that at the same time creates a world outside of the observer. This should not be taken as an indictment of the whole field. Indeed, there are salutary examples – especially in the anthropology of science (Haraway 1989; Hayden 2003) – of rich readings of scientific texts. It is just that there are too few of them. Deleuze's fold provides an argument for and encouragement to provide many more such analyses.

On to biogeography. There is no one discipline of biogeography, though the generic topic is the study of the distribution of life, past and present. This is an interesting field for our purposes because of the multiple ways in which the binaries discussed above play out. At the scale of biogeographical analysis, one needs to recognize the role of living beings in terraforming the globe (Margulis and Olendzenski 1992): we wouldn't be breathing oxygen if it weren't for organic life. Biogeographical analysis also raises the complex issue of whether one can use species distribution as markers for geological change (the upthrust of the Rockies changed patterns of speciation) – can one infer geology from species or is it more possible to infer species from geology? A number of technical fixes have been proposed to handle this inherent circularity: each has philosophically rich implications begging for STS interpretation.

Let us look at some ways in which the kinds of arguments I have developed about the fold are playing out in this arena. In *Frontiers of Biogeography* there is a wonderful paper by Humphries and Ebach entitled 'Biogeography on a Dynamic Earth' (2004). We begin with a rather opaque assertion near the beginning of the paper:

> The consequence of not making a distinction between cladograms and trees has many other implications ... The old chestnuts of dispersal versus vicariance and the role of ancestors in determining centers of origin are good cases in point. Also, the postmodern versions of historical biogeography are creating a milieu of mixed messages that consider life to be separate from geophysical history. Old oppositions constantly resurface in different guises and so we argue for the theory, methods and implementation of historical biogeography from a pattern perspective that sees cladograms as different from phylogenetic trees; that ancestors and centers of origin cannot be ascertained; and that the distinction between dispersal and vicariance cannot be justified. (Humphries and Ebach 2004: 68)

Trees are the problematic branching structures which that many representations of the development of life on earth – the 'tree of life' representations many of us saw in school, which started with single-celled organisms and branched and branched until humans, horses and other contemporary species emerged, are an example. A cladogram is a way of representing species change; it maps 'acquired characteristics' that are sufficiently stable to mark a branching point. A cladogram can be represented in tree form *inter alia*, but the question of whether or not it actually depicts historical pathways rather than a map of acquired characteristics is at stake here. There have been furious arguments about whether the cladogram should be seen as a historical representation or a logical presentation whose historicity is an open question. From the inside of the science (tree or cladogram), we get from Humphries and Ebach a denial of origins – the question of 'dispersal' (species leaving from a common centre of origin) and vicariance (dispersed species getting cut off by climatic or tectonic events) is made moot.

Along the way, precisely the same denial of origins, as we have seen, is associated with structuralist history (Foucault, Serres) is now associated with the history of species. Without some way of folding our sociocultural present into our science and our history, we would have a resonance without a cause. With a fold, it would be surprising if there were no such resonances. (If you want to see the assertion without the theory, look at Holton's classic *Thematic Origins of Scientific Thought* (1988), which poses the resonances without any possible explication of them.)

Dispersal from centres of origin is the general line of argument that species start somewhere and then spread out in linear fashion, fighting other species until they find a suitable niche they can settle into. The move has generally been painted as from centres in the tropics to ecosystems beyond; in some marine molluscs, for example, one finds the older species further out from the tropics, suggesting dispersion. Its rival, vicariance biogeography, argues that speciation occurs when major tectonic or climatic events cause the separation of similar taxa for long enough for them to evolve into separate species. One can imagine this most clearly for parts of Arizona, where there are multiple species of hummingbirds in the high mountains. Before the area between the mountains was desert, these would have been able to interbreed; more recently, they have been trapped in their island forests and have lost the ability to interbreed. Similarly, there are 'tropical' fish trapped in the Gulf of California by a swatch of cold water that prevents connection with sister species elsewhere.

For Humphries and Ebach, phylogenetic (or vicariance) biogeography 'views the worlds as static, but cladistic biogeography ... allows us to assess relationships of areas on a dynamic Earth' (Humphries and Ebach 2004: 68). What they mean by this is that the tectonic events and their climatic consequences are seen as external to and acting on life. For those in science studies, they want to use precisely the kind of arguments deployed against technological determinism: there is no 'impact' of the world on life because you cannot separate the two, just as you cannot separate technology and society. In order to move forward, they argue, we need to switch foreground and background, as well as the inside and the outside of life: 'For almost two centuries, biogeographers in the dispersalist tradition have considered life and Earth as separate entities. Dispersal from centers of origin and geophysical stasis, or at best, gradual change, was the enduring paradigm from the mid-nineteenth century until well into the twentieth century '(ibid.: 69)

Or again: 'What is common to both panbiogeography and cladistic biogeography is the notion that life and Earth have evolved together' (ibid.: 73). In this process, insides and outsides get defined in new ways; we no longer have the rock and the range as stable frontier akin to Bentley's skin: 'An "area" in area cladistics is not the inorganic substrate of soil, rock, and sea, but rather the endemic biota. A case of two conflicting patterns may sometimes be resolved if the same area, over time is treated as two different areas. This is known as "time slicing"' (ibid.: 79).

The phrasing is clumsy but the analysis is clear: both inside/outside and foreground/backround need to be rethought if the story of life on earth is to be told. A Deleuzian reading of such texts brings to the fore the radical similarities of our natural entities: people, organisms, flora, fauna, planets are all built up of monads within monads. The world is in us and we are in the world. Rather than being surprised at tripping over parallels between people, peoples and areas over the question of whether they are identical with themselves over time, we can invert the issue and take the parallels as evidence for a richly multiple ontology.

The biogeographical arguments here are very similar to the historiographical arguments above. Whence all this similarity? A range of possible theoretical responses to such replication of patterns are available. There is the argument by ideal origins, which, of course, I reject. A pure form of this is that we are dealing with, for example, the history of the Hegelian subject: thought drives all practice and thus a single set of concerns appears quite naturally in all spheres of human activity. There is the somewhat more attractive but equally flawed argument by materiality. This can range from a base/superstructure model, where there is determination in the last instance by the mode of production to the more pragmatist position that it's all work, and that work practices tend to travel well from one sphere of activity to the next. The material argument, for example, would say that at the limit there is a single political economy (defined broadly as the relationship between ourselves and nature) or there is a single history of bureaucracy, which we dress up into a series of separate internalist histories (especially when we make the false divide between scientific, philosophical and other work). Indeed, if we see both the history of science and the history of society as the history of the organization of information in order to act on the world (namely, the history of bureaucracy), then much becomes clear.

However, these forms of the ideal and material solutions to similarity reify a divide between us and the world that Deleuze and others (Bergson, Whitehead) see as problematic. For it still puts us in the position of being outside nature and interacting with it – either through our consciousness or through our practices.

Deleuze on Leibniz on the Fold

The greatest value that I see in Deleuze's account of Leibniz is precisely his ability to sketch a position where there need be no great divide between writing grounded history, philosophizing about the nature of reality and doing scientific work. According to Deleuze, Leibniz gives 'a conception of the object which is not only temporal but also qualitative, in so far as sounds, colours are flexible and are comprised in modulation. It is a mannerist object – not an essentialist one – which becomes event' (Deleuze 1988).

This reading renders the pyramid an event as much as a party or a primrose – and an event that is to be understood by its variable folding into other events. Isabelle Stengers in her account of Whitehead (Stengers 2002) sketches a similarly rich account of an event ontology. The object as mannerist event is not an object that can exist separately in the world (with a well-defined inside or outside) and it is inherently multiple – the object 'does not exist outside of its metamorphoses or in the declension of its profiles; perspectivism guarantees the truth of relativism (and not the relativity of truth)' (Deleuze 1988: 30). We thus escape the essentialism of Aristotle (with his essences or categories) or Descartes (with his reductions to the machinic). For Leibniz, 'the world itself is an event – and as a predicate it must be taken as a background to (incorporated in) the subject' (Ibid.: 72). For Leibniz, then, there is no contradiction between singularity and multiplicity (Ibid.: 80) – through the dissolution of the object into the event, the world can be both radically singular and one (the world is folded into me) and multiple (any event is apprehended through its set of 'accidental' predicates).

Deleuze's Leibniz motivates the breakdown of what ANT theorists call the human/non-human divide: 'The great difference is not then between the organic and the inorganic, but rather traverses both by distinguishing between what is individual and what is a mass or crowd phenomenon, what is absolute form and what is figure or massive, molar structure' (Deleuze 1988:139).

Through ignoring such divides, Deleuze brings together a new form of architecture characterized by folds (the baroque), a new economic system also characterized by folds (capitalism) and a new apprehension of microscopic life folded into our bodies (ibid.: 180). For him, the baroque instantiates a new kind of narrative, one in which 'description replaces the object, the concept becomes narrative and the subject a point of view, the subject of enunciation' (ibid.: 174).

Bringing Science Studies into the Fold

For the field of science studies, then, what can emerge from this multiple, porous apprehension of science, nature and the world is a new way of reading scientific texts. Serres once made the observation that scientific disciplines were very good at doing what they did – defining and then manipulating their objects to achieve a partial vision of the world. The role of the philosopher of science, he said, was to hold open the spaces between disciplines to create a productive tension in our reading of science, literature and myth that constantly situated and relativized our certainties. Within STS, we have done in general just a part of that work. Hampering us has been a general invocation not to enter as actors into the worlds we describe. The principle of symmetry has been conjured into a claim that we should not make judgements about the scientific programmes we analyse; all we can do is use the same sets of causal factors to explain success and failure. This frankly silly position (and let me not talk about 'supersymmetry') has set us up within science studies as the guardians of the singular, the outside and the background. Similarly, tricks of the trade like anthropological strangeness do us a disservice when they keep us ever poised at the threshold of a text, refusing to analyse it in its own terms. This means that, at the very time we are bringing in the whole mix of ways in which science gets developed, we risk losing the ability to read.

I was stirred when I first read actor–network theory (ANT). It provides a theoretical language that simultaneously denies insides or outsides for scientific practice: the work of being a scientist is precisely the work of bringing science into the world and the world into the scientific laboratory. It denies that external categories like 'society' exist outside of the natural world and have an impact only on our apprehension. Society is comprised of microbes, scallops, people, practices and technology; and each apparently separable unit (the scientific truth, the technical artefact, the social fact) has the others folded into it at some point – it is a question of the levels of granularity one is choosing to apply. There is no point at which you can make that final cut that will separate off us and our thoughts from the world and history. To extend actor–network theory, which has a rather bland 'it's turtles all the way down approach' to scale, we can follow Jacques Revel's beautiful analysis of his data on the visible and the invisible in colonial practice: 'the change in the scale of observation revealed not just familiar objects in miniature but different configurations of the social' (Revel 1995). ANT also provides a good – though less robust – language for multiplicity. Hence the resonance for Latour with ethno-psychiatry, a discipline that seeks to change the individual not by going back to their origin (childhood trauma) but by changing the self in the present through changing the sets of current connections with the world, talking in clinical sessions about one's current interpenetration with the world (sport, politics and so forth) instead. The ANT self is defined by its connections in the moment, just as any technical artefact or natural object is multiple over time. Hennion's (1997) and Latour's previously mentioned work on the centrality of intermediation – its existence before objects – is a development of Leibniz's phenomenology. However, ANT has singularly (or multiply ...) failed to carry through on its early promise. From being a place where one can see the world in new ways, it has become a machine for turning out safe analysis.

I often think of science studies as being predicated on a set of not particularly useful negative commandments. Thou shalt not judge a work of science. Thou shalt not talk ontology, for this is the work of our subjects. Thou shalt not talk about large-scale social effects, for each scientific practice is peculiar unto itself.

These negative commandments are all problematic and are all interrelated. I really don't know how one can talk about the nature of reality without talking about the state of society; about the state of nature without talking about the social. Deleuze in *Le Pli* – and elsewhere in his work – offers ways of weaving together scientific, philosophical and historical discourse. And he does it in a resolutely 'irreductionist' (Latour 1988; Smith 2006) way – no one of the trio is determined by the other two, either in the first or in the last instance.

We are in a historical period (we always already have been, perhaps) where it is crucial to interpret and engage science: as Latour points out, every newspaper swarms with politically charged questions relating to non-human entities created in the laboratory. This means that STS must develop ways of engaging politically with science. Yet, as a field STS has been notoriously bad at doing just that: the political content of science has been eviscerated in the same movement as the philosophical content has been eviscerated. Let's bring it all back into the fold.

Notes

1. Citations are to the French edition and translations are my own.
2. For an illustration, see the animated sequence at http://physiol.umin.jp/hrm/chp1_e/1-101_e.html
3. See text by Clay Shirkey at http://www.shirky.com/writings/ontology_overrated.html.

Bibliography

Bentley, A.F. (1941). 'The Human Skin: Philosophy's Last Line of Defense', *Philosophy of Science* 8, 1–19.

Bohm, D. (2002). *Wholeness and the Implicate Order*. London and New York, Routledge.

Boland, R. J. and U. Schultze (1995). 'From Work to Activity: Technology and the Narrative of Progress' in G. Orlikowski, G. Walsham, M.R. Jones and J.I. DeGross (eds) *Information Technology and Changes in Organizational Work*. London, Chapman Hall, pp. 308–24.

Bowker, G.C. (1994). *Science on the Run: Information Management and Industrial Geophysics at Schlumberger, 1920–1940*. Cambridge, MA, MIT Press.

———. (2006). *Memory Practices in the Sciences*. Cambridge, MA, MIT Press.

Deleuze, G. (1986). *Foucault*. Paris, Editions de Minuit.

———. (1988). *Le pli: Leibniz et le Baroque*. Paris, Editions de Minuit.

———. (1996). *Proust et les signes*. Paris, Presses universitaires de France.

Fukuyama, F. (1992). *The End of History and the Last Man*. New York, Free Press.

Goff, J.L. and J.-C. Schmitt. (1999). *Dictionnaire raisonné de l'occident médiéval*. Paris, Fayard.

Gunderson, L.H. and C.S. Holling. (2002). *Panarchy: Understanding Transformations in Human and Natural Systems*. Washington, DC, Island Press.

Halbwachs, M. (1968). *La Mémoire collective*. Paris, Presses universitaires de France.

Haraway, D. (1989). *Primate Visions: Gender, Race, and Nature in the World of Modern Science*. New York, Routledge.

———. (1997). *Modest_Witness@Second_Millennium. FemaleMan_Meets_OncoMouse: Feminism and Technoscience*. New York, Routledge.

Hayden, C. (2003). *When Nature Goes Public: the Making and Unmaking of Bioprospecting in Mexico*. Princeton, Princeton University Press.

Holton, G. (1988). *Thematic Origins of Scientific Thought: From Kepler to Einstein*. Cambridge, MA, Harvard University Press.

Hennion, A. (1997). 'Baroque and Rock: Music, Mediators and Musical Taste', *Poetics* 24, 415–35.

Humphries, C.J. and M.C. Ebach. (2004). 'Biogeography on a Dynamic Earth', in M.V. Lomolino and L.R. Heaney (eds) *Frontiers of Biogeography: New Directions in the Geography of Nature*. Sunderland, MA, Sinauer Associates, pp. 67–86.

Hutchins, E. (1995). *Cognition in the Wild*. Cambridge, MA, MIT Press.

Jonasse, R. (2001). *Making Sense: Geographic Information Technologies and the Control of Heterogeneity*. San Diego, Department of Communication, UCSD.

Jones, G.S. (1972). 'History: The Poverty of Empiricism', in R. Blackburn (ed.) *Ideology and the Social Sciences*. London, Fontana, pp. 95–116.

Khoshafian, S. (1993). *Object-oriented Databases*. New York, John Wiley.

Latour, B. (1987). *Science in Action: How to Follow Scientists and Engineers Through Society*. Milton Keynes, Open University Press.

———. (1988). *The Pasteurization of France*. Cambridge, MA, Harvard University Press.

———. (1993). *We Have Never Been Modern*. Cambridge, MA, Harvard University Press.

Lévi-Strauss, C. (1971). *L'Homme nu*. Paris, Plon.

Linde, C. (1993). *Life Stories: The Creation of Coherence*. New York, Oxford University Press.

Margulis, L. (1992). *Diversity of Life: the Five Kingdoms*. Hillside, NJ, Enslow.

Margulis, L. and L. Olendzenski. (1992). *Environmental Evolution: Effects of the Origin and Evolution of Life on Planet Earth*. Cambridge, MA, MIT Press.

Martin, R.M. (1988). *Metaphysical Foundations: Mereology and Metalogic*. München, Philosophia.

Revel, J. (1995). 'Microanalysis and the Construction of the Social', in J. Revel and L. Hunt (eds) *Histories: French Constructions of the Past*. New York, The New Press, pp. 492–502.

Serres, M. (1968). *Le Système de Leibniz et ses modèles mathématiques*. Paris, Presses universitaires de France.

———. (1980). *Le Parasite*. Paris, B. Grasset.

———. (1985). *Les Cinq Sens*. Paris, Grasset.

Serres, M. and B. Latour. (1995). *Conversations on Science, Culture and Time.* Ann Arbor, MI, University of Michigan Press.

Shapin, S. and S. Schaffer. (1985). *Leviathan and the Air-pump: Hobbes, Boyle, and the Experimental Life.* Princeton, NJ, Princeton University Press.

Smith, B.H. (2006). *Scandalous Knowledge: Science, Truth and the Human.* Edinburgh, Edinburgh University Press.

Staley, J.T. and A.-L. Reysenbach. (2002). *Biodiversity of Microbial Life: Foundation of Earth's Biosphere.* New York, Wiley.

Star, S.L. (1991a). 'The Sociology of the Invisible: the Primacy of Work in the Writings of Anselm Strauss', in D. Maines (ed.) *Social Organization and Social Process: Essays in Honor of Anselm Strauss.* Hawthorne, NY, Aldine de Gruyter, pp. 265–83.

———. (1991b). 'Invisible Work and Silenced Dialogues in Representing Knowledge', in I.V.Eriksson, B.A. Kitchenham and K.G. Tijdens (eds) *Women, Work and Computerization: Understanding and Overcoming Bias in Work and Education.* Amsterdam, North Holland, pp. 81–92.

Stengers, I. (2002). *Penser avec Whitehead: une libre et sauvage création de concepts.* Paris, Seuil.

Tort, P. (1989). *La Raison classificatoire: les complexes discursifs – Quinze Etude.* Paris, Aubier.

Vann, K. (2001). *The Duplicity of Practice.* San Diego, Department of Communication, UCSD.

Veyne, P. (1971). *Comment on écrit l'histoire; augmenté de Foucault révolutionne l'histoire.* Paris, Éditions du Seuil.

Waldron, A. (1989). *The Great Wall of China: From History to Myth.* Cambridge, UK and New York, Cambridge University Press.

Watson-Verran, H. and D. Turnbull. (1995). 'Science and Other Indigenous Knowledge Systems', in S. Jasanoff, G. Markle, J. Petersen and T. Pinch (eds) *Handbook of Science and Technology Studies,* Thousand Oaks, CA, Sage Publications, pp. 115–39.

Yates, J. (1989). *Control Through Communication: The Rise of System in American Management.* Baltimore, MD, Johns Hopkins University Press.

Yerushalmi, Y.H. (1996). *Zakhor: Jewish History and Jewish Memory.* Seattle, WA, University of Washington Press.

Zimmer, C. (2000). *Parasite Rex: Inside the Bizarre World of Nature's Most Dangerous Creatures.* New York, Free Press.

Chapter 6

Every Thing Thinks: Sub-representative Differences in Digital Video Codecs

Adrian Mackenzie

> Every body, every thing, thinks and is a thought to the extent that, reduced to its intensive reasons, it expresses an Idea the actualisation of which it determines. (Deleuze 2001: 254)

What would it mean for anthropologies of technology if we took seriously Deleuze's claim that every thing thinks? This claim is hard to stomach for several reasons. Technologies are generally seen as the expression of much highly organized thinking (scientific, design, engineering, artistic, financial, political, etc). Moreover, many technologies today are very much designed to think in specific, albeit limited, ways. This is particularly the case in so-called 'intelligent systems', but it holds for any designed or made thing. Smartness, intelligence, sophistication, cleverness: are not all of these prized qualities in technology actually expressions of thought and of much mental effort? What does the affirmation '*every* thing thinks and is a thought' add? Somewhat counter-intuitively, or at least, contrary to common sense, Deleuze's more specific claim that things determine the actualization of ideas, I would argue, points in a different direction, towards a much more problematic mode of existence of things. Intelligent or smart technologies may, in their ingenuity, render this problematic mode less visible. They may detract from the problem ideas, from the singular problem-setting imperatives, actually at work in technologies.

From Deleuze's perspective, thinking would not merely refer to things through the mental work of developing concepts that represent them. It would actually be part of things. In other words, Deleuze can be read as bringing a radical constructivism to the fore. Some years after Gilles Deleuze published *Difference and Repetition* in 1968, the US Patent Office awarded the early patents on compressed digital image transmission. Two lineages of patents began to emerge. Both sought to isolate and intensify certain forms of repetition in moving images. In the first, the patents described some ideas for the application of signal processing techniques known as fast Fourier transforms (FFT) to video images (Means 1974; Speiser 1975). These transforms extracted more compact encodings of video images and sounds so that information networks and transmission systems could store and move them more easily. In the second, the patents described how to computationally predict future video images from past images (Haskell and Limb 1972; Haskell and Puri 1990). The 'motion estimation' or 'motion compensation' techniques analysed patterns of movement of objects and figures across successive video or television frames. The transformations of images through the fast Fourier transform (later by the discrete cosine transform (DCT)) and motion estimation still stand at the very centre of compression techniques used in digital media technologies (used in JPEG images, MPEG, DVDs, etc.), and in digital video *codecs,* such as MPEG-2 in particular. The entwining of these two lineages turns out to be important in very many domains of contemporary culture.

Just as Deleuze's *Difference and Repetition* concerned the rumble of *sub-representative* differences in philosophical thought, these patent lineages expressed the envelopment of sub-representative differences in technologically mediated perception. Just as Deleuze's book advanced an account of how identities and sameness stem from 'a more profound game of difference and repetition' (Deleuze 2001: xvix), these patents announced an alteration in the micro-perceptual and infrastructural supports of representation and repetition of images and sound. The image or the frame, which in media such as photography, film and even television could still be regarded as the support of representation, began to dissolve or at least redistribute itself into thresholds of brightness, colour and motion vectors, into what, in Deleuze's terms, could be understood as differentials. Today, the tremendous architectural complexity of a single technological instance, the MPEG-2 video codec, might serve as an index to certain spatio-temporal dynamics (problems or desires, depending on your viewpoint) in media technologies.

What is at stake in developing a sub-representative account of media technologies, or in developing radical constructivist accounts of technology more generally? The 'sub-representative', in Deleuze's thought, refers to that aspect of things that cannot be consciously thought or reduced to the presence of an object to a subject mediated by a concept or category. While the sub-representative cannot be identified, measured or calculated as such, it is felt and, in some cases, felt intensely. So, while the proliferation of digital video might on the one hand be seen as a paroxysm of representation, an unbounded expansion of the power to represent, from the sub-representative standpoint, it could also be seen as 'imbued with a presentiment of groundlesness' (Deleuze 2001: 276), alterity and differences. In making sense of this claim, the discussion that follows makes a big jump from Deleuze's account of ideas, difference and repetition to a specific but hardly very well-known technology. One can easily speak of technology in general without acknowledging the specificity of a technology. Deleuze's claim that 'every thing thinks', however, might bring to an anthropology of technology a refreshed conception of specificity and the grounds of specificity.

Stated rather baldly, specificity and actual differentiations, as seen in sensibilities, social institutions, global organizations, local workarounds, civic epistemologies, political economies and forms of personhood associated with audio-visual media, would all represent affirmations of the problem-setting imperatives that move through codecs. As Constantin Boundas writes, in his commentary on *Difference and Repetition*, 'Just like Kant, Deleuze believes that Ideas are problem-setting imperatives. But unlike Kant, Deleuze believes that the ability of a problem to be solved must be made to depend on the form that the problem takes' (Boundas 1996: 88).

The growth of video material culture can be seen, then, as an affirmation of a problem idea. In the labyrinthine plenitude of digital video, one path leads to a key technical component: *codecs.* Software and hardware codecs transform images and sound. Transformed images move through communication networks much more quickly than uncompressed audio-visual materials. Without codecs, an hour of raw video footage would need 165 CD-ROMs or take roughly twenty-four hours to move across a standard computer network (10 Mbit/sec Ethernet). Instead of 165 CDs, we take a single DVD on which a film has been encoded by a codec. We play it on a DVD player that also has a codec, usually implemented in hardware. Instead of 32 Mbyte/sec, between 1 and 10 MByte/sec streams from the DVD into the player and then on to the television screen.

The economic and technical value of codecs can hardly be overstated. DVD, the transmission formats for satellite and cable digital television (DVB and ATSC), HDTV as well as many Internet streaming formats such as RealMedia and Windows Media, third-generation mobile phones and voice-over-IP (VoIP) all depend on video and audio codecs. They form a primary technical component of contemporary audio-visual culture in many of its most global dimensions. Physically, codecs take many forms, in software and hardware. Today, codecs nestle in set-top boxes, mobile phones, camcorders, video cameras and webcams, personal computers, media players and other gizmos. Codecs perform encoding and decoding on a digital data stream or signal, mainly in the interest of finding what is different in a signal and what is mere repetition. They scale, reorder, decompose and reconstitute perceptible images and sounds. They only move the differences that matter through information networks and electronic media. This performance of difference and repetition of video comes at a cost. Enormous complication must be compressed in the codec itself.

Much is at stake in this infrastructure and image logistics from the perspective of cultural studies of technology and media. On the one hand, codecs analyse, compress and transmit images that fascinate, bore, fixate, horrify and entertain billions of spectators. Most of these images are repetitive or clichéd. They reinforce or shore existing orderings of difference, identity and power. There are many reruns of old television series or Hollywood classics. Youtube.com, a video upload site, currently offers 13,500 wedding videos. Yet the spatio-temporal dynamism of these images matters deeply. They open new patterns of circulation and permit new symbolic differences to appear. To understand that circulation matters deeply, we could think of something we don't want to see, for instance, the many executions of hostages (Daniel Perl, Nick Berg and others) in Jihadist videos since 2002. Islamist and 'shock-site' web servers streamed these videos across the Internet using the low-bit-rate Windows Media Video codec, a proprietary variant of the industry-standard MPEG-4. The shock of such events – the sight of a beheading, the sight of a journalist pleading for her life – depends on circulation through online and broadcast media. A video beheading lies at the outer limit of the visual pleasures and excitations attached to video cultures. Would a beheading, a corporeal event that takes video material culture to its limits, occur without codecs and networked media? We need to understand how media infrastructures such as codecs envelop differences such that differences, modifications and variations in sensibility, subjectivities and institutions arise.

From Eye to Infrastructure: Envisioning Centres of Calculation

One way to glimpse the different path opened up by Deleuze's account of sub-representative differences is to contrast it with the now canonical approaches to technology developed by science studies in the 1980s and 1990s. From a science and technology studies (STS or SCOT) perspective, we could say that the compression techniques presented in 1970s patents began to anchor centres of calculation in the midst of the chaotic, surging flows of cinematic, televisual and video images. A centre of calculation is, according to Bruno Latour, 'any site where inscriptions are combined and make possible a type of calculation. It can be a laboratory, a statistical institution, the files of a geographer, a data bank, and so forth. This expression locates in specific sites an ability to calculate that is too often placed in the mind' (Latour 1999: 304).

A codec is a site where calculations are done on images. Why should images need calculation? Ostensibly, the patents addressed the logistics of moving images. They proposed different ways of doing this: an 'apparatus capable of performing a discrete cosine transform with lightweight, low-cost, high-speed hardware suitable for real-time television image-processing' (Means 1974: 1); 'a linear transform device capable of the rapid generation of linear transforms of a spatial or temporal signal' (Speiser 1975: 2); or a 'system for encoding a present frame of video signals comprising means for dividing the picture elements of the present frame into moving and nonmoving regions' (Haskell and Limb 1972).

Moving images and, to a lesser extent, sound indeed pose logistical problems. People in electronic media cultures have constantly imagined images circulating everywhere. Millions of images flicker across TV and cinema screens. Analogue television broadcasting solved the logistics problem in a Fordist fashion: images produced in studios passed through electromagnetic waves transmitted from central stations to many identical receivers. With few exceptions, the central platform of studio and transmitter came under state and/or large corporate ownership and control. As many studies of television and radio have shown, the forms of identity, representation and consumption associated with broadcast television align closely with political, economic and cultural forms of nationstates. However, despite all its recent attempts to offer interactivity, broadcast television cannot keep up with the information age's kaleidoscopic imagining of images flowing in many directions at once.

The calculations that codecs perform on images are not purely logistical. Put differently, any change in image logistics does not only concern media infrastructures. It affects embodied habits of perception, sometimes at a micro-perceptual level (sensations of brightness, colour and movement alter), sometimes at the level of media-historical habits (where, when and how images are made and seen), sometimes at the level of affect and temporality. Today, the calculations done by codecs take many forms and occur on many platforms: set-top boxes, mobile phones, cameras, computers, media players and other gizmos cradle codecs.

Those arrangements are indeed fascinating. However, here I want to concentrate on the problem of how to connect embodied vision and media infrastructure. To comprehend this connection, I suggest, we need to shift attention from centres of calculation to *centres of envelopment,* a concept that Gilles Deleuze proposes late in *Difference and Repetition.* Centres of envelopment, according to Deleuze, interiorize differences in situations already structured by settled orders of representation, resemblance, extension and qualities perceived by human subjects. Such centres crop up in 'complex systems' where series of differences come into relation 'to the extent that every phenomenon finds its reason in a difference of intensity which frames it, as though this constituted the boundaries between which it flashes, we claim that complex systems increasingly tend to interiorise their constitutive differences: the centres of envelopment carry out this interiorisation of the individuating factors' (Deleuze 2001: 256).

Here, I shall not be able to develop a full explanation of this concept. It comes late in *Difference and Repetition,* after long consideration of the constitution of time, memory, habit, difference and intensity. Importantly, the notion of a centre of envelopment is only one of a series of notions Deleuze developed to describe how differences relate to differences. Instead, I concentrate on the basic idea that a centre of envelopment interiorizes differences since this is almost a refrain in Deleuze's thought. Sub-representative differences are essentially interior since they differ intensively rather than extensively. If we take the rather risky step of understanding codecs as a centre of envelopment, we could ask how codecs interiorize 'constitutive differences' in ways that lead to a proliferation of moving images. The functioning of codecs, their capacity to compress and move moving images in space and time and to generate sensations of qualities of colour, depth, light and form, certainly relies on calculation. However, calculation itself must derive in principle, for Deleuze, from 'a difference of intensity'. By tracking how intensities (or differences of difference) matter in codecs, we might begin to get a feel for how video material cultures eventuate. In substituting envelopment for calculation, Deleuze allows us to shift focus from extension to intensity, and thereby to describe how altered feelings, expectations and transformed sensibilities occur in and around technologies. If every thing expresses an Idea, if 'our experience of things, if you will, *can be conceptual*' (Henare et al. 2007: 13), if every thing is a partial traversal of a field of affirmation or 'solutions' to the problem-setting imperative of an idea, then feelings, sensibilities and expectations of 'more to come' might be seen as symptoms of the impersonal individuations set in train by differences.

Spatio-temporal Dynamisms and Differences in Repetition

The concept of centre of envelopment addresses something quite elementary: the persistence of intensities, singularities and purely spatial dynamisms in worlds largely organized by systems that order, calculate and represent. A centre of envelopment is like a microstructural black hole peppering the fabric of the everyday, constantly interiorizing elementary differences that give rise to phenomena, and serving as a site of actualization for problem-setting ideas. According to Deleuze, 'disparity ... difference or intensity ... is the sufficient reason of all phenomena' (Deleuze 2001: 222). The problem in developing this philosophical perspective into an analytical engine for an anthropology of technology is that the technology of video codecs seems thoroughly actual. How do the densely interwoven calculations that codecs perform on images and sounds participate in what Deleuze called the 'asymmetrical synthesis of the sensible' (Deleuze 2001: 222), or the ordinary everyday sensibilities of electronic media as they play out on information networks today?

Spatio-temporal dynamisms are critical here. As mentioned above, the primary characteristic of codecs is to allow digital images to move around in many forms. Video iPods, digital cameras, mobile phones, media players, DVDs, satellite broadcasts or Internet media encode and decode video streams. Starting from the 1970s patent lineages, several decades of heavily funded public and private research have gone into liberating moving images from the bulkiness of film stock and projects and the fixity of television transmitters and television sets. Tens of thousands of patents litter the wake of that research begun in the early 1970s. Today, probably the most widespread video or moving image codec is MPEG-2, which stands at the confluence of several lineages of image encoding. DVDs, for instance, consist of MPEG-2 files. The International Standards Organization (ISO) formalized the MPEG-2 (aka H.262) encoding and decoding procedures as a standard (ISO/IEC 13818–1, 1999) in the early-1990s. The standard calls itself a 'transport system' (ISO/IEC 13818–1, 1999). The standards documents grew from the work of several thousand engineers and software designers meeting in Europe and North America. Other engineers implement the arrangements described in the documents in software or hardware codecs (coder–decoder). Video codecs for different standards (MPEG-1, MPEG-2, MPEG-4, H.261, H.263, the important H.264, theora, dirac, DivX, MJPEG, WMV, RealVideo, etch) litter electronic media networks.

Any codec that implements the MPEG-2 coding standard incarnates extraordinarily complicated calculations. Algorithmically it must draw on several distinct compression techniques: converting signals from time domain to frequency domain using discrete cosine transforms, quantisation, Huffman and run length encoding, motion compensation, timing and multiplexing mechanisms, retrieval and sequencing techniques. The standard borrows from the earlier, lower-resolution video standard, MPEG-1 (ISO/IEC 11172–1, 1993), and from other image standards, such as JPEG. Legally, it imposes relations with many intellectual property claims (700 patents held by entertainment, telecommunications, government, academic and military owners administered through the MPEG-LA patent pool). Finally, it competes with many other codecs (e.g. Chinas AVC – Advanced Video Codec – versus the increasingly popular H.264 versus other versions such as Microsoft Windows VC-1 – Windows Media 9).

At the intersection of technical, legal and economic forces, codecs display a mosaic, composite character. They make sensible compromises between extensity and quality, between how images circulate (online, in media materials and transmission formats) and the sensations (brightness, detail, luminance, etc.) associated with them. Their composite character reflects a constant and dynamic negotiation between the political economy of telecommunications and the media-historical perceptual habits of visual cultures. Telecommunications and media companies create cable, satellite and wireless network bandwidth as a market commodity. They sell bandwidth to anyone who will pay for information and images to move. At a very deep level, the architecture of an MPEG-2 codec reflects the assumption that all movement costs something in time, computation or bandwidth. Reducing the cost of movement means that more people can pay for that movement. If a codec compresses images, it makes their movement more likely. However, any reduction in image size has to take into account human eyes, just as every reduction in the space between airline seats should take into account the postures and shapes of passengers' bodies. Eyes and ears do not have universal, timeless physiological properties. They have media-historical habits. Electronically mediated visual culture shapes eyes and ears and creates perceptual habits at many levels. For instance, the conventions of the rectangular 4:3 ratio TV screen, the 16:9 ratio cinema screen, the number of scan lines or the colour models of PAL/NTSC television broadcasts go deep into visual habits. Sensations of colour, texture, brightness and level of detail all feed into habits of viewing. The video codecs behind DVDs, High Definition Television, mobile TV for 3G cellular telephones, RealPlayer or satellite digital video broadcast attempts to take those expectations into account and meld them with the limited channel capacities of networks, broadcast spectrum or cables.

This outline of the situation of MPEG-2 codecs suggests that the spatio–temporal dynamisms found in a centre of envelopment link very different scales, levels and orders of movement and difference. Codecs make trade-offs between micro-perceptual sensations of brightness, colour, resolution and movement in trying to meet constraints concerning the cost of bandwidth on satellite or cable infrastructures (for instance, SkyChannel, a UK-based digital satellite TV broadcaster, uses MPEG-2 compression to transmit many channels from one satellite). An analysis of trade-offs between image quality and media logistics could determine how the codec mangles or blends different interests (Pickering 1995). Undoubtedly interesting, such an analysis would show how social relations have been displaced into the codecs. At base, it would indicate how '[m]en [*sic*] and things exchange properties and replace one another' (Latour 1996: 61). The sheer volume of moving images, their extension, repetition and multiplication, would correlate with changes in their quality.

This approach to technological projects as a series of compromises or exchanges can yield rich results. However, any analysis of trade-offs between extension (or distribution) and image quality remains fundamentally conservative in relation to differences. Here Deleuze's thought opens a very different path. Deleuze links repetition to irreducible differences. For every instance of repetition, Deleuze suggests, we need to look for the hidden repetition or resonances between differences. The system of resonances between differences comprises an idea. 'Ideas have the power to affirm divergence; they establish a kind of resonance between divergent series' (Deleuze 2001: 278). It is 'pure movement' (ibid.: 24) or the dynamic of an idea in process of becoming, organizing and unfolding a time and space in which repetition occurs. 'Every thing thinks' (ibid.: 254) because every thing, no matter how banal, ordinary, repetitive or singular, occurs within spatio-temporal processes that unfurl from movement inherent to an Idea. Movements of becoming 'are' the mode of existence of the dissimilar, the different or the unequal (Deleuze 2001: 128).

Transformation into Tendencies

In contrast to the notion of *problematization* that Paul Rabinow has distilled from the work of Michel Foucault (Rabinow 2003), the problem imperative of an idea is not primarily concerned with truth and falsehood. It is based on a distribution of the singular and the ordinary. This can be illustrated by one of the two main ways in which video codecs handle differences and repetition in images. It comes from the first lineage of patents. This lineage treats human vision as a sub-representative process that detects differences in brightness, illumination and shadow rather than seeing things. For the purposes of image transmission, variations in brightness and colour count more than the typical analysis of representations in terms of figures, figure–ground relations, forms and contents.

In MPEG-2, the discrete cosine transform (DCT) treats a video frame (or field) as a spatially extended distribution of brightness and colour. This treatment bears no resemblance to the figures and forms found in a given frame. It slices each frame into three separate planes: one of luminance (brightness) and two of chrominance (colour). The most detailed spatial calculation done by the codec on video images begins by analysing luminance of different areas in the picture. (It handles colour planes as larger blotches or patches.) The transform coding process seeks to elicit forms of repetition or redundancy from the variations of luminance across the plane of the image: the luminance or chrominance of a pixel in an image mostly exhibits a high level of correlation with neighbouring pixels. The colour and brightness of a pixel usually predicts those of its neighbours. So a pixel belongs horizontally and vertically in a *sequence*, a well-defined order of increasing or decreasing values of luminance or chrominance. The codec defines a transformation that summarizes or *contracts* the relations between the individual adjacent pixels as a *series* (the sum of the sequence) or as a periodic function that expresses variations of luminance and chrominance. It transforms a spatially extended distribution of pixels into a function, which can then be expressed as a series (for reasons explained below). The transformation pivots on the fact that the information content of an individual pixel is relatively small because by far the majority of adjacent pixels in a given image are identical. Tendencies or variations of colour or brightness across the plane of the image have more value than any particular element of the image.

The discrete cosine transform treats each image as a set of periodic signals or waveforms. Once transformed into a periodic signal, it can be broken down into a series of component cosine waves of different frequencies. For a given signal, some of the component waves contribute more energy to the overall signal than others. The transform coding selects only the most energetic or high-amplitude components and discards the rest. In more technical terms, the transform coding extracts the components of the signal with greatest *spectral density*. It reduces repetition (and hence storage space or transmission) by extracting the differentiating or individuating traits of luminance and chrominance in the image. When MPEG-encoded images are displayed on some screen, the decoding process reconstructs an image from the series. It assigns values of luminance and chrominance to pixels on the screen on the basis of the coefficients in the series. It unfolds the displayed image by putting parts of it back together.

A great deal more could be said about the provenance and development of the discrete cosine transform. It has a rich and continuing history of development coming out of natural sciences and communications engineering. The important point is, that via such techniques, codecs in a certain sense perceive the image. The spatio-temporal dynamisms they introduce in electronic media cultures concern how far this movement of perception can go. These dynamisms do not only *exchange* properties between human and things. Deleuze's thought diverges from any exchange-based account of these dynamisms. Deleuze writes: 'Every spatio-temporal dynamism is accompanied by the emergence of an elementary consciousness which itself traces directions, doubles movements and migrations, and is born on the threshold of the condensed singularities of the body or object whose consciousness it is' (Deleuze 2001: 220).

In its analysis of spectral density and selection of the most energetic component of the signal, the codec isolates tendencies or emphases. Perhaps this is something that also occurs in bodily perception. It treats what can be seen in an image as composed of tendencies and emphases that can be seized in a contractile movement. The codec embodies an 'elementary consciousness', an awareness of transitions, of variable distributions of light and colour. It contracts variations of luminance and chrominance distributed across the plane of the image. Transform compression addresses the differences that spatially extend patches of brightness and colour in the image. This redundancy organizes the image as an extended field of sensations of light and colour.

The transform compression turns spatially distributed or extended repetition into transient differences expressed by coefficients of the different component frequencies. It synthesizes space and time differently. The movement of contraction and the elementary consciousness it presupposes no longer occur only in the bodies of seeing subjects, but also in the technical apparatus of the codec and hence in assorted media technologies. Already here, we see how a thing expresses an idea, if an dea can be understood as a problem-setting system of differential elements (eyes, infrastructures, screens, images, calculations, etc.) that form centres of envelopment around singularities.

Intensities and Differences in Sameness

By rendering an image as a set of (digitized or 'quantized') waveforms, MPEG-2 deeply fissures the objecthood of visual representation. Digital signal processing intimately concerns sensation rather than representation (of objects or figures). It processes brightness and colour, without concern for form or figure in an image. In its analysis of spectral density and selection of the most energetic component of the signal, it isolates tendencies or emphases. It treats an image as composed of tendencies and emphases that can be seized in a contractile movement and summed as a series.

The codec connects two apparently ontologically distant entities, eyes and media infrastructures. They articulate embodied sensations of light and colour with the economically valuable markets for bandwidth of information and communication infrastructures. However, everything discussed so far concerns the extension of images in the world, and how to move images around more often, in greater numbers. Where are the interiorized differences characteristic of centres of envelopment?

One could say that codecs devote themselves to the reduction of difference (and this would be very much in line with critical theory and phenomenology's general treatment of media). They multiply the repetition of the same. In extending the reach of images, they constitute an *extensity* for video. In particular, we could say that the compression of the image responds to a demand for sameness (Terranova 2004: 136). The demand is for the same qualities of brightness and chrominance wherever the image goes. Yet something more is at stake in codec than massive reproduction of sameness. In Deleuze's account of the synthesis of the sensible, perceptions of extension and quality derive from differences, and particularly from *intensity*: 'The extension and "extensity" (the result of the process of extending) of phenomena in space–time and qualities of sensation (*qualia* such as 'redness') attached to phenomena flow from a 'deeper disparateness' or 'difference in intensity' (Deleuze 2001: 236).

Basically, intensities bind differences to each other. When differences come in relation to each other, intensities arise. Deleuze defines intensity as 'a difference which refers to other differences' (ibid.: 117). Intensities differ from representations. Such second-order differences can be physical: differences such as pressure, temperature and density 'drive fluxes of matter or energy' (de Landa 2002: 159); they can also be biological, psychic, social, aesthetic or philosophical, or some mixture of these. Intensities inhabit sensation.

An account of technology in terms of intensities would be somewhat novel. Intensities arise when differences come into relation. Just as the DCT turns the spatial distribution of brightness and colour of an image into a series of frequency components, in *Difference and Repetition* Deleuze developed a series-based explanation of intensity. Deleuze abstractly understands differences as series: 'The first characteristic seems to us to be organisation in series. A system must be constituted on the basis of two or more series, each series being defined by the differences between the terms which compose it' (Deleuze 2001: 117).

According to the quasi-mathematical notion of series, each term in a series differs from the preceding and following terms: A' – A'' – A''' – A'''' – etc. Deleuze says that intensity arise from two or more series of differences in relation. Within the codec, DCT literally generates one set of series or one set of differential relations. What is the other series?

Motion Estimation: 'the Embedding of Presents within Themselves'

MPEG video never flickers. The second lineage of patents, starting in the 1970s, took movement as its problem. Using techniques they inherit from that lineage, MPEG codecs perform a second major calculation called *motion compensation*. Transform coding treats individual pictures themselves as spatial distributions of luminance and chrominance values to be reorganized in series. Motion compensation, in contrast, treats the relation between successive pictures in terms of vectors of movement. MPEG video never flickers because it calculates and predicts transitions between pictures. It dismantles the temporally discrete recording, storing and transmitting of pictures put in place by filmstrip or television. Film frames and analogue video 'fields' have fixed boundaries. They leave the habits of human perception to bridge between frames. In video codecs, calculation fills in the gap between frames. Again, the technical apparatus of the codec takes on part of the work of embodied perception in the interests of changing the relation between body and media infrastructure.

In order to do this, components of the codec involved in motion estimation assume that nothing much happens between successive frames apart from spatial transformations (translation, rotation, skewing, etch) of parts (macro-blocks) of the image. In the process of encoding a video sequence, the MPEG-2 codec analyses each picture in relation to a previous and a future reference picture. It calculates and transmits a series of motion vectors describing how different parts of the frame move in relation to their position in the reference pictures. Motion compensation does not distinguish types of movement. Film scholars often distinguish camera movements (pans, tilts, zoom in) and then ascribe different motivations to movement. MPEG-2 decomposes every movement into the directions and rate of movement of macro-blocks. A typical PAL DVD image contains roughly 800 macro-blocks. At 30 frames/sec, the codec tracks the movement of roughly 24,000 macro-blocks. The 'pictures' streamed on the Internet, downlinked via satellite or burned on DVD mostly comprise long series of vectors describing blocks in motion. Decoding the MPEG stream means turning these vectors back into patterns of blocks moving around in frames on the screen. The decoding side of a codec frenetically recomposes images from blocks moving in all directions.

We could view motion compensation as a second series of differences. Motion compensation alters the temporality of moving images. We have already seen that, in a first centre of envelopment, transform coding contracts the image into a series of differences of brightness and colour. In motion compensation, the image or picture itself is no longer the elementary component of motion perception. Motion compensation reorganizes the picture into series of motion vectors describing relative movements of blocks in time.

Images in Overflow

What happens when these two series come into communication in the codec? According to Deleuze, when series of differences communicate with each other, spatio-temporal dynamisms emerge: 'Once communication between heterogeneous series is established, all sorts of consequences follow within the system. Something passes between the borders, events explode, phenomena flash, like thunder and lightning. Spatio-temporal dynamisms fill the system, expressing simultaneously the resonance of the coupled series and the amplitude of the forced movement which exceeds them' (Deleuze 2001: 118).

As centres of calculation, codecs repeat and render images with a strong concern for accurate repetition. The MPEG-2 bitstream and most other contemporary video codecs put two different series in relation. First they generate a series of values that express a key frame, and then they generate a series of values that express how elements of that key frame move. However, as centres of envelopment, they also do something that lies at the heart of technological repetition. Deleuze refers to this in terms of events, explosions, excess and flashes. Where do we see these? In a sense, the consequences can be seen everywhere today in the growth of video material culture. A preliminary analysis of the spatio-temporal dynamisms could examine ways in which images have become extended and new qualities of images emerge as they proliferate.

The first facet concerns the spatio-temporal dynamism of calculation done within the codecs themselves. The actual ratio of transform-coded or 'intra-frame' pictures (I-pictures) and 'inter-frame' pictures (P- and B-pictures) in a given MPEG-2 bitstream varies. It depends on where the encoding is done, the bandwidth of the expected transmission channel and the size of the display screen. In an MPEG-2 datastream, the *Group of Pictures* (GOP) structure defines the precise mixture of different picturetypes at encoding time. A GOP usually has twelve or fifteen pictures in a sequence such as I_BB_P_BB_P_BB_P_BB_P_BB_. The order of the pictures in a GOP does not correspond to their viewing sequence. A dozen or so block motion-compensation frames follow one transform-coded I-picture. The ratio of different picture types in a bitstream directly affects the encoding time, the transmission time and the decoding time. Calculating motion compensation is much slower than the highly optimized block transforms. Yet motion vectors take much less time to transmit than transform-coded pictures. (Some DVD players offer an option that displays the bit rate of the images on screen. Variations in this bit rate indicate different ratios of DCT and motion estimation being done by the player's codec.)

Even the drastically simplified description I have given should indicate that both transform coding and motion compensation entail much comparison, sorting and shifting of numerical values around in arrays. The time and cost of this calculation can be high, and every reduction of them matters. Practically, codecs must make direct trade-offs between computational time and bandwidth/storage space. A highly compressed image will take longer to generate but require less storage space or network bandwidth. The trade-offs made in encoding sometimes result in artefacts visible on screen, such as blocking and mosaic effects. At times, motion prediction cannot work smoothly. A change in camera shot or an edit breaks the flow of movement between adjacent frames. In that case, the codec falls back on transform coding. A momentary but visible splintering of the images – so called 'motion blocking' – occurs. Whereas film flickers, digital video 'motion-blocks'. Motion blocking appears as horizontal and vertical edges where the MPEG-2 motion compensation algorithm has sliced the image into macro-blocks. (We could also look at 'reordering delay', a 'delay in the decoding process that is caused by frame reordering' [ISO/IEC 13818–1, 1999]). All of these fringe effects and the technical design processes that trade-off between different forms of compression and rendering of images come from the space-time of calculation itself.

The second facet concerns the spatio-temporal dynamism of video materials on contemporary mediascapes. If we ignore all the physicalities of spectatorship, video encoded and decoded by codecs probably looks much the same. Much effort has gone into making them look the same or almost the same. In fact, researchers and standards organizations such as the ITU (International Telecommunications Union) have developed complicated testing regimes that distinguish objective versus subjective video quality. Are these visible artefacts, themselves the effect of technical compromises made in the name of cost, the only sign of an event for a codec, the only phenomena that flash and explode? Videos might look the same but circulate very differently. As Deleuze writes, 'difference pursues its subterranean life while its image reflected by the surface is scattered' (Deleuze 2001: 240). The production of sameness seeks to cement the relation between eye and screen image through intensified sensation. It does detailed work on framing, resolution, brightness, colour and movement. Yet this sameness envelops a very different relation to infrastructure. It generates powerful spatio-temporal dynamics in the relation between eye and infrastructure. Video churns on the Internet. Video streams and broadcasts surge (satellite TV, digital high definition, Web). New ecologies of spectatorship, consumption and occasionally citizenship both mimic and differ from cinema and television spectatorship. As formats, platforms and products mushroom and new forms of making, viewing and moving populate flows of images.

Finally, the two lineages of patents I mentioned at the beginning of the chapter suggest another spatio-temporal dynamism or 'forced movement' around codecs. Almost 700 patents apply to the MPEG-2 standard (http://www.mpegla.com/m2/m2–patentlist.cfm). Tremendously tight intellectual property arrangements bind the codecs. Each aspect of the codec calculations we have been discussing undergoes intensive variation as slight reductions in computational costs and detours around existing intellectual properties are sought out.

Centres of envelopment interiorize differences: 'we claim that complex systems increasingly tend to interiorise their constitutive differences', writes Deleuze (2001: 256), and at the same time give rise to spatio-temporal dynamisms, patterns of extension and qualitative differences on multiple scales. As things that think, how do codecs interiorize 'constitutive differences'? The MPEG-2 codec, I have suggested, can be understood as a composite process of change in the relations between eye and media infrastructure. An intensity inhabits these relations. Deleuze describes an eye as 'bound light': 'An animal forms an eye for itself by causing scattered and luminous excitations to be reproduced on a privileged surface of its body. The eye binds light, it is itself a bound light' (ibid.: 96).

The codec also binds light, but not just to 'privileged surfaces' of the body. It binds light to the movement of images in media infrastructures. It causes some forms of luminance, chrominance and movement to be reproduced on 'privileged surfaces' within media technologies. In order to do this, different temporal syntheses must come into relation: the habit-based contraction of perception in DCT; the embedding of presents in each other in motion estimation.

The relation between transform coding and motion estimation, between the technical treatments of luminance–chrominance and movement, is however unstable. It envelops or comprehends some aspects of the movement of images, but not all of them: 'each intensity *clearly* expresses only certain relations or certain degrees of variation. Those that it expresses clearly are precisely those on which it is focused when it has the *enveloping* role. In its role as the *enveloped*, it still expresses all relations and all degrees, but confusedly' (Deleuze 2001: 252).

The MPEG-2 codec clearly expresses only certain relations between eye and infrastructure. It focuses on brightness, colour and movement of images as framed by the history of photography, cinema and television but puts these in relation to the political economy of telecommunications (with its constraints on bandwidth, memory, processing power, etc.). A primary intensity comes from the light-binding relations between eye and infrastructure. This relation frames the dynamics. The two lineages of patents – transform coding and motion estimation – represent different treatments of repetition in digital video. When transform coding and motion estimation come together in MPEG-2, different spatio-temporal dynamics and forms of interiorization result. Interiorization occurs when the spatio-temporal phenomena begin to cover over the intensities that gave rise to them. A proliferation of patents around codecs occurs. At the same, the trade-offs made between computation and bandwidth in the MPEG-2 codecs mean that the physical forms of codecs proliferate in chips, software, gadgets and boxes. Finally, the circulation of video itself changes. New ecologies of images burgeon.

The codecs, I have been arguing, envelop relations between eye and infrastructure. Does it make any difference that the streaming digital video of decapitation never flickers? It could be argued that the intensive paths generated by codecs in the extension of images make no difference to the viewers. In other words, viewers might see straight through the codecs. They might not be seen. For viewers, however, execution and hostage videos and vast pools of pornography open up alongside the streams of wedding, baptism and graduation videos. Viewers may not be highly conscious of how brightness, chrominance and movement have been minutely altered by the codec. These differences can be easily cancelled out or remain almost imperceptible. This does not mean that they make no difference. On the contrary, the proliferation of video materials and the degrees of variation opening up around video streams today suggest that viewers are caught up in the spatio-temporal dynamisms of video material culture.

The Idea that Thinks the Thing

How far have we come in thinking technologies as thinking things? No thing is a pure expression of an idea. An idea cannot be thought as such. Moreover, things inhabit already structured worlds. Hence, Deleuze's notion of a centre of envelopment, a site for actualization of an idea in worlds organized by systems of extension and quality, reflects the need to accommodate in any radically constructivist account of things the existing orders of representation, personhood and identity. Certainly, nothing of what I have said of codecs belongs solely to advanced technologies. It is not as if codecs think more than, say, photographs, rock carvings or oilpaintings. The singularity of codecs, the different worlds to be found in them, come from the intensities they put in relation. Those differences actualize in the spatio-temporal, scale-transforming dynamics of video material culture.

Why privilege the two patent lineages and the series of transformations they produce in the moving images as the locus of differences? The patent lineages stand in the discussion above as centres of envelopment, as tendencies that attract much mental effort to create new variants, versions and modifications as well as property claims. Centres of envelopment mark the surface of things with zones of infolding and ingression, with regions where an idea or a problem is in actualization, where 'intensive reasons' play out. Crucially, to make sense of Deleuze's claim that every thing thinks, we cannot treat an idea as a concept, a construct that represents or holds difference in identity. It is a system of 'positive, differential multiplicity' (Deleuze 2001: 288) that needs to be thought from a radical constructivist viewpoint. The effort to think oriented by a radically constructivist alignment is particular relevant to composite things that exist in a flux of documents, international standards, versions, software and hardware implementations and diverse and constantly decentred applications.

Bibliography

Boundas, C. (1996) 'Deleuze–Bergson: An Ontology of the Virtual', in P. Patton (ed.) *Deleuze : A Critical Reader*. Oxford and Cambridge, MA, Blackwell, pp. 51–77.

de Landa, M. (2002). *Intensive Science and Virtual Philosophy*. London and New York, Continuum.

Deleuze, G. (2001). *Difference and Repetition*. London and New York, Continuum.

Haskell, B.G. and J.O. Limb. (1972). 'Predictive Video Encoding Using Measured Subject Velocity', USPTO 3,632,865, USA, Bell Telephone Laboratories.

Haskell, B.G. and A. Puri. (1990). 'Conditional Motion Compensated Interpolation of Digital Motion Video', USPTO 4,958,226, USA, AT&T Bell Telephone Laboratories.

Henare, A., M. Holbraad and S. Wastel (eds). (2007). *Thinking Through Things: Theorising Artefacts Ethnographically*. New York, Routledge.

ISO/IEC 13818–1. (1999). "Information Technology – Generic Coding of Moving Pictures and Associated Audio Information: Systems, Pictures and Associated Audio Information: Video." International Organization for Standardization, Geneva, Switzerland. Available at http://www.iso.org/iso/catalogue_detail?csnumber=44169

Latour, B. (1996). *Aramis, or the Love of Technology*. Cambridge, MA, and London, Harvard University Press.

———. (1999). Pandora's Hope: Essays on the Reality of Science Studies. Cambridge, MA, Harvard University Press.

Means, R. (1974). 'Discrete Cosine Signal Transform Processor', 3,920,974, USA, Secretary of the Navy, United States Government.

Pickering, A. (1995). *The Mangle of Practice: Time, Agency, and Science*. Chicago, University of Chicago Press.

Rabinow, P. (2003). *Anthropos Today. Reflections on Modern Equipment*. Princeton and Oxford, Princeton University Press.

Speiser, J.M. (1975). 'Serial Access Linear Transform', USPTO 3,900,721, USA, Secretary of the Navy, US Government.

Terranova, T. (2004). *Network Culture. Politics for the Information Age*. London, Pluto Press.

Chapter 7
Cybernetics as Nomad Science
Andrew Pickering

> There is a kind of science, or treatment of science, that seems very difficult to classify, whose history is even difficult to follow ... [I]t uses a hydraulic model ... inseparable from flows, and flux is reality itself ... The model in question is one of becoming and heterogeneity, as opposed to the stable, the eternal, the identical, the constant. (Deleuze and Guattari 1987: 361)

A Thousand Plateaus makes a tantalizing distinction between what Deleuze and Guattari call *royal* and *nomad* science. The royal sciences are integral to the established state, while the nomad sciences sweep in from the steppes to undermine and destabilize any settled order. I like the sound of these nomads, but just what is the contrast here, and where can it take us? Deleuze and Guattari are, as usual, not entirely clear. I can think of two readings of their story. In the first, the royal/nomad distinction refers in a generalized way to two *phases* of scientific practice. Royal science is finished science, cold, rigid, formalized and finalized, like the state itself – a given repository on which projects of government can draw. Nomad science is instead science in action, research science developing in unforeseeable ways – warm and lively, always liable to upset existing arrangements and to suggest new ones. This would be Bruno Latour's (1993, 2004) reading, I think, and would feed nicely into his notion of a *Politics of Nature* as a rather conservative transformation of the present politico-scientific order (Pickering 2009a).

I am tempted by a more radical reading of Deleuze and Guattari. Their point might be that there are two *kinds* of science. The royal sciences would then include classically modern sciences like physics and sociology, which have, indeed, been enfolded in projects of state formation and governance since their inception – the very name of the Royal Society of London points us in that direction. But what, then, of nomad science? What could count as examples of this? My idea is that the sciences of complexity, emergence and becoming might fit Deleuze and Guattari's description, but, rather than staying at the level of generalities, we need a concrete example to examine, and I focus here on just one such science: cybernetics, especially as it developed in Britain after the Second World War.[1]

There are many different stories about what cybernetics is (or was) and many different political appraisals of it, so let me start by emphasizing that I am interested in *one specific strand* of its history, a strand that took the brain as its primary referent, and which can be defined by its specific conception of the brain and its function. The British cyberneticians, in particular, were concerned with the brain not immediately as cognitive but as embodied and performative – as integral to action in the world. And, beyond this, the cybernetic brain was understood as an organ of adaptation, as central to our ability to cope with situations we have not encountered before. One characteristic activity of the first generation of cyberneticians, including Grey Walter and Ross Ashby, was thus the construction of electromechanical adaptive systems – the 'tortoise' and the 'homeostat', respectively – understood as scientifically illuminating the mechanisms of the brain. Later work in cybernetics focused less on the construction of physical models of the brain and more on questions of identity and social relations as conceptualized around a notion of the adaptation – here I think of the work of Gregory Bateson, Stafford Beer and Gordon Pask.[2]

I need to say more about the substance of cybernetics, but let me start with its nomadism. Why call cybernetics a nomad science? First, because the cyberneticians were *literally* nomads, wandering around outside established social institutions and career structures for much of their lives. Almost all of the early achievements in British cybernetics were made on an amateur, hobbyist basis. Grey Walter built his first tortoises at home in 1948; likewise Ross Ashby and his homeostat (a least in the apocryphal version of the story). Ashby referred to his cybernetic work up to about 1950 as his hobby, and the entire development of his cybernetics is recorded in a set of private notebooks he kept from 1928 onwards, while working as a research pathologist in mental hospitals. Beer and Pask's visionary work on biological computers in the late 1950s and early 1960s was a spare-time activity for Beer (who ran one of the world's largest industrial OR and cybernetics groups for a living), while Pask's institutional base was his private research and consulting firm, System Research, located in the basement of his family home.[3]

So cybernetics lived outside the realms of established society, and one corollary of this was its odd mode of transmission. If the royal sciences have their established modes of propagation – undergraduate degrees and postgraduate training – cybernetics advanced instead in a series of chance encounters, often going via popular and semi-popular books. Norbert Wiener's 1948 book, *Cybernetics,* both put the word 'cybernetics' into circulation and convinced many people that they, too, were cyberneticians. In Britain, its appearance led directly to the formation of the so-called Ratio Club, the first self-conscious grouping of British cyberneticians, which characteristically took the form of an informal and private dining club. In robotics, the cybernetic approach of Walter and Ashby was eclipsed by symbolic AI in the early 1960s, only to come back in the 1980s with the situated robotics of Rodney Brooks, now at MIT – and Brooks had read Walter's book, *The Living Brain* (1953), as a schoolboy in Australia. In another field entirely, it was a turning point in his musical career when Brian Eno's mother-in-law lent him a copy of Stafford Beer's book, *Brain of the Firm* (1972), in 1974. He visited Beer several times, and at one point Beer suggested that Eno was the inheritor of the cybernetic mantle (which Eno politely declined).

Sociologically, then, cybernetics wandered around as it evolved, and I should emphasize that an undisciplined wandering of its subject matter was a corollary of that. If PhD programmes keep the royal sciences focused and on the rails, chance encounters maintained the openness of cybernetics. Beer's *Brain of the Firm* is a dense book on the cybernetics of management, and music appears nowhere in it, but no-one had the power to stop Eno developing Beer's cybernetics however he liked. Ashby's first book, *Design for a Brain* (1952), was all about building synthetic brains, but Christopher Alexander made it the basis for his first book on architecture, *Notes on the Synthesis of Form* (1964). A quick glance at *Naked Lunch* (1959) reveals that William Burroughs was an attentive reader of *The Living Brain*, but Burroughs took cybernetics in directions that would have occurred to no-one else (see also Geiger 2003).

So cybernetics was strikingly nomadic in at least three interconnected ways: it grew outside the usual institutions of support; it lacked systematic modes of transmission; and it could thus mutate wildly in its development. Deleuze and Guattari (1987: 366) speak of the nomad sciences as carried by families and lineages and of the 'secret power' of 'agnatic solidarity' that can 'rise up at any point' (ibid.: 363) – more prosaically one might think of social movements, cults and gurus. But we have not got to the heart of the matter. Why did cybernetics live outside the law? In what sense did it promise to destabilize the state? I need to talk about the connection between sociology and ontology.

Ontology: very crudely, the royal sciences assume that the world is a *knowable* place, and that our relation to it is a cognitive one that goes through knowledge. Our understanding of the hidden structures of the world enables us to submit it to our will in a process that Heidegger (1977) called *enframing*. Historically, this picture has a lot going for it, and one can see that such sciences would hang together nicely with the ambitions of the state. Cybernetics, instead, envisaged a world that was in the end *unknowable*, but to which we can indeed adapt performatively: as I said, cybernetics was a science of adaptation (and *revealing*, to borrow another term from Heidegger). And now I want to distinguish two lines of development of this cybernetic ontology, which map on to a more familiar distinction between what are often called first- and second-order cybernetics.

British cybernetics was the science of the adaptive brain, in two guises, the normal and the pathological, the sane and the mad. Cybernetics emerged, that is, from the matrix of psychiatry, and in its earliest phase the cybernetics of Walter and Ashby ratified, so to speak, the existing psychiatric sociotechnical status quo. The period from the 1930s to the 1950s was the age of the 'great and desperate' psychiatric cures – chemical and electrical shock therapies and lobotomy – and Walter and Ashby used their electromechanical models both to show how an adaptive brain could become mad (as maladaptation) and how the great and desperate cures might undo that.

I described the cyberneticians earlier as nomads, but here we find them acting just like royal scientists. What should we make of this? Part of the solution to this puzzle would be to see Walter and Ashby's work as bifurcated between the nomad and the royal. The radical aspect of their cybernetics – electromechanical robots as brain science – evolved, as I said, outside any established social framework, while their understanding of psychiatric therapy remained tied to the traditional institutions where they in fact made their living.[4] The complication here, as just noted, is that this bifurcation was by no means complete: Walter and Ashby read their cybernetics into a form of psychiatric practice, which they treated as simply given. We could, then, take this as an index of the effectiveness of institutions in repelling the nomad – the disruptive aspects of Walter and Ashby's cybernetics were left largely outside the gates.[5] We could follow Latour (1993) here, and speak of a certain institutional *purification* of their practice. And it is illuminating in the present context to focus on one aspect of that purification.

I have already noted that a concern with adaptation was the hallmark of British cybernetics, but the institutional framework of British psychiatry, from which Ashby and Walter's cybernetics emerged and to which it returned, was anything but adaptive. It was highly asymmetric and hierarchical, seeking to enforce social relations in which the psychiatrists were the only genuine agents, and the patients were literally patients, with no real agency of their own, entirely subject to the psychiatrist's will.[6] As Ashby's horrifying notion of 'blitz therapy' – the use of hypnosis, LSD and electroshock treatment in combination with one another – made clear, the idea was that the patient should adapt to the psychiatrist and not the reverse.

With this in mind, it is interesting to turn to the other line of cybernetic psychiatry that emerged in the 1950s, which undid this asymmetry in taking the concern with adaptation beyond the social circumscription that marked Ashby and Walter's cybernetics. The expatriate and highly nomadic Englishman, Gregory Bateson, one of the founding members of the Macy cybernetics conferences in the US, was the key figure here. Bateson understood madness along much the same lines as the other cyberneticians, though he focused on communication patterns as the site of 'double binds' rather than on brain mechanisms (Bateson et al. 1956), but he stepped outside the orbit of Walter and Ashby's models in postulating a further level of adaptability in the human brain. Walter and Ashby understood madness as a jammed cybernetic mechanism that could only be unjammed from the outside, by ECT or whatever (this is how their technical cybernetics was inserted into established psychiatric practice). But Bateson (1961) redescribed psychosis as an 'inner voyage' comparable to an initiation ceremony, in which some 'endogenous dynamics' might sometimes serve to undo doublebinds and even lead to inner enlightenment. Under this description, the great and desperate cures of psychiatry appeared as completely misconceived, serving only to block the adaptive inner voyage and leaving patients trapped in their double binds. The prescription instead would be to care for schizophrenics, to help them see such voyages through to their conclusion.

The person who took this reasoning to the limit and symmetrized it even further was the Scottish psychiatrist R.D. Laing. During the 1960s he arrived at the conclusion that in modernity we are all mad, in the sense of being cut off from our own inner lives, and therefore the sane can learn from the mad, understood as explorers of inner space. 'We need a place where people ... can find their way *further* into inner space and time and back again' (Laing 1967: 128). Laing and his Philadelphia Association put this idea into practice at Kingsley Hall in London between 1965 and 1970, and, in the 1970s, in a series of communities in Archway, North London.[7] At Kingsley Hall, psychiatrists and schizophrenics, as well as artists and dancers, lived symmetrically together, the sane providing a support community for the mad, reciprocally adapting to their often bizarre behaviours rather than prescribing electroshock treatment, and, at the same time, becoming something new themselves (even, at Archway, sometimes entering into their own inner voyages).

And this is the point I wanted to arrive at. Kingsley Hall is the best exemplification I can come up with of a destabilizing nomad science in action. Taken to the limit, the cybernetic ontology of unknowability and adaptation hung together at Kingsley Hall with a radical transformation of social relations and institutional forms. And the socially disruptive force of cybernetics as nomad science is thematized here by the fact that Kingsley Hall grew out of David Cooper's earlier Villa 21 project, which had aimed to implement symmetric relations between doctors and patients *within* an established mental hospital (Cooper 1967). The institutional frictions between Villa 21 and the rest of the hospital fed directly into the decision of the Philadelphia Foundation to operate entirely *outside* the established mental health system in England. We can thus see that cybernetics was a *different kind* of science from the royal sciences of discipline and governance, and that, as elaborated by Bateson and Laing, it invited a *different kind* of social organization – a self-organizing and adaptive institutional form quite different from the state form of hierarchical command and control.

So this cybernetic anti-psychiatry is my way of putting flesh on the radical reading of Deleuze and Guattari's idea of nomad science – my way of thinking through what they could possibly have had in mind – and I want to close with a few brief remarks on it. First, we can see that in this version cybernetics had a radical political edge, entailing the abandonment of a well-entrenched institution of state governance. Secondly, we might note that the influence of Kingsley Hall extended well beyond psychiatry. The Kingsley Hall community was itself a key element of the 1960s counterculture in Britain, with all its well-known challenges to established forms of life. The Philadelphia Association sponsored the Dialectics of Liberation Congress held at the Roundhouse in London over three weeks in 1967, which brought together many of the luminaries of the counterculture in Europe and the US, including Allen Ginsberg, Gregory Bateson, Emmett Grogan, Simon Vinkenoog, Julian Beck, Michael X, Stokely Carmichael, Alexander Trocchi, Herbert Marcuse and Timothy Leary. Kingsley Hall was also the model for the anti-University of London – a radical and anti-hierarchical formation that seems to have foundered when the students decided to charge the lecturers for the privilege of teaching them (Green 1988).

But, thirdly, I want to mention Deleuze and Guattari's idea that the royal and the nomad sciences have a mutually constitutive relation. History, according to Deleuze and Guattari, has the quality of an *interplay* between the state and the nomad. The nomad supplies a transformative dynamic, upsetting state formations, which are then reconstituted on a new basis, only to be nomadically disrupted again, and so on. The state adapts to the nomad.[8] That kind of interplay has been, at best, only partial in the post-war history of cybernetics. If robotics is different since the work of Walter and Ashby, psychiatry is not. Ashby and Walter themselves domesticated their cybernetics to their institutional milieu, while Kingsley Hall had little effect on psychiatric practice more broadly. The only institutionalized change since the 1950s has been the rise of pharmaceuticals instead of ECT and lobotomy as our chosen means of blocking inner voyages at the expense of reciprocal adaptation. Over the last forty years, brute exclusion and forgetting rather than interplay have become our rule for coping with the nomads at the level of the state. Maybe that has something to do with the grimness of the world we now find ourselves in.

Notes

1. The following discussion of cybernetics is taken from a book currently in press, *The Cybernetic Brain: Sketches of Another Future,* and fuller documentation and analysis can be found there (Pickering forthcoming). Deleuze and Guattari's discussion of nomad science is to be found on pp. 361–74 of *A Thousand Plateaus,* in chapter 12, '1227: Treatise on Nomadology – the War Machine.'. Deleuze and Guattari's examples of nomad science cluster around civil engineering (building cathedrals and bridges), thematizing an informal, not codified, relation to the world that develops *in situ,* in the hands of engineers/scientists who constitute a mobile and quasi-autonomous 'band' undisciplined by the state (Deleuze and Guattari 1987: 364), and I associate these examples with Latour's finished-science/science-in-action pairing. Latour's proposals for a 'politics of nature' then hinge on incorporating science in action into the political process (without letting go of finished science). But Deleuze and Guattari also make an ontologically based distinction between sciences of laminar and turbulent flows (1987: 361–64), the latter being less useful to projects of governmentality. I want to put some flesh on this second notion here. A related ontological question that Deleuze and Guattari touch upon is whether we should think of matter as inherently formless, a blank slate upon which we write our designs, or whether science and engineering are better seen in terms of adaptive attempts to enrol the tendencies of matter. Deleuze and Guattari, of course, favour the latter, on which see also de Landa (2002). Deleuze and Guattari ascribe a 'hylomorphic' model of matter to the royal sciences (1987: 369), while I associate a sort of 'hylozoism' with cybernetics (Pickering 2009b). I should note that it is possible to tilt the balance in favour of my reading of Deleuze and Guattari (and away from the Latourian one). Deleuze and Guattari discuss methods of stone-cutting such that the accumulation of stones produces the kind of arch that can support a cathedral, without any 'over-arching' [*sic*] geometrical vision of the arch. This invites a connection to the mathematics of fractals, cellular automata, simulations of non-linear systems – the unknowable (see below). (Of course, they are also talking about the difficulty of abstracting unformalized knowledge from the workers and hence subjecting them to state control. See also Linebaugh (1992).)
2. For more on these individuals, see Pickering (2002, 2004a, b, c, 2007).
3. 'It is not that the ambulant sciences are more saturated with irrational procedures, with mystery and magic ... Rather, what becomes apparent ... is that the ambulant or nomad sciences do not destine science to take on an autonomous power, or even to have an autonomous development. They do not have the means for that because they subordinate all their operations to the sensible conditions of intuition and

construction – *following* the flow of matter, *drawing and linking up* smooth space. Everything is situated in an objective zone of fluctuation that is coextensive with reality itself ... [T]he experimentation would be open-air, and the construction at ground level' (Deleuze and Guattari 1987: 373–74).

4. Almost all of Walter's working life was spent at the Burden Neurological Institute, where he became one of the world's leaders in EEG research. Ashby worked at a series of mental hospitals in England, before starting a new career at Heinz von Foerster's Biological Computing Laboratory at the University of Illinois in 1960, at the age of fifty-seven.
5. 'Whenever this primacy [of the 'man of the State'] is taken for granted, nomad science is portrayed as a prescientific or parascientific or subscientific agency' (Deleuze and Guattari 1987: 367).
6. Laing (1985) later recalled that in the early days of his career psychiatrists were strongly discouraged from even speaking to schizophrenics.
7. There are no very good scholarly sources on Kingsley Hall. The only book-length account, Barnes and Berke (1971), is very much focused on the experience of its authors. Sigal (1976) is a wonderful fictional account. The Archway communities are better documented. See, for example, Burns (2002) and Peter Robinson's documentary film, *Asylum* (1972). Guattari himself worked at a similarly radical institution, the psychiatric clinic La Borde, south of Paris (Guattari 1984: 52). 'The aim at la Borde was to abolish the hierarchy between doctor and patient in favour of an interactive group dynamic that would bring the experiences of both to full expression in such a way as to produce a collective critique of the power relations of society as a whole' (Massumi's foreword to Deleuze and Guattari 1987: x).
8. For example:

   '[A]mbulant procedures and processes are necessarily tied to a striated space – always formalised by royal science – which deprives them of their model, submits them to its own model, and allows them to exist only in the capacity of "technologies" or "applied science." ... There is a type of ambulant scientist whom State scientists are forever fighting or integrating or allying with, even going so far as to propose a minor position for them within the legal system of science and technology ... [T]he ambulant sciences quickly overstep the bounds of calculation; they inhabit that 'more' that exceeds the space of reproduction and soon run into problems that are insurmountable from that point of view; they eventually resolve those problems by means of a real-life operation ... Only royal science, in contrast, has at its disposal a metric power that can define a conceptual apparatus or an autonomy of science ... That is why it is necessary to couple ambulant spaces with a space of homogeneity, without which the laws of physics would depend on particular points in space ... This is somewhat like intuition and intelligence in Bergson, where only intelligence has the scientific means to solve formally the problems posed by intuition, problems that intuition would be content to entrust to the qualitative activities of a humanity engaged in *following* matter.' (Deleuze and Guattari 1987: 372–74, emphasis in original)

Bibliography

Alexander, C. (1964). *Notes on the Synthesis of Form.* Cambridge, MA, Harvard University Press.

Ashby, W.R. (1960) [1952]. *Design for a Brain,* 2nd edn. London, Chapman and Hall.

Barnes, M. and J. Berke. (1971). *Two Accounts of a Journey Through Madness.* New York, Harcourt Brace Jovanovich.

Bateson, G., D. Jackson, J. Haley and J. Weakland. (1956). 'Towards a Theory of Schizophrenia', *Behavioral Science* 1, 251–64. Reprinted in G. Bateson (1972). *Steps to an Ecology of Mind.* New York, Ballantine, pp. 201–27.

Bateson, G. (ed.). (1961). *Perceval's Narrative: A Patient's Account of His Psychosis, 1830–1832.* Stanford, CA, Stanford University Press.

Beer, S. (1972). *Brain of the Firm.* London, Penguin.

Burns, D. (2002). 'The David Burns Manuscript', unpublished. Available at laingsociety.org/colloquia/thercommuns/dburns1.htm#cit

Burroughs, W.S. (2001) [1959]. *Naked Lunch, the Restored Text,* edited by J. Grauerholz and B. Miles. New York, Grove Press.

Cooper, D. (1970) [1967]. *Psychiatry and Anti-Psychiatry.* London, Tavistock Institute, reprinted by Paladin.
de Landa, M. (2002). *Intensive Science and Virtual Philosophy.* London, Continuum Books.
Deleuze, G. and F. Guattari. (1987). *A Thousand Plateaus: Capitalism and Schizophrenia.* Minneapolis, University of Minnesota Press.
Geiger, J. (2003). *Chapel of Extreme Experience: A Short History of Stroboscopic Light and the Dream Machine.* New York, Soft Skull Press.
Green, J. (1988). *Days in the Life: Voices from the English Underground, 1961–1971.* London, Heinemann.
Guattari, F. (1984) [1973]. 'Mary Barnes, or Oedipus in Anti-Psychiatry', *Le Nouvel Observateur* 28 May 1973; translated in F. Guattari, *Molecular Revolution: Psychiatry and Politics.* Harmondsworth, Middlesex, Penguin, pp. 51–59.
Heidegger, M. (1977). 'The Question Concerning Technology', in *The Question Concerning Technology and Other Essays.* New York, Harper and Row, pp. 3–35.
Laing, R. D. (1967). *The Politics of Experience.* New York, Pantheon.
———. (1985). *Wisdom, Madness and Folly: The Making of a Psychiatrist.* New York, McGraw-Hill.
Latour, B. (1993). *We Have Never Been Modern.* Cambridge, MA, Harvard University Press.
———. (2004). *Politics of Nature: How to Bring the Sciences into Democracy,* Cambridge, MA, Harvard University Press.
Linebaugh, P. (1992). 'Ships and Chips: Technological Repression and the Origin of the Wage', in P. Linebaugh, *The London Hanged: Crime and Civil Society in the Eighteenth Century.* Cambridge, Cambridge University Press, pp. 371–401.
Pickering, A. (2002). 'Cybernetics and the Mangle: Ashby, Beer and Pask', *Social Studies of Science* 32, 413–37.
———. (2004a). 'The Science of the Unknowable: Stafford Beer's Cybernetic Informatics', in R. Espejo (ed.) *Tribute to Stafford Beer, Kybernetes* (special issue) 33, 499–521.
———. (2004b). 'Mit der Schildkröte gegen die Moderne: Gehirn, Technologie und Unterhaltung bei Grey Walter', in H. Schmidgen, P. Geimer and S. Dierig (eds) *Kultur im Experiment.* Berlin, Kulturverlag Kadmos, pp. 102–19.
———. (2004c). 'Cybernetics and Madness: From Electroshock to the Psychedelic 60s', paper presented in science studies seminar series, Simpson Center for the Humanities, University of Washington, Seattle, 7 May 2004.
———. (2007). 'Science as Theatre: Gordon Pask, Cybernetics and the Arts', *Cybernetics and Human Knowing,* 14 (4), 43–57.
———. (2009a). 'The Politics of Theory: Producing Another World, with Some Thoughts on Latour', *Journal of Cultural Economy* 2, 199–214.
———. (2009b). 'Beyond Design: Cybernetics, Biological Computers and Hylozoism', *Synthese* 168, 469–91.
———. (forthcoming). *The Cybernetic Brain: Sketches of Another Future.* Chicago, University of Chicago Press.
Sigal, C. (1976). *Zone of the Interior.* New York, Thomas Y. Cromwell.
Walter, W.G. (1953). *The Living Brain.* London, Duckworth.
Wiener, N. (1948). *Cybernetics, or Control and Communication in the Animal and the Machine,* Cambridge, MA, MIT Press.

Part III
Minor Assemblages

Chapter 8

Cinematics of Scientific Images: Ecological Movement-Images

Erich W. Schienke

With the cinema, it is the world which becomes its own image, and not an image which becomes the world. (*Cinema I*)

Theory of Cinema I and II as Analytical Device for Science Studies

Deleuze's work on cinema spans two volumes (*Cinema I – the Movement-image* (1986) and *Cinema II – the Time-image* (1989)) and the greater breadth and depth of film history. *Cinema I* and *II* theorize the construction of images as the realization of sign, framing, continuity, disjuncture, closure and reflexivity. Film, in this case, is Deleuze's empirical landscape, into which he brings and works through aspects of Bergsonianism and Piercean semiotics. What is important to understand about the term *image* as used in *Cinema* is that it refers not so much to the visual elements (scopic) of an image, but to what the image is as a plane of immanence, which has a system of relations (is bounded) and which communicates (is connected to other planes).

While film is the central topic of *Cinema* I and II, the theories within open up a new means for theorizing and engaging the construction of scientific images as partial and fractured, but which, when assembled together like a film, tell a story (as scientific theory) about the world. The (experience of) sense that is made by how our theories fit the world do so only at a particular scale, framed in a particular way, capturing only particularities of the whole. Within scientific enquiry, as within film production, there is always something left out of our frame of enquiry and beyond the threshold of a narrative that can distinguish between 'real' data and 'outliers' and can suture together one data set with the next in a way that communicates a story about the world. In the communication of science or the production of a scientific product, as in film editing, there is always something left on the cutting room floor, significant moments of the enquiry (or filming) that never make it to the final edit.

Film provides a useful analogy to science in thinking through styles of enquiry and presentation of scientific theory amidst the international diversity of styles and matters of content, where both film and science are formed by and represent the political and cultural contexts within which they operate. Deleuze does not argue for a deeper meaning or underlying communication in film images, in that a film's meaning cannot be separated from its distinctly representational forms (as *movement-image* or as *time-image*). The film hides no meaning, as the meaning is always at the face, or surface. It is in this way that the narratives of science, the stories we tell about meaning in the world, cannot be separated from the models, diagrams, equations, sets, etc. That is, the language and form of film and the language and form of science are always communicating at the surface (face) – and without hidden meaning, i.e. prima facie. Granted, there may be metaphorical or encrypted content that needs unpacking, but there is no part of the communication that is hidden. A different meaning or theory requires a reframing and is a different communication, a different image that gets committed to memory.

Deleuze performs his analysis of film as a *pragmatic vitalist*, at least where the theoretical genealogy of *Cinema* extends from Bergson(ianism). Where other thinkers (Husserl, Bergson, Sartre, Merleau-Ponty) tended to treat cinematic images as moving stills, Deleuze brought to cinematic theory a very vivid sense of the organic image – of not just movement and action, but of vitality itself (Deleuze 1986: 5). In science studies literature, most theoretical treatments of scientific images rarely move beyond a scopic analysis – beyond prioritizing the visual in analysis (Crary 1990; Lynch and Woolgar 1990; Galison 1997, Jones and Galison 1998; Dumit 2003) – to thinking of the scientific image as categorical assemblages that perform an *ordering of things* that both communicates and remembers. Pulling together the conceptions of *movement-images* and *time-images* from Bergson's *Matter and Memory* (1991) and ordering them according to Peirce's three modes of being (firstness, secondness, and thirdness), Deleuze produces a robust theory of images through this analysis of cinema. Deleuze's theory on cinema is particularly rich for an analysis of not just the visual elements of scientific images, but of how scientific images (movement and time) of the world come together, are constrained, are ordered and shift over a duration and across milieux.

This chapter primarily uses the *movement-image (Cinema* I) as a device for categorizing and analysing the contemporary production of *scientific images* – images not just as visual articulations, but as regimes of ordering and communicating. A complete articulation of *scientific images* would also require further investigation of the *time-image (Cinema* II) and the forms it takes. Specifically, theorizing the production of ecological images, with concern for scale and sets (data, classes, categories), is richly informed by the organic qualities of the *movement-image*. There are also scientific images that are necessarily produced in a virtual mode; in that one looks to perceive and recognize a trace of a thing/object for which we have no memory, a virtual image needs to be created. This is the *time-image*, which emerged in post Second World War film. Atomic diagrams, such as Feynman Diagrams and bubble chamber traces, are an almost literal example of the *time-image* in its most basic crystalline (circuit) form. The *time-image* is equally worthy of deeper analysis, but requires delving further into arrays of organic circuits (*machinic-phylum*) and cybernetics. This chapter will mainly investigate how an analysis of *movement-images* can be applied to analysing ecological sets.

The set of ecological images and supporting analysis in this chapter emerge from sixteen months of fieldwork and laboratory studies conducted within the Chinese Academy of Science's (CAS) Research Centre for Eco-Environmental Science (RCEES) in Beijing, China. Throughout the duration of the fieldwork, I investigated how researchers, particularly within the Geographic Information Systems (GIS) lab of the RCEES, constructed their ecological investigations, data sets, maps and models for the support of ecological management by local municipalities and provincial-level planning and for assessment by the Chinese central government. Extending from my own training with GIS and the history of cartography, I was able to work extensively with ecologists in the construction and communication of their research, mainly as scientific papers for peer-reviewed journals. The images covered in this chapter come from the paper by Rongbo Xiao et al., entitled 'Land Surface Temperature Variation and Major Factors in Beijing, China' (Xiao et al. 2008). This paper used a variety of remote-sensing techniques (interpretation of data from satellites) to look at the effects of building clustering, building height, impervious surfaces and a variety of other factors that can have a significant impact on a city's urban heat island (UHI) effect, an effect that can cause an increase of +0–15°C in urban areas compared with outlying green regions. This sort of analysis is very useful in considering future urban planning and ways to address the problem under current circumstances. What I refer to as *ecological movement images* in this chapter are those images that are considered to be scientific communications, such as maps, models, posters, videos, parts of reports and other forms of communication used by both scientists and decision makers.

What emerges overall from combining an analysis of scientific representation with Deleuze's treatise on film is a strategic way for science studies to access and analyse the languages and modes of scientific representation, particularly in how scientific representations are sutured together and move from one representation to the next in a manner that makes sense – as a mode of representing and remembering at the same time.

Movement-images

The set of things that have appearance are classified as 'image', where *image* (not pose) entails *movement*. Movement is generative of image, in that appearance is generated through a trace (in memory). Without movement, there is no trace and thus no appearance, and no 'image'. In other words, without movement, there is no foundation for the persistence of vision, which is a form of memory (*durée*). The connection from one image to another image, in Deleuze's terms, is complete, 'where every image acts on others and reacts to others, on "all their facets at once" and by all their elements'. The trace and the perception of it, then, are the movement-image.[1] Preceded by a necessary explanation of framing (set), the four progressions or dimensions of movement-images

– namely, *perception-images, affect-images, action-images,* and *relation-images* – will be used to analyse how these progressions inform an understanding of the production and circulation of specific scientific images, namely ecological movement-images.[2]

> Framing is the art of choosing the parts of all kinds which become part of a set. This set is a closed system, relatively and artificially closed. The closed system determined by the frame can be considered in relation to the data that it communicates to the spectators: it is 'informatic', and saturated or rarefied. Considered in itself and as limitation, it is geometric or dynamic-physical. Considered in the nature of its parts, it is still geometric or physical and dynamic. It is an optical system when it is considered in relation to the point of view, to the angle of framing: it is then pragmatically justified, or lays claim to a higher justification. Finally, it determines an out-of-field, sometimes in the form of a larger set which extends it, sometimes in the form of a whole into which it is integrated. (Deleuze 1986: 18)

Movement-image is used here to evaluate how ecological images are constructed as scientific images, and how 'moving', either from one pixel to the next or from one scale to the next, requires a logic of intervals. Movement, in this regard, can take a variety of different forms, such as movement across a frame (between pixels); zooming in and out of a frame (between scales); movements in relationships between elements, sets and actions within and outside of the frame (between types); movement of the entire frame itself in some way to create a trace and exhibit difference from one aspect to the next (across map types). However, using each of these different kinds of movements in a communication requires a different logic of intervals, in that a logic of intervals between pixels is not the same as a logic of intervals between scales. However, the main factor that determines types of movement, and thus the appropriate logic of intervals, is how the frame itself is constructed. The *art of framing,* as Deleuze refers to it, provides a way to discuss processes of constructing closed sets within ecological contexts. Qualifying the closure of a set directly relates to how ecological analyses develop their units and various subsets for descriptive analysis and define bounded limits to constructing categories. Often these sets are articulated within ecological fields of study as maps, diagrams and data sets. (Maps are the presence of a visual frame. Diagrams are the presence of movement between frames, i.e. processes. Data sets are the 'informatic' frame, i.e. elements of communication.) Articulating each of these framings as scientific objects requires that they be understood as 'closed' for the sake of description, albeit arbitrarily, as such sets are never completely or entirely closed. For knowledge claims to be made and decisions to be actualized, choices are made about a cut, about what to leave out of the ecological set (frame) and about what constitutes such a set as a 'whole' or bounded. The scientific *movement-image,* then, is an 'image' that can be used to further articulate something about the world and how it is ordered, and thus governed. There are, of course, political consequences of the constructions of scientific sets (as categories of organization) and, as such, it is important to critically interrogate and analyse how and why a particular set is 'closed'.

Framing and Closing Sets

> The whole and the 'wholes' must not be confused with sets. Sets are closed, and everything which is closed is artificially closed. Sets are always sets of parts. But a whole is not closed, it is open; and it has no parts except in a very special sense, since it cannot be divided without changing qualitatively at each stage of the division ... The whole is not a closed set, but on the contrary that by virtue of which the set is never absolutely closed, never completely sheltered, that which keeps it open somewhere as if by the finest thread which attaches it to the rest of the universe. (Deleuze 1986: 10)

The first step is the closure of a set, which, as Deleuze points out, is always an artificial closure. As to the closure of ecological sets (databases, maps, scales, etc.) and subsets (data ranges, map layers, pixels, etc.), the closure (boundedness) of a set is performed as a *strategic closure* within a wider context of possible sets and subsets. The closure of a set is not arbitrary, but rather is produced or arrived at through a variety of possible contingencies and limits of description. It is the goals of a description (as content), such as a description of ecological conditions, which pre-construct how parts, sets, subsets are assembled and bounded (closed). 'The divisibility of content means that the parts belong to various sets, which constantly subdivide into sub-sets or are themselves the sub-set of a larger set, on to infinity. This is why content is defined both by the tendency to constitute closed systems and by the fact that this tendency never reaches completion' (Deleuze 1986: 16).

Categories of classification require that a part can belong to more than one set. For example, a small stand of trees (as a part to be looked for as a cluster of pixels) could be categorized as a kind of broader land-use/land-cover set, as a subset of a vegetation index (NDVI), as a resource for fuel wood or food, as a pixel of particular colour and size, as an ecological construction project, as an invasive species, as shelter for other species, as shade, etc. That is, the stand of trees as part can be any of these things. The part, in this case the stand of trees, when placed into a set ('forest', 'recovered farmland' or unit of 'ecological construction') is a move towards closure of the set. However, as the stand of trees could easily reside in multiple sets depending upon the context of the enquiry or probe, closure is never complete or absolute in that any given part would only belong to one and only one set. There must be a way out of the set/subset so that the trees can be re-categorized according to the context of a different enquiry – of a qualitatively different framing of analysis.

> Every closed system also communicates. There is always a thread to link ... any set whatever to a larger set. This is the first sense of what we call the out-of-

> field: when a set is framed, therefore seen, there is always a larger set, or another set with which the first forms a larger one, and which can in turn be seen, on condition that it gives rise to a new out-of-field, etc. (Deleuze 1986: 16)

It is those sets, which are framed, that are put into motion through a particular enquiry. That is, the set does not come together in a coherent manner unless it is framed. The frame of the inquiry is synonymous with the bounds of a particular inquiry. In constructing an ecological inquiry, for example, the 'stand of trees' is only described as such because the bounds of the enquiry, the framing, determine it as such. It is a stand of trees; or a single tree; or the sub-parts of a tree, such as bark, stems, leaves, roots; or a grouping of different types of plant cells, chemical processes, nitrogen-fixating nodules and microbes in the soil; or, moving 'up' in scale or context: it is a small part of a larger stand of trees; or a forest; or a continental carbon stand. The framing of the ecological analysis, thus, determines how that stand of trees is categorized into a set within a continuum of categorical possibilities. The framing of trees, in this example, would be predominantly determined by the use of scale[3] and political-economic evaluations of ecological services.

There are various forms of 'out-of-field' that could be applied in these instances. A map, for example, produces a very distinct notion of what is in the frame and what is out of field, namely, that which lies within and beyond the borders of the map. In addition, thematic layers on a map[4] only communicate specific kinds of information. Thematic map layers that are selectively not shown still reside in the same spatial frame of the map, but effectively, they are also out of field in that they are rendered invisible. Ecological scale, such as the difference between trees and forest, also determines what is out of field. Thus, ecological sets at a finer or coarser granularity (smaller or larger scale) are also out of field of the frame, which is again, determined by the context of the enquiry.

> We know the insoluble contradictions we fall into when we treat the set of all sets as a whole.[5] It is not because the notion of the whole is devoid of sense; but it is not a set and does not have parts. It is rather that which prevents each set, however big it is, from closing in on itself, and that which forces it to extend itself into a larger set. The whole is therefore like thread which traverses sets and gives each one the possibility, which is necessarily realized, of communicating with another, to infinity. (Deleuze 1986: 10)

A description of all relationships, as a whole, can never reach closure itself. It is the 'whole' then, which forces open the sets, allowing for relationships between the parts. Even if the relationships between sets are out of the field of the frame of enquiry, the relationships still persist. As in the example of the stand of trees, the set can never be fully closed. That is, the stand of trees can be drawn in relation to the local soil microbes, the forest, the continental carbon pool, the solar influx of energy, etc., all at the same time. It is the thread connecting these relationships between ecological sets (a categorical continuum) that force them to remain incomplete and with permeable boundaries.

> Hence, it is perhaps not sufficient to distinguish ... a concrete space from an imaginary space in the out-of-field, when it thus ceases to be out-of-field. In itself, or as such, the out-of-field already has two qualitatively different aspects: a relative aspect by means of which a closed system refers in space to a set which is not seen, and which can in turn be seen, even if this gives rise to a new unseen set, on to infinity; and an absolute aspect by which the closed system opens on to a duration which is immanent to the whole universe, which is no longer a set and does not belong to the order of the visible. (Deleuze 1986: 17)

The first aspect of being out of field, the relative aspect, can be seen in the example of the thematic map layers (see Figure 8). Making the layer visible, such as a land-use layer, brings about a qualitatively different analysis, a qualitatively different framing. What this layer (set) may mean in relationship to other visible sets, such as water basins (as in flood plains), takes on a different relationship. What is out of field and not framed, then, is qualitatively different from what it was before that layer was made visible. Say, for example, we had a map of the area downstream from the world's largest dam; the Three Gorges Dam. Then, we add the layer predicting the possible flood zone if that dam were to burst. What has been strategically left out of field is a wide variety of land-use could be at risk and threaten life concerns. Again, add to it a layer visualizing the dwellings of residents within that flood zone, and what becomes apparent that is out of the field is a plan for evacuating those residents quickly, such as in the case of an earthquake. This is the relative aspect. The absolute, as opposed to the relative, aspect of out of field cannot ever be rendered visible. In this case, there would be no 'act of God'[6] map layer that could be seen. All of the forces that have persisted through time to configure the landscape in a particular way, to make it favourable or unfavourable to damming, to arrive at a particular configuration at a particular time, cannot be visualized. Yet it is all of these forces that could bring us to the moment of rupture, of release, of destruction and of regeneration that would follow.

Perception–images

'From the point of view which occupies us for the moment, we go from total, objective perception which is indistinguishable from the thing, to a subjective perception, which is distinguished from it by simple elimination or subtraction' (Deleuze 1986: 64). As a 'long shot',[7] the perception–image, as it relates here, would be described as an image of or at the level of systems – a system-wide view. Deleuze, drawing on Bergsonianism, infers that 'a subjective perception is one in which the images vary in relation to a central and privileged image' (Deleuze 1986: 74).

'[O]bjective perception is one where, as in things, all the images vary in relation to one another, on all their facets and in all their parts' (Deleuze 1986: 76). A connected network of sensors measuring all significant aspects of an ecosystem, for example, would constitute a strongly objective perception – a *network perception.* In defining objectivity as 'see[ing] without boundaries or distances', Deleuze describes a form of networked perception where every significant relationship within the system is connected and measured in relation to every other. In other words, networked perception is the complete subtraction of the other, or rather, the negation of the subject–object dichotomy – as there is no observation that exists outside the network. For example, the *subject-perception* of 'seeing as the state' is centralized and hierarchical, while the *object-perception* of the network (across an ecological network) is decentralized and 'panarchical'.[8] The 'ecosystem' is a *perception-image* – the central and privileged image that constitutes the sum of networked (ecological) relationships.

While affection-images equate to a Peircean firstness, action-images to a secondness, relation-images to a thirdness, Deleuze considers perception-images to inhabit a zeroness, i.e. in that they determine all that is able to be perceived by a particular framing or enquiry (Fig. 8.1).

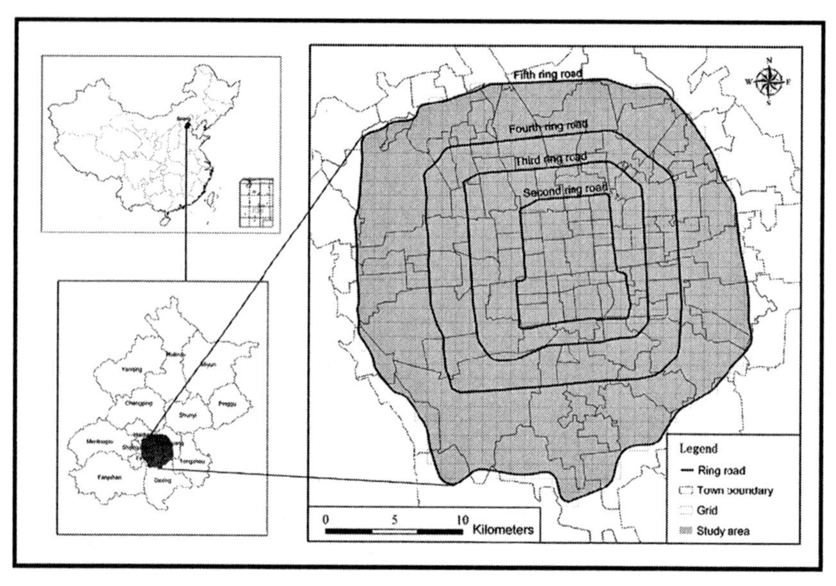

Figure 8.1. Perception-Image – locating the enquiry, in this case within Beijing's Fifth Ring Road. Image courtesy of Xiao Rongbo.

Affection-images

At the face of enquiry, the close-up image reveals the expression of the object of enquiry – what is revealed is pure affect without spatial or structural referent. The affection-image is as close as one can get to a plane of immanence – where it falls within the quality of a sense, i.e. it is not perceived or recognized. It is a pre-categorical sensation of an object/place. Equating the affection-image to Peirce's firstness, Deleuze explains that:

> it is not a sensation, a feeling, an idea, but the quality of a possible sensation, feeling, or idea. Firstness is thus the category of the Possible: it gives a proper consistency to the possible, it expresses possible without actualizing it, whilst making it a complete mode. Now, this is exactly what the affection-image is: it is quality or power, it is potentiality considered for itself as expressed. (Deleuze 1986: 98)

In a close-up – an affection-image – we are not observing what the image is, but rather how the image is. An expression of how an image is cannot be reduced further, nor can it be divided, without changing qualitatively, i.e. it is 'dividual'. 'The affect is impersonal and is distinct from every individuated state of things: it is none the less singular, and can enter into singular combinations or conjunctions with other affects' (Deleuze 1986: 98).

The affection-image is the possibility of expressions, or, rather, the set of possible expressions. Within scientific enquiry, for example, the affection-image is the set of possible outcomes, at its most reductive, of any applied test/observation. Take, for example, remote-sensing imagery, where the affection-image (close-up) is the possibility for a pixel within the image (one pixel out of, say, ten million) to be expressed at a certain wavelength of reflectance (of the sun reflecting back from the surface of the earth to a sensor), i.e. at a certain colour. In this regard, scientific affection-images could be more generally construed as those of 'pure signal', i.e. data from measurements preceding any interpretive or comparative context (Fig. 8.2). The wavelength of the pixel will also indicate, within a more zoomed-out perception-image, the 'treeness' or 'waterness' of the landscape across that particular area, but only in relation to other pixels. The irreducibility of the pixel, in this case, is that it cannot express both 'treeness' and 'waterness' at the same time, a situation that produces significant

Figure 8.2. Affection-Image – realm of possible expression of a pixel wavelength.

fuzziness when trying to determine boundaries between ecosystems – a problem in the fractal nature of things. Of course, a change in pixel size (where the large pixel can be divided further into a smaller area, say, from 0.5 km to 0.5 m) will open up the possibility for a finer-grain analysis, so we could instead observe a clump of trees between a river and a field where such a distinction could not be made before. Nevertheless, each of those pixels can still only express a given wavelength (affect). That is, in Deleuze's terms, the pixel (or vector) is the construction of 'any-space-whatevers', or rather, the *genetic element* of a remotely sensed image.

Action-images

Affection-images bring us the immediate sensations of the object, but the object cannot be understood as a body (figure) without differentiation through time, in relation to other objects in the foreground. The action-image also produces a means for analysing the difference between the figure of enquiry and the background. The object then becomes the body from which various affection-images emanate. In an expression of Peirce's secondness, Deleuze describes action-images as images where:

> [q]ualities and powers are no longer displayed in any-space-whatevers, no longer inhabit originary worlds, but are actualized directly in determinate, geographical, historical and social space-times. Affects and impulses now only appear as embodied in behaviour, in the form of emotions or passions which order and disorder it. This is Realism ... What constitutes realism is simply this: milieux and modes of behaviour, milieux which actualise and modes of behaviour which embody. The action-image is the relation between the two and all the varieties of this relation. (Deleuze 1986: 141)

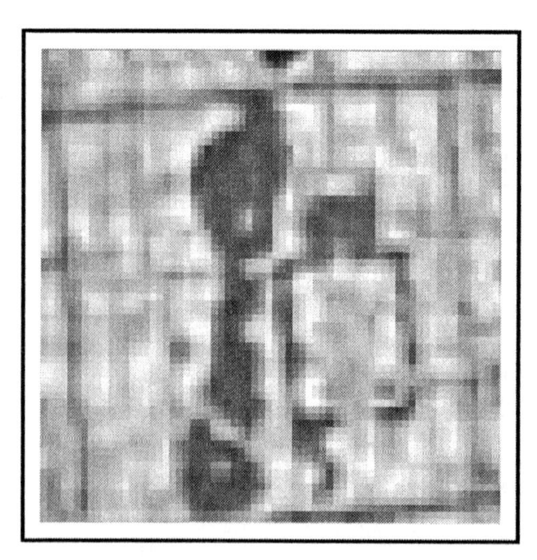

Figure 8.3. Action-Image – pixel differentiation across movement of space and affect. (Forbidden City and Beihai park to the left.)
Image courtesy of Xiao Rongbo.

In ecological terms, the action-image extends the interpretation of a quality or power expressed into a categorical set that identifies the kind of body the expression emanates from. For example, wetness, blueness, coolness, flatness, flow and high reflectance are typical expressions of a body of water. The body of water also fluctuates according to seasons and the status of the surrounding elements of the ecosystem. The ecological action-image, then, is one that expresses, through observation and enquiry, the *status* of the ecosystem or ecological body (Fig. 8.3).

Mental Images (Relation-Images)

'The affection-image undoubtedly already contains something mental (a pure consciousness). And the action-image also implies it in the end of the action (conception), in the choice of means (judgement), in the set of implications (reasoning)' (Deleuze 1986: 198). Perceptions-images (mental images), effectively equating to a Peircean 'zeroness' (Bogue 2003), describe the necessary assumption of an observer in the expression of any given movement-image. Affection-images, exhibiting Peircean firstness, express the primacy of images, as irreducible qualities or powers. Action-images, as secondness, are extensions of the affection-images to wider physical, social and historical contexts. The relation-image, as thirdness, is the ordering of affection-images, action-images and other relation-images according to relationships between sets and parts; in other terms, it is the construction of the mental image that is communicated and remembered.

> When we speak of mental image, we mean something else: it is an image which takes as objects *of* thought, objects which have their own existence outside perception. *It is an image which takes as its object, relations,* symbolic acts, intellectual feelings. It can be, but is not necessarily, more difficult than the other images. It will necessarily have a new, direct, relationship with thought, a relationship which is completely distinct from that of the other images. (Deleuze 1986: 198)

The mental image is not just an impression of an object or a system, it is perception of the relationship of various sets within a system, which leads to the characterization (framing) of the set of relationships as a particular system. Taking this a step further, while the genetic element of the affection-image would be the pixel in this case, the genetic elements of the relation-image are the relations themselves. Relation-images (Fig. 8.4), like action-images, come from and circulate within wider milieux, a wider set of current and historical relations, where, at its limits, it is the complete genealogical image of a system.

Pushing relation-images to the limits of 'renderability' is where we find scientific images at their most theorized, at their most constructed. It is also where we are able to say, 'Here, these sets of questions and concerns are environmental, these are political, these are social, while these are historical.' Through the construction and performance of relation-images (thirdness) one is able to distinguish between systems or modes of

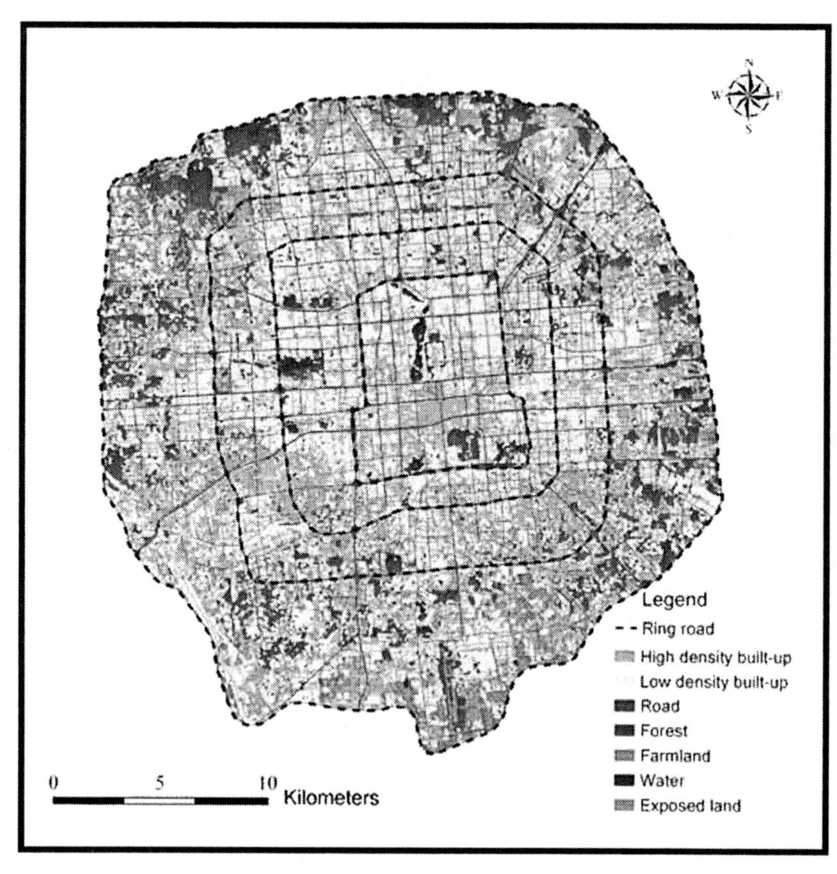

Figure 8.4. Relation-Image – correlation of actions with the system.
Image courtesy of Xiao Rongbo and Li Weifeng.

enquiry. As will be discussed in the next section, within systems ecology, the relation-image is an image of the various ecological systems and subsystems, and allows for the ability to analyse from one scale to the next, from the microbes to the trees to the economics of forestry.

Ecological Movement-Images for Governance

Contemporary ecological governance requires the production and circulation of images for use by the state, the decision-making community and stakeholders. The typical land-use map is an excellent example of a relation-image, one that relates across features (thematic layers), resides in a common frame (scale) and communicates the memory of a place where one has not been before (the cartographic communication). For example, in a conversation with Xuehua Liu, a professor of ecology and spatial analysis at Tsinghua University, I learned that she uses maps to develop a mental image of a small region before conducting fieldwork, such as in her investigations of the Giant Panda habitat. In this example, it is the map that is used to generate a memory of a place and various features which she had not seen previously.

> For example, before I go to the field I always first do geographical research on the region. Such as administrative boundaries, local government and management issues. You will see, generally, which village is located in which area, how many rivers there are there and what their names are, and the general vegetation types. After that, if you go to the field then you will just feel something. Or, I should say, originally it's a kind of image and now it's real to you. You will feel that. (From an interview with Professor Xuehua Liu at Tsinghua University)

How ecological images are produced, communicated, circulated and embodied is crucial to understanding their role in ecological governance. As such, ecological images cannot be considered to be solely the results of ecological science, but, rather, as the product of ecological governance, which includes the constraints by science, politics, society and geography. The application of Deleuze's movement-image to the construction of ecological images allows us to bring in all the various dimensions that force a certain framing of ecological analysis – a forcing that can leave many significant issues out of field and without memory, and thus unconsidered.

Framing Ecological Relation-Images

Individual actors working in some capacity of decision-making for the state work between two positions – through *subjective centralized perceptions,* on the one hand, and through *objective networked perceptions,* on the other. The subjective centralized perceptions of the actors, such as determining priorities for local and regional development in China, would mostly be derived from the sociocultural and political milieux they inhabit (their habitus), while the objective networked perceptions of the state (in the form of tables, indexes, matrices, spreadsheets and maps), such as natural resources and ecosystem services, are gathered from a physical network of instrumentation and data-processing centers.[9] So far, these are only perception-images, the 'things' of the decision-making process.

Taking form as models, scenarios, decision support systems, forecasts, etc., relation-images used in ecological decision-making are formed through building a theory of relations between various scientifically, politically, culturally and economically derived framings of the environment and its life's ystems. Here, relation-images are considered as an assemblage of affection-images (intensities of the status of various aspects of a landscape, such as wetness, dryness, slope, etc.), action-images (intensities put into wider context, such as 'treeness' or 'habitatness') and other relation-images (such as sub-models, environmental and political imaginaries, such as China's 'well-off society', or ecological protection).

In terms of framing an ecological movement-image, the most significant factors are scale and data availability. But, in considering the map as communication, aesthetics (style) and translation from numerical data to visual data (opsign) require the application of an art. Remote-sensing interpretation is an excellent example of how data are finessed into a communication, both through proper framing and interpretation of data/information taken of the earth's surface from either satellite or fly-over and through a sense for visual coherence and aesthetics. In remote sensing work, the construction of a visual end product (the map) by the specialist is usually considered to be a less than obvious task; as there is definitely an art to the construction of a map as communication. That is, beginning with the same data, a remote-sensing specialist could not easily reconstruct a perfect replication of another rendered remotely-sensed image (an image which they happen to see by sight and is not already loaded on to a computer system) by following a simple set of predetermined (obvious) procedures. A remote-sensing specialist needs to develop a feel for the technological process, for the landscape phenomena, for the meaning of spectral signatures and for their colour separations.

The development of this 'tacit' sense for the desired end product and how to arrive at it are very much a crafted skill, one that takes time and practice, the development of which resembles what Deleuze refers to as the *art of framing*. A cartographic specialist I worked with wanted to produce a map, using the same data set, as well done as one produced by someone working out of the central government mapping bureau. She indicated to me that, even after two weeks of working on the image, she still needed more practice to match the skill of the one produced by the state scientist. This is one example of the *art of framing*, of choosing the parts of all kinds that become part of a set, where the set (or set of subsets) is the map, in this case, the map of Beijing land use based on remotely sensed data. Here, the remote-sensing specialist was personally motivated to develop a 'beautiful' object of image/analysis, partly for her own personal aesthetic satisfaction and partly to communicate the quality of her abilities. While there are a process and a science in producing a map from remote-sensing data, the results of an aestheticized treatment do improve the map as communication. Thus, applying the *art of framing* is significant in this case because it does not necessarily affect the determination of scale, as will be discussed below, but it does affect the quality and power of the map as a *scientific movement-image*. The quality/art of the map does matter in what and how well the map communicates, which is a crucial factor in moving the map from the scientific community to the decision-making community.

Scale in Ecological Movement-Images for Governance

Just as the conception of site is immanent to ethnography, so the conception of scale is immanent to ecology. While conducting ethnographic research among ecological scientists in China, I was able to study how various constructions of scale were formed and closed. Studying the construction of scale within the geographical and ecological research of the scientists, I addressed questions such as: what scale dominates an analysis when and for what reasons; and when is there a strong difference in practice between techno-scientific, geographical, and social scales in the formation of analyses for ecological decision-making. Going back to the earlier discussion of framing and sets, scale could well be considered one of the more significant thresholds at which a set is closed.

Scale can be understood as a discursive formation at a *threshold of epistemologization* – a concept developed by Foucault in *The Archaeology of Knowledge*. In the *Archaeology*, Foucault articulates how specific categorical systems are at, near or just on the other side of certain thresholds of epistemologization, where their assemblage constitutes a specific form or mode of epistemic production. In this regard, a specific scale would be the threshold between that which is outside the epistemic unit (set) and that which is inside. As Foucault suggests, '[w]hen in the operation of a discursive formation, a group of statements is articulated, claims to validate (even unsuccessfully) norms of verification and coherence, and when it exceeds a dominant function (as a model, a critique, or a verification) over knowledge, we will say that the discursive formation crosses a threshold of epistemologization' (Foucault 1972: 187).

When a threshold 'spillover' happens, that is, when a scale no longer suffices to properly categorize or contain the phenomenon of informatic interest and when the phenomenon is effectively out of frame, a finer or coarser resolution is required. In that they are constructed, physical and geographical scales are thresholds of analytical framing and, thus, are also thresholds of scientists', decision-makers' and other frame-workers' capacity to strategize and mobilize actions (based on their image of the situation). How such thresholds are arrived at is a critical question for analysis of how ecological informatics is constructed by and as a system of governance.

There are various 'forcings' that determine scale (as threshold) in spatial analysis of ecological phenomena. Though constructed by other political and social forces, the technology itself, such as the satellite infrastructure, obviously plays a strong role in determining scale, and thus what can be seen and known. In using data from the existing remote-sensing infrastructure, when considering the analytical frame as forced by the techno-scientific infrastructure of remote-sensing imagery (satellite), the difference between scales is discrete (quantized), i.e. it is not continuous. That is, an analysis is determined by the forced pixel size (depending on the spectral band desired), such as 20 m x 20 m, 15 m x 15 m, 10 m x 10 m, or 2.5 m x 2.5 m per image pixel. With these systems, there are no pixel ranges in between these set standards; thus, the jump between scales is discrete (quantized), not continuous. As mentioned, when working with ecological phenomena for governance, the differences between scales of analysis are often continuous, or at least not explicitly quantized in the manner forced by the technology.[10]

Before producing an analysis, remote-sensing specialists will spend a considerable amount of time determining what the best pixel-to-phenomena ratio is. For example, a remote-sensing specialist in the lab in which I conducted participant observations spent much time in trial-and-error figuring out the best pixel size and cluster size for analysing ecological construction sites specific to certain regions of western China. In this case, the ecological construction sites were constituted as a farmer planting a stand of trees paid for by the provincial government to improve local ecosystems. Local decision makers, however, have thousands of these types of small-scale projects they have funded and they have a difficult time in determining whether or not the funds have been put to use and whether or not the farmer or landscaping corporation has followed through on the construction, perhaps with follow-up maintenance. Local officials would like to know how to survey these kinds of construction sites without having to send someone out into the field, which is cost-prohibitive and often non-conclusive due to the possibility of misleading data given when on the ground. The amount of pixels in the cluster the ecologist is trying optimize is very specific. The cluster size, then, is framed by multiple factors such as: technological limits (pixel size); the fact that the sizes of the ecological construction projects themselves are very specific to China; the lack of resources to monitor these projects *in situ*; and the possibility of corruption even if they were able to be monitored locally. Finally, the scientist determined that, with SPOT imagery (2.5 m/pixel), a cluster of 6 x 6 2.5 m pixels (15 m x 15 m) is the optimum grouping for analysing whether or not the ecological construction site exists. The analysis was made even richer by including the TM (elevation) data,[11] which allowed for a forced three-dimensional (not true 3–D) visualization of the landscape. Though it does not change the map qualitatively, this last step (last framing) was taken to improve 'readability' of the map for use by decision makers.

On the one hand, in this example, the scale of the frame construction is fixed by the pixel size of the remote-sensing image type chosen. On the other hand, the scale of the phenomena is determined more by land-use issues, such as tree stands or irrigation projects. The scale of governance is also a component in the construction of this

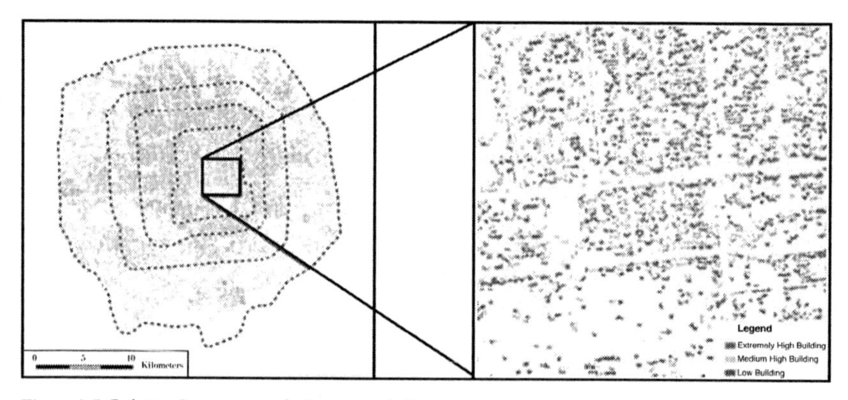

Figure 8.5. Relation-Image – correlation across scales.
Image courtesy of Xiao Rongbo.

frame. Township and city governments are not likely to have the resources to hire remote-sensing specialists or to develop remote-sensing expertise at the local level. Such requests would, most likely, need to originate from the provincial or central government, usually those giving the funds for the construction of the ecological projects in the first place. The construction of scale,[12] in this case, is configured through imbrications between the scalar rigidness of the remote-sensing imagery, the ecological-economic scale of the land-use phenomena and the information needs of the scale of governance. Each of these contingencies regulates how an analytical set is determined and how an analytical system for decision making is closed.

Scale is a structural relationship that allows for various sets (data sets) to be brought together (federated) for generating images of an ecological system under enquiry. Understanding relationships between scales of analysis is also crucial to understanding how regional choices will affect local situations, and vice versa. The scale of governance determines and is determined by the scale of the *informatic* (Fig. 8.5). That is, capacity to govern is directly dependent upon the resolution of the relationships (in bringing data together). The higher the resolution, the finer the detail of phenomena, which results in the need for scalar-specific governance (micromanagement). Of course, micromanagement of higher-resolution (small-scale) phenomena is not always efficient and often requires unrealistic amounts of labour to administer and govern. Such small-scale governance (micromanagement) can also hinder larger-scale flows of capital and resources and inhibit the growth of local economies. In such cases, the scale of the *informatic* (data set) is often predetermined by the desired scale and capacity of governance. In other words, the scale and capacity of governance will play largely in how an

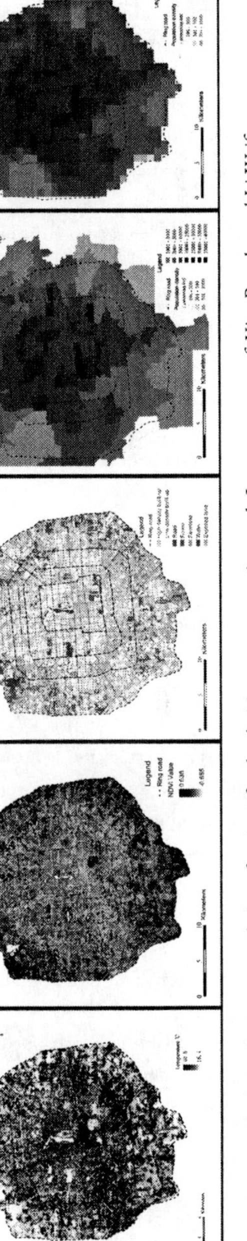

Figure 8.6. A set of relation-images – correlation of actions for the decision system at a given scale.Images courtesy of Xiao Rongbo and Li Weifeng.

ecological enquiry is framed (zeroness) and built up through three types of the *movement-image* (firstness, secondness, and thirdness).

Ecological governance, as in the management of ecosystem resources and services, requires relationships to be revealed and managed across a wide field of interests, from forest-wide data to the diets of local residents.[13] Finding a common scale for analysis is not always possible or even necessarily germane to the analysis. The point here is that scale is one of the primary means by which ecological images are related and assembled. A relation-image becomes the effective means for interfacing with the system, in this case, the relevant local aspects of the ecosystem. The relation-image of the ecosystem, at a particular scale, becomes a basis for planning and taking action. The kind of scientific image that is produced for ecosystem management, then, is an *image of relationships* between the various subsystems (those parts and sets within the frame of analysis). Relation-images are mental images (which have the appearance) of the system that will be acted upon by various actors and institutions of governance. As such, investigating ecological images as relation-images is crucial in thinking about how ecological images are formed and put into practice by scientists, decision makers and stakeholders (Fig. 8.6).

'Cinematics' and Scientific Images

This cursory treatment of Deleuze's work on cinema only looked at the construction of ecological movement-images. Closer consideration of scientific images from qualitatively different disciplines, such as particle physics, would be found in Deleuze's *Cinema* II: *the Time-image,* which explores fragmentary or solitary virtual images, rather than images of systems and intervals. An extended 'director's cut' of this chapter would include further examples of scientific movement-images, such as the type of 'image' created through ethnographic enquiry and representation,[14] as well as various examples of scientific time-images, such as traces from a bubble chamber or Feynman diagrams, or even ecological time-images that represent relationships as ecological circuits (flow charts) as opposed to movements (across space and scale). Certainly, at the very least, a robust theory of scientific images drawn from *Cinema* I and II would require this further analysis. What is significant here is not that filmic and scientific images have structural similarities, or that science is somehow a film, but that the theory of images Deleuze generates through looking at film is highly applicable to an analysis of both filmic and scientific images as stylized and politicized images of the world that are immanently both objective and subjective in their construction and communication.

Cinematic images, categorized by Deleuze into movement-images and time-images, are visual (opsign) and auditory (sonsign) in their expression and interface and only ever communicate at the surface. However, much predetermines the construction and communication of a relation-image. The application of Deleuze's cinematic images to an analysis of scientific images, here, is not centred only on the visual elements of the communication. Using the framework of 'cinematic theory' when discussing scientific images requires analysing scientific images as specific formalized expressions that take on qualities of movement-images and/or the time-images. Thinking through ecological maps as movement-images, for example, provides a way to understand their ordering and closure, and is analytically effective not because it relies on the eye/optics, but because it relies on framing and piecing together. The analysis above is centred on those elements of image construction necessary to produce systems that signify and communicate ecological relationships, be they forced by natural, technological, social or political factors. Moving forward along this trajectory will require further analysis of scientific images, specifically in drawing out a further comparison with time-images – another form of image assemblage. 'Cinematic analysis' not only provides a means of analysing scientific communications in terms of textual communications, but also allows for an analysis as to how many small instants (within a series of experiments, for example) are combined (through a logic of associations or theory of intervals) to generate an actionable 'image' of the open system, be it a cell or the solar system.[15]

It is important to remember that scientific images, like cinema, are always produced in the context of wider milieux, and as such need to be evaluated according to their structural and interpretative relationships within a given milieu. In evaluating scientific images in terms of their effects in the world, differences between what constitutes a scientific image and what an image of governance, for example, begin to break down and become less empirically relevant. Investigating questions concerning the assemblage and consequences of scientific images is central to a critical and deeper analysis of scientific communication more broadly – as systems, as experiments, as facts, as data sets and as actionable understandings of the world.

Notes

1. Material traces within science, according to Hans-Jorg Rheinberger, are generated through the production of experimental systems. 'Experimental systems are the units within which the signifiers of science are generated. They display their dynamics in a space of representation in which graphemes, material traces, are produced, are articulated and disconnected, and are placed, displaced, and replaced' (Rheinberger 1998: 287). Rheinberger's articulation of material traces elicits a sense of not just a writing on but an etching into the world. The traces that are produced within an experimental system are then interpreted, through a diagrammatic process (a framing of the movement-image), as signifiers of significant relations between poles such as 'known and unknown' or 'subjective and objective'. The 'surprises' are made less surprising through a process of articulation and re-articulation, sometimes displacing and sometime transforming previous diagrams (movement-images).
2. Deleuze categorizes six types of movement-images in total. The two that I do not cover here are the impulse-images and the reflection-images. While they could find applicability here, they do not correspond to a discussion of 'zeroness', and Peirce's 'firstness', 'secondness' or 'thirdness'. Impulse-images and reflection-images are useful in better understanding the transitions between affection-images and action-images, and action-images and relation-images, respectively.

3. Deleuze's articulation of *dividual* describes further a quality of framing and/or closure of sets that is helpful in understanding how scale (scaleset) is a feature of closure. Dividual, as Deleuze explains, is 'that which is neither indivisible nor divisible, but is divided (or brought together) by changing qualitatively. This is the state of the entity, that is to say of that which is expressed in an expression' (Deleuze 1986: 14). Scale (the scaleset) is ecological enquiry at a particular resolution and at a particular resolve. Arguably, one cannot see a tree when one is working at the level of microbes, for the tree is out of frame of the scale of the microbes, though they are still connected. Scale cannot simply be divided or split off, but, rather, a qualitative difference in systematic characteristics needs to be generated. Different scales of systems and sets can exhibit and express radically different features. This is, perhaps, best understood as a qualitative difference in analytical features – a clash of scales. One cannot readily analyse in the same frame simultaneously at two different scales. A healthy forest requires the death of trees and undergrowth (usually in the form of fire), or the successful growth of a parasite (success at one scale) can ultimately kill the host (failure at another scale). This is not to say that there is always an incongruity between the two, but, rather, an incompatibility of relevant features of systematic success or failure. Further, both scale and the concept of dividuality are derivative of Deleuze's broader use of immanence – of existing or operating within and of pervading and sustaining the universe. A system existing or operating within a particular plane of immanence is a system that exhibits an emergent vital coherency on that plane or at that scale.
4. Thematic layers, in cartographic terms, are descriptive layers on maps, such as roads, rivers, vegetation, etc.
5. See Gödel's (1992) incompleteness theorems.
6. 'Acts of God' are critiqued more as failures of a social system in an editorial by Donald Kennedy, editor-in-chief of *Science*.
   We know with confidence what has made the Gulf and other oceans warmer than they had been before: the emission of carbon dioxide and other greenhouse gases from human industrial activity, to which the United States has been a major contributor. That's a worldwide event, affecting all oceans. When Katrina hit the shore at an upgraded intensity, it encountered a wetland whose abuse had reduced its capacity to buffer the storm, and some defective levees gave way. Not only is the New Orleans damage not an act of God; it shouldn't even be called a 'natural' disaster. These terms are excuses we use to let ourselves off the hook. (Kennedy 2006)
7. 'These three kinds of spatially determined shots can be made to correspond to these three kinds of varieties; the long shot would be primarily a perception image; the medium shot an action-image; the close-up an affection image' (Deleuze 1986: 70).
8. Panarchy is a concept developed by Crawford Holling to describe how abrupt change (crisis) within an ecological system presents opportunities for adaptation, i.e. learning.
   Panarchy focuses on ecological and social systems that change abruptly. Panarchy is the process by which they grow, adapt, transform, and, in the end, collapse. These stages occur at different scales. The back loop of such changes is a critical time and presents critical opportunities for experiment and learning. It is when uncertainties arise and when resilience is tested and established. (Holling 2004)
9. States, ultimately, are run by individual actors, which is one reason why a state could never achieve fully objective and rational direction, even if it could achieve highly effective networked perception.
10. In China's ecological planning and construction community, the SPOT imagery format is used most often.
11. Landsat TM (Thematic Mapper) data and elevation data (DEM) were used to produce the appearance of 3–D of the SPOT data.
12. I use the 'frame/scale' pairing to describe when the closing of a frame is intrinsically co-constructed according to the spatial or social scale; which, in the content of the three main case studies here, is most of the time. However, frame and scale are less linked when discussing issues such as global scale, or regional scale, institutional scale, etc., when the construction of the 'frame' of analysis is not based on scale in the less stringent policy-derived framing of sustainable development, poverty alleviation, diversity protection, etc.
13. Methods of participatory geographic information systems/science (PGISs) are increasingly being employed by NGOs and researchers to empower communities in achieving sustainable, grass-roots

development. Instead of gathering data in a top-down manner for a typical GIS, a PGIS actively solicits and incorporates the knowledge of local residents to improve environmental management through a sense of entitlement. A PGIS can incorporate all of this information generated through local knowledge (of farmers, for example) into a layered map that facilitates conversation and generates actionable knowledge. In implementing participatory GIS, working at the appropriate 'scale of practice' is essential to maintain coherency between the scale of social practice and the scale of spatial analysis. That is, cartographic scale needs to be in the range that is visible and relevant to social practices. If the scale is too small, say 1:50,000, then the framing of finer community-level details is lost and there is a disjuncture in the participants' connection to place – where place is the locus of an activity. For example, in the 2004 PPGIS conference summary, 'the international track underlined the critical importance of scale in our work, pointing out that women withdraw from discussions when the scale is less than 1:5,000, because that is beyond the scope of their daily activity field' (Craig and Ramasubramanian 2004). Poole discusses how aerial photography at scales of 1:20,000 and 1:60,000 are useful for wider area decision-making, but he also confirms that '1:5,000 is optimum as a stimulus for local discussions' (Poole 1995: 15). Thus, for the participatory management of local forestry, 1:5,000 would be an appropriate target scale for GIS layers.

14. Early ethnography and early film have very close ties in both content and style.
15 Deleuze, in extending Bergson's views, argues that 'the modern scientific revolution has consisted in relating movement not to privileged instances, but to any-instant-whatever. Although movement was still recomposed, *it was no longer recomposed from formal transcendental elements (poses), but from immanent material elements (sections)*' (Deleuze 1986: 4). Historically, the scientific goal for universalizing the construction of the ecological frame to 'any-instant-whatever' has presented a deep structural problem for the construction and analysis of ecological movement (or complex movement more generally, i.e. movement across systems, framesets and framescales); where preferring 'any-instant-whatever' does not capture the continual movement/dynamics of the system.

Bibliography

Bergson, H. (1991 [1896]). *Matter and Memory*. New York, Zone Books.

Bogue, R. (2003). *Deleuze on Cinema*. London, Routledge.

Craig, W.J. and L. Ramasubramanian. (2004). 'An Overview of the 3rd Annual PPGIS Conference', 3rd Annual PPGIS Conference.

Crary, J. (1990). *Techniques of the Observer: On Vision and Modernity in the Nineteenth Century*. Cambridge, MA, MIT Press.

Deleuze, G. (1986). *Cinema 1: The Movement-image*. Minneapolis, University of Minnesota Press.

———. (1989). *Cinema 2: The Time-image*. Minneapolis, University of Minnesota Press.

Dumit, J. (2003). *Picturing Personhood: Brain Scans and Biomedical Identity*. Princeton, Princeton University Press.

Foucault, M. (1972). *The Archaeology of Knowledge*. New York, Pantheon Books.

Galison, P. (1997). *Image and Logic*. Chicago, University of Chicago Press.

Gödel, K. (1992[1931]). *On Formally Undecidable Propositions on* Principia Mathematica *and Related Systems*. Translated by B.Meltzer. New York, Dover Publications.

Holling, C.S. (2004). 'From Complex Regions to Complex Worlds', *Ecology and Society* 9 (1). Available at http://www.ecologyandsociety.org/vol9/iss1/art11/.

Jones, C.A., and P. Galison. (1998). *Picturing Science, Producing Art*. New York, Routledge.

Kennedy, D. (2006). 'Acts of God?', *Science* 311, 303.

Lynch, M. and S. Woolgar. (1990). *Representation in Scientific Practice*. Cambridge, MA, MIT Press.

Poole, P. (1995). *Indigenous Peoples, Mapping and Biodiversity Conservation: An Analysis of Current Activities and Opportunities for Applying Geomatics Technologies*, Peoples, Forests, Reefs Program Discussion Paper Series, Washington D.C, World Wildlife Fund.

Rheinberger, H.J. (1998). 'Experimental Systems, Graphematic Spaces', in T. Lenoir (ed.) *Inscribing Science: Scientific Texts and the Materiality of Communication*, Stanford, Stanford University Press, pp. 285–303.
Xiao, R., Q. Weng, Z. Ouyang, W. Li, E.W. Schienke and Z. Zhang. (2008). 'Land Surface Temperature Variation and Major Factors in Beijing, China', *Photogrammetric Engineering and Remote Sensing* 74 (4), 451–61.

Chapter 9

Social Movements and the Politics of the Virtual: Deleuzian Strategies

Arturo Escobar and Michal Osterweil

Introduction: Deleuze and the Politics of Theory

Gilles Deleuze has undoubtedly become one of the most influential critical thinkers of our time. As was the case with Foucault in the 1970s–1990s, many critical currents today are in dialogue with Deleuze and Deleuze and Guattari. What are the features of Deleuze's theory that many intellectuals, activists and academics find so appealing? A Deleuzian reading of this question suggests that Deleuze's work opened up again the field of the virtual to other thoughts and other theoretical and political projects, and that this reawakening of the virtual has found tremendous resonance in the dreams and desires of many social actors; to paraphrase a fashionable formula, the appeal of Deleuze's work would lie in enabling 'other theories and theory otherwise'. That these other theories could be linked to the construction of other possible worlds is of particular concern to social movement activists and their theorists. The idea is also found in Deleuze and Guattari, most clearly stated in *What is Philosophy?*. '*We lack resistance to the present,*' they write (Deleuze and Guattari 1994: 108, original italics), in discussing the link between philosophy and capitalism peculiar to the age of the total market. What they intuit is an absolute deterritorialization of philosophy, one that brings about a novel conjunction of philosophy with the present – a renewal of political philosophy. This political philosophy takes place around the concept of utopia: 'it is with utopia that philosophy becomes political and takes the criticism of its time to its highest point' (ibid.: 99). The goal?: To '*summon forth a new earth, a new people*' (ibid.: 99, see also 110). This would constitute a reterritorialization of philosophy in the future. Is Zapatismo, for instance – with its dream of 'a world where many worlds might fit' – an attempt at a reterritorialization of this kind? Other social movements? Can these be seen as 'Deleuzian movements'? Or, more appropriately, what kinds of conversations can be established between contemporary social movement practices and Deleuze's work that could further illuminate one and the other? These are some of the questions to be discussed in this chapter.

As this volume's editors state, one could argue that there is a Deleuzian turn in contemporary studies of techno-science, culture and politics in disciplines such as geography and anthropology and fields such as cultural studies, and it is time to explore in a more systematic fashion this turn's epistemological, ontological and methodological consequences, as well as its political and practical implications. This is a complex issue, given that there is always a tight connection between social reality, the theoretical frameworks we use to interpret it and the sense of politics and hope that emerges from such an understanding. In other words, our hopes and politics are largely the result of the particular framework through which we analyse the real. Marxism was paradigmatic in this sense, to the extent that Marxist analyses of capitalism made revolutionary politics and hopes inevitable; for post-structuralism, politics and hope lay in the transformation of discourse and the regimes of truth-telling, since this is what the theory highlights as key to the production of the real. These frameworks have been extremely productive, and continue to be important, but there are other articulations of the connection between theory, politics and hope that are emerging.

This connection is brought to the fore with acuity in times of heightened struggle and crisis, as Deleuze and Guattari implicitly suggest in their discussion of the de-/re-territorialization of philosophy mentioned above. We might be going through one such period at present, in which we see both dire social processes – unprecedented in their destructiveness and ideological reach – and an eagerness for novel approaches that are more radically contextualizing and relational (Grossberg 2006). As is often the case, the more well-known theories are no longer able to say something radical about the contemporary situation because the languages at their disposal do not allow it (Santos 2002). Most discussions of capitalism and modernity, for instance, reveal that the languages of contemporary theory are capitalocentric (e.g. Gibson-Graham 1996, 2006), Eurocentric (e.g. Mignolo 2000) and/or globalocentric (e.g. Harcourt and Escobar 2005), thus making place, other knowledges and diverse forms of economy invisible or inconsequential.

It would thus seem, then, as if the world of today demanded a new social theory altogether – as if from the depths of the social an urge to revamp social theory were springing more intensely than is usually the case. There is one crucial difference as compared with similar theoretical moments of the recent past: the cohort of those interested in the production of new theories has expanded well beyond the usual suspects in the (largely northern) academies. Today this urge is being heeded by a growing number of researchers, activists and intellectuals, both within and outside of the academy. In this sense, the complex conversations that are beginning to happen among many kinds of knowledge producers worldwide are in and of themselves a hopeful condition of theory at present. A second feature is that this urge addresses not only the need to transform the contents of theory but its very form; as Walter Mignolo (2001) likes to put it, the aim is to transform both the content and the form of the conversation – to which we need to add changing the place of the conversation (Osterweil 2005a). In the last instance, what is at stake is the transformation of our understanding of the world in ways that allows us to contribute to the creation of different worlds. More philosophically, this means that one main feature of the current wave of critical theory is that it is concerned not only with questions of epistemology (the conditions of knowledge, as was still the case with post-structuralism) but with ontology, that is, with basic questions about the nature of the world. In other words, today's critical theories are fuelled by a fundamental scrutiny of the kinds of entities that theories assume to exist and, concomitantly, the construction of theories based on different ontological commitments.

Our emphasis on theoretical orientations that take ontological questions seriously also means that this is largely a theoretical chapter. We say 'largely' because the argument we make resonates with our experience of looking at and working with some social movements, but the chapter is not based on ethnographic research. Consider it a middle way in between theory and ethnography. If what we are arguing makes sense, we hope it will foster ethnographic studies to examine some of the claims we make more closely.

The first part of the chapter delves into the ontological turn from a Deleuzian perspective; it does so by discussing the concept of 'flat ontologies' currently being developed in various fields, particularly by one of the most astute among Deleuze's commentators, in these authors' view, Manuel de Landa. The second part starts the discussion about social movements proper; it reviews theories of networks and self-organization and their applications to social movements. The third part, finally, presents some preliminary ideas for discussing social movements that emerge more directly from Deleuze's and Deleuze and Guattari's work, particularly their conceptualization of the dynamics between the smooth and the striated, territorialization and de-territorialization and the politics of the virtual.

The 'Ontological Turn' in Social Theory: Assemblages and 'Flat' Alternatives[1]

The various waves of social constructionism, deconstruction and discursive approaches of the past few decades included a critique of realism as an epistemological stance. Some of the most interesting social theory trends at present entail, implicitly or explicitly, a return to realism. Since this is not a return to the naive realisms of the past (particularly the Cartesian versions, or the realism of essences or transcendent entities, which Deleuze criticized with particular care), these tendencies might be called neo-realist. Other viable metaphors for the emerging social theories are 'biological sociologies', a term applied to the phenomenological biology of Maturana and Varela in particular (1980), or new materialist sociologies (e.g. actor–network theories). Deleuze has inspired some of these developments; in this section, we shall focus on the reconstruction of Deleuze's ontology by the Mexican theorist Manuel de Landa (2002) and de Landa's own form of neo-realism (2006). Deleuze, in de Landa's view and unlike many constructivists, is committed to a view of reality as autonomous (mind-independent); his starting point is that reality is the result of dynamic processes in the organization of matter and energy that leads to the production of life forms (morphogenesis); things come into being through dynamic processes of matter and energy driven by intensive differences; these processes are largely self-organizing. Deleuze's morphogenetic account, in other words, makes visible the form-generating processes that are immanent to the material world; it amounts to 'an ontology of processes and an epistemology of problems' (de Landa 2002:6).[2]

A central aspect in de Landa's social ontology arises from Deleuze's concept of the virtual. There are three ontological dimensions in the Deleuzian world: the virtual, the intensive and the actual (de Landa 2002: 61–88). The larger field of virtuality is not opposed to the real but to the actual. This is a very different way to think about the relation between the possible and the real – here, the possible is not thought about in terms of a set of predefined forms that must retain their identity throughout any process of change, thus already prefiguring the end result (this is one of the most fatal modernist assumptions, since it precludes real difference). The possible does not necessarily resemble the real, as in the notion of 'realization'. In the actualization of the virtual, the logic of resemblance no longer rules, rather that of a genuine creation through differentiation. The actualization of the virtual in space and time entails the transformation of intensive differences into extensive (readily visible) forms through historical processes involving interacting parts and emerging wholes; this leads to what de Landa calls 'a *flat ontology*, one made exclusively of unique, singular individuals, different in spatio-temporal scale but not in ontological status' (2002: 47). 'The existence of the virtual is manifested ... in the cases where an assemblage meshes differences as such, without canceling them through homogenization ... Conversely, allowing differences in intensity to be cancelled or eliminating differences through uniformization, effectively hides the virtual and makes the disappearance of process under product seem less problematic' (ibid.: 65). This concealment is the result of human action – hence the need to investigate the unactualized tendencies of the virtual wherever they are expressed.[3]

In other words, differences have morphogenetic effects; they display the full potential of matter and energy for self-organization and result in heterogeneous assemblages. Intensive individuation processes occur through self-organization; differences in intensity drive fluxes of matter and energy; they can be amplified through positive feedback, mutually stimulating coupling and autocatalysis; individuals possess an openness and capacity to affect and be affected and to form assemblages with other individuals (organic or not), further differentiating differences through meshing (ibid.: 161). One consequence is that in a flat ontology 'there is no room for totalities, such as "society" or "science" in general' (ibid.: 178).

Based on a careful reconstruction of Deleuze's concepts, de Landa goes on to propose his own approach to 'social ontology' as a way to rethink the main questions of classical and contemporary sociology (including notions of structure and process, individuals and organizations, essences and totalities, the nationstate, scale, markets and networks). His goal is to offer an alternative foundation for social theory (an alternative 'ontological classification' for social scientists). His starting point is the realist stance of asserting the autonomy of social entities from the conceptions we have of them. This does not mean that social science models do not affect the entities being studied; this was one of post-structuralism's stronger arguments. It means that the focus of realist social ontology is a different one; the focus is on the objective, albeit historical, processes of assembly through which a wide range of social entities, from persons to nationstates, come into being. The main objects of study are 'assemblages', defined as wholes whose properties emerge from the interactions between parts; they can be any entity: interpersonal networks, cities, markets, nationstates, etc. The goal of this notion is to convey a sense of the irreducible social complexity in the world.

Assemblage theory differentiates itself from theories that rely on concepts of totality and essences and that assume the existence of seamless webs or wholes. It is an alternative to the organic totalities postulated by classical social science. Assemblages are wholes characterized by relations of exteriority; the whole cannot be explained by the properties of components but by the actual exercise of the components' capacities. Rather than emphasizing the creation of wholes out of the synthesis of parts from logically necessary relations – the organismic metaphor – assemblage theory asserts that the relations among parts are 'contingently obligatory', as, for instance, in the co-evolution of species. In this way, ecosystems may be seen as assemblages of thousands of different plant and animal species; co-evolution results from the symbiosis of species and the relations of exteriority obtained among self-sufficient components. Assemblage theory thus does not presuppose essential and enduring identities, or 'natural kinds', to which pre-given individuals would belong. This idea resonates with trends in evolutionary theory that rely on a view of speciation in terms of historical individuation of species and individuals, thus avoiding taxonomic essentialism; in some of these views, evolution is due as much to natural selection as to self-organization (e.g. Kaufman 1995).[4]

A particularly salient problem for social theory is the causal mechanisms that account for the emergence of wholes from the interaction between parts; this impinges, for instance, on the question of the micro and the macro. Conventional approaches assume two levels (micro, macro) or a nested series of levels (the proverbial Russian doll). The alternative approach is to show, through bottom-up analysis, how, at each scale, the properties of the whole emerge from the interactions between parts, bearing in mind that the more simple entities are themselves assemblages of sorts. Through their participation in networks, elements (such as individuals) can become components of various assemblages operating at different levels. This means that most social entities exist in a wide range of scales, making the situation much more complex that in conventional notions of scale:

> Similar complexities arise at larger scales. Interpersonal networks may give rise to larger assemblages like the coalitions of communities that form the backbone of many social justice movements. Institutional organizations, in

> turn, tend to form larger assemblages such as the hierarchies of government organizations that operate at a national, provincial, and local levels ... A social movement, when it has grown and endured for some time, tends to give rise to one or more organizations to stabilize it and perform specialized functions ... That is, social movements are a hybrid of interpersonal networks and institutional organizations. And similarly for government hierarchies which, at each jurisdictional scale, must form networks with non-governmental organizations in order to be able to implement centrally-decided policies. All of these larger assemblages exist as part of populations: populations of interpersonal networks, organizations, coalitions, and government hierarchies. (de Landa 2006: 33)

The processes of assembly through which physical, biological or social entities come into being are recurrent. This means that assemblages exist in populations that are generated by the repeated occurrence of the same processes. There is recurrence of the same assembly process at a given spatial scale, and recurrence at successive scales, leading to a different conceptualization of the link between the micro and the macro levels of social reality. For de Landa, the question becomes: How can we bridge the level of individual persons and that of the largest social entities (such as territorial states) through an embedding of assemblages in a succession of micro and macro scales? (2006: 34–38). For the case of markets, for instance, the issue is to show how differently scaled assemblages operate, with some being component parts of others, which, in turn, become part of even larger ones. In his historical work on the development of markets, de Landa (1997) shows how larger entities emerged from the assembly of smaller ones, including town, regional, provincial, national and world markets, following the Braudelian explanation.

Wholes exercise causal capacity when they interact with one another. Groups structured by networks may interact to form coalitions (or hierarchies). These larger assemblages are also emergent wholes – the effect of their interactions goes beyond the interaction of the individuals, with a sort of redundant causality. In sum:

> social assemblages larger than individual persons have an objective existence because they can causally affect the people that are their component parts, limiting them and enabling them, and because they can causally affect other assemblages at their own scale. The fact that in order to exercise their causal capacities, internally as well as externally, these assemblages must use people as a medium of interaction does not compromise their ontological autonomy any more than the fact that people must use some of their bodily parts (their hand or their feet, for example) to interact with the material world compromises their own relative autonomy from their anatomical components. (de Landa 2006: 38)

To sum up:

> The ontological status of any assemblage, inorganic, organic or social, is that of a unique, singular, historically contingent, individual. Although the term 'individual' has come to refer to individual persons, in its ontological sense it cannot be limited to that scale of reality ... Larger social assemblages should be given the ontological status of individual entities: individual networks and coalitions; individual organizations and governments; individual cities and nation states. This ontological maneuver allows us to assert that all these individual entities have an objective existence independently of our minds (and of our conceptions of them) without any commitment to essences or reified generalities. On the other hand, for the maneuver to work the part-to-whole relation that replaces essences must be carefully elucidated. The autonomy of wholes relative to their parts is guaranteed by the fact that they can causally affect those parts *in both a limiting and an enabling way,* and by the fact that they can interact with each other in a way not reducible to their parts, that is, in such a way that an explanation of the interaction that includes the details of the component parts would be redundant. Finally, the ontological status of assemblages is two-sided: as actual entities all the differently-scaled social assemblages are individual singularities, but the possibilities open to them at any given time are constrained by a distribution of universal singularities, the diagram of the assemblage, *which is not actual but virtual.* (ibid.: 40, emphasis added)[5]

Flat ontology and assemblage theory are implicated in an important reformulation of the concept of scale in geography. The past two decades in this field have seen a lot of interesting thinking on this concept, intended to move away from the vertical hierarchies associated with established theories and towards conceptions that link vertical with horizontally networked models (e.g. scalar structuration, glocalization). Building on the insights of flat ontology, these latter conceptions have recently been critiqued for remaining trapped in a foundational hierarchy and verticality, with concomitant problems such as lingering micro–macro distinctions and global–local binaries. According to these critics, these problems cannot be solved just by appealing to a network model; as they see it, the challenge is not to replace one 'ontological–epistemological nexus (verticality) with another (horizontality)' but to bypass altogether the reliance on 'any transcendent pre-determination' (Marston et al. 2005: 422). This is achieved thanks to a flat (as opposed to horizontal) ontology that discards 'the centering essentialism that infuses not only the up–down vertical imaginary but also the radiating (out from here) spatiality of horizontality' (ibid.: 422). For these authors, flat ontology refers to complex, emergent spatial relations, self-organization and ontogenesis.

The resulting conceptualization also seeks to move away from the 'liberalist trajectories' that fetishize flows, freedom of movement and 'absolute deterritorialization' that are present in some theories inspired by Deleuze and actor–network theories. In contradistinction, this geographical application of flat ontology emphasizes the assemblages constructed out of differential relations and emergent events, and how these result in both *systemic orderings* (including hierarchies) and *open-ended events*. One conclusion is that 'overcoming the limits of globalizing ontologies requires sustained attention to the intimate and divergent relations between bodies, objects, orders, and spaces'; for this, they propose to invent 'new spatial concepts that linger upon the singularities and materialities of space', avoiding the predetermination of both hierarchies and boundlessness (Marston et al. 2005: 424). In this flat alternative, 'sites' are reconceptualized as contexts for event relations in terms of people's activities. Sites become 'an emergent property of its interacting human and non-human inhabitants'; they are manifolds that do not precede the interactive processes that assemble them, calling for 'a processual thought aimed at the related effects and affects of its *n*-connections. That is, we can talk about the existence of a given site only insofar as we can follow the interactive practices through their localized connections' (ibid.: 425).

It follows that processes of localization should not be seen as the imprint of the global on the local, but *as the actualization of a particular connective process, out of a field of virtuality.* Indeed, what exists is always a manifold of interacting sites that emerge within unfolding event relations, which include, of course, relations of force from inside and outside the site. This site approach is of relevance to ethnography and anthropology as much as it is to geography. It is important to emphasize that these recent flat frameworks and site approaches provide an alternative to much established scalar, state-centric, capitalocentric and globalocentric thinking, with their emphasis on 'larger forces', hierarchies, determination and unchanging structures. These newer visions see entities as made up of always unfolding intermeshed sites. To paraphrase a well-known work (Gibson-Graham 1996), flat approaches spell out 'the end of globalization (as we knew it)'. To the disempowering of place and social agency embedded in globalocentric thinking, these approaches respond with a new plethora of political possibilities. Some of these possibilities are being tapped into by social movements, and even by individuals seeking to become new kinds of subjects of place and space.

Social Movements, Networks and Self-organization

Social Movement and Network Theories

It is possible to differentiate between two kinds of theories of networks (Escobar 2000). In the first type, the concept of network fits into an existing social theory. In the second, social theory is re/constructed on the basis of the concept of network. Castells' application of networks to contemporary society is the most well-known case among the first set. Central to Castells' theory of the network society is a distinction between the space of flows (the spatial structures relating to flows of information, symbols, capital, etc.) and the space of places (1996: 415–29). The former is composed of nodes and hubs hierarchically organized according to the importance of the functions they perform. For Castells, places have to network or perish. This structural schizophrenia can only be avoided by building bridges between the two spatial logics, that of flows and that of places (ibid.: 428). From his globalocentric perspective, power resides in flows and strategic nodes, while the structural meaning of people and places disappears. However, as we know, many social movements affirm the centrality of place in the constitution of societies today.[6]

Most theories of social movement networks belong in the first class – that is, they assume a particular social theory into which networks fit. Examples are theories of social networks of activism, such as Smith et al. (1997), and Keck and Sikkink (1998); broader attempts at theorizing social movement networks (e.g. Alvarez 1998, forthcoming; Diani and McAdam 2003); and ethnographic studies of particular social movement networks, such as those associated with Zapatista networks (e.g. Leyva Solano 2002, 2003, Olesen 2005) or anti-globalization movements (e.g., Juris 2005, 2008; Osterweil 2005a,b; King 2006,). A general model for network-centred social movement research has been proposed by Diani, McAdam and co-workers (Diani and McAdam 2003), focusing on network configuration and the research requirements for mapping networks.

Actor-network theories (ANT) are the most well-known example of the second type; ANT 'aims at accounting for the very essence of societies and natures. It does not wish to add social networks to social theory but to rebuild social theory out of networks' (Latour 1997: 1). The theory of actor networks asserts that the real is an effect of networks. Reality arises in the bringing together of heterogeneous social, technical and textual materials into patterned networks. No matter how seamless it might look at times, reality is the end product of actor networks, which have put it together after a lot of work. As Latour explains, reality comes into being after a lot of 'dissemination, heterogeneity and careful plaiting of weak ties ... through netting, lacing, weaving, twisting of ties that are weak by themselves'; concomitantly, analysis must start 'from irreducible, incommensurable, unconnected localities, which then, at a great price, some times end up into provisionally commensurable connections' (Latour 1997: 2; see also Law 2000 [1992]).[7] Accounts of networks derived from complexity theory are also part of the second type. However, there are very few studies to this date that systematically apply notions of complexity to social movements, although interest is rising in connection with anti-globalization mobilizations, as discussed in the next section (e.g. Escobar 2000, 2004; Chesters 2003; Summer and Halpin 2005; Chesters and Welsh 2006; Peltonen 2006).

Some recent conceptualizations suggest bridges between social movements and flat ontologies. In revisiting prevailing notions of both 'social movements' and 'political contestation', Sonia Alvarez (forthcoming) moves in the direction of a flat ontology. Her call is for a reconceptualization of social movements as expansive, heterogeneous and polycentric discursive fields of action that extend well beyond a distinct set of civil society organizations. These fields are constructed, continuously reinvented and shaped by distinctive political cultures and power distributions. Movement fields configure alternative publics in which dominant cultural-political meanings are refashioned and contested; the publics can be seen as parallel discursive arenas, where subaltern groups reinvent their own discourses, identities and interests. The fields are potentially contentious in two ways: they create and sustain alternative discourses, identities and challenges in conflict with dominant meanings and practices; and they maintain an internal contestation about their agendas in ways that enable them to respond adequately to their own ethico-political principles. It is easy to see how the concept of a social movement field and the double contestation that structures it may apply to anti-globalization movements, for instance, to the extent that their networks can be seen as apparatuses for the production of alternative discourses and practices, on the one hand, and as enacting forms of cultural politics that find articulation in dispersed networks on the other (Alvarez et al. 1998; Osterweil 2004). Alvarez's work also calls attention to the impact of differential access to cultural, political and material resources on local network nodes.

One final example of social movement network research concerns the ensemble of networks that have emerged throughout the years around the Zapatista movement in Chiapas. What is interesting in the analysis of this case provided by Mexican anthropologist Xochitl Leyva Solano (2002, 2003) is that she treats neo-Zapatismo as precisely that: an ensemble of articulated networks emerging from broad political contexts, many of them with deep historical roots in the region and the nation. In her network ethnography, Leyva Solano distinguishes among six interrelated but distinct neo-Zapatista networks: those based on historical agrarian and peasant demands; democratic-electoral and citizen-based networks; *Indianista*-autonomist networks, focused on indigenous peoples; women's rights networks; alternative revolutionary networks, promoting anti-state ideology and radical change; and international solidarity networks. All of these networks are both socio-political and cybernetic at the same time; after 1994, they became articulated around the armed Zapatista movement (EZLN) but are in no way restricted to it: they all emerged from local historical and regional conditions (and global in the case of the sixth network); they share moral grammars (e.g. concerning rights, citizenship, land, autonomy, etc.) and construct cognitive frameworks through which they have an impact on power relations, institutional politics and daily life; and, of course, they are characterized by tensions and conflicts within themselves and with other networks. The image of 'Zapatismo' that emerges from this conceptualization is complex – a manifold of actualizations out of a field of virtuality, to appeal to Deleuze's and de Landa's concepts. Each network may be considered as an assemblage in itself and in relation to other assemblages; each represents a multilayered entanglement with a host of actors, organizations, the natural environment, political and institutional terrains and cultural-discursive fields, which may be properly seen as a result of assemblage processes.

Networks and Complexity

Self-organization, assemblage theory and autopoiesis constitute relatively new forms of thinking about the organization of the living. The application of these concepts to social questions is just beginning. De Landa has applied complexity theory to social processes, highlighting the extent to which, over the past few hundred years, economic and social life has been organized on a logic of order, centralization and hierarchy building. Complexity applications and flat ontologies aim at making visible a different logic of social organization; this attempt resonates in two domains that are particularly pertinent to this chapter: digital technologies (cyberspace, as the universe of digital networks, interactions and interfaces) and social movement networks. To start with cyberspace, the argument is that, while modern media operate on the basis of a top-down, action-reaction model of information, the model enabled by new information and communication technologies (ICTs) is based on a novel framework of interaction – a relational model, where all receivers are also potentially emitters, a space of truly dialogical interaction (best expressed in some instances of net.art). As a space for inter-cultural exchange and for the construction of shared artistic and political strategies, cyberspace potentially affords unprecedented opportunities to build shared visions. (However, a number of events and activities not limited to cyberspace, such as the World Social Forum process, as well as a number of '*encuentros*' (or gatherings) can also be seen partly as a result of this dynamic.) The 'fluid architecture' of cyberspace thus enables a micro-politics of local knowledge production, which in turn emphasizes the 'molecular' nature of cyberspace. This micro-politics – as opposed to state-centric, goal-oriented macro-politics – consists in large part of practices of mixing, reusing and recombining knowledge and information.[8]

This vision resonates in turn with the principles of complexity and self-organization, which emphasize bottom-up processes in which agents working at one (local) scale give rise to sophistication and complexity at another level. Emergence happens when the actions of multiple agents interacting dynamically and following local rules rather than top-down commands result in some kind of visible macro-behaviour or structure. These systems may be adaptive in that they learn over time, responding more effectively to the changing environment. The network concept is central to these approaches. Physical and natural scientists are currently busy mapping networks of all kinds, and trying to ascertain network structures, topologies and mechanisms of operation. Social scientists have also been joining the bandwagon for complex networks research.[9]

De Landa (1997, 2003, n.d.) has introduced a useful distinction between two general network types: hierarchies and self-organizing meshworks. This distinction underlies two alternative life philosophies. Hierarchies entail a large degree of centralized control, ranks, overt planning, homogenization and particular goals and rules of behaviour; they operate under linear time and conform to tree-like structures. The military, capitalist enterprises and most bureaucratic organizations have largely operated on this basis. Meshworks, on the contrary, are based on decentralized decision-making, self-organization, heterogeneity and diversity. They develop through their encounter with their environments, while conserving their basic organization (their autopoiesis). Other metaphors used to describe these phenomena are tree-like structures or 'strata' (for hierarchies) and 'rhizomes' or 'self-consistent aggregates' for meshworks (from Deleuze and Guattari 1987). Like Deleuze and Guattari, de Landa is clear that these two principles of organization are found mixed in most real-life examples. They can also give rise to one another, for example, when social movement meshworks develop hierarchies; or the Internet, which can be said to be a hybrid of meshwork and hierarchy components, with a tendency for the elements of command and control to increase. The reverse could be said about the global economy, since today's corporations are seeking to evolve towards a networked form with flexible command structures.

Most social movements are a mixture of hierarchies and self-organization. Unfortunately, most network approaches and flat ontologies do not explicitly address the power dimension. However, it might be possible to differentiate between two kinds of networks: subaltern actor networks (SANs) and dominant actor networks (DANs; see Escobar 2000, 2008). Most theories reviewed so far do not make this distinction, for good reason, since SANs and DANs overlap and often co-produce each other, yet they can be analytically differentiated on political grounds and in terms of contrasting goals, practices, modes of agency, mechanisms of emergence and hierarchy and scale. It is undeniable that social movement networks constitute a wave of confrontational engagement at many levels, so their oppositional character is hard to deny; it is important, however, to avoid falling back into modernist notions of opposition – that is, into representations of discrete entities that are independent of their own enactment and self-production (King 2005). In other words, in characterizing networks as either dominant or oppositional, it is important to remain within a 'flat' terrain. A simple criterion is to say that DANs are networks in which elements of hierarchy predominate over those of self-organization; conversely, SANs are those in which the opposite is true.

Building on the field of biological computing, Terranova adds useful elements to the conceptualization of networks as self-organizing systems that engender emergent behaviour. For her, networks can be thought of in terms of 'abstract machines of soft control – a diagram of power that takes as its operational field *the productive capacities of the hyperconnected many*' (Terranova 2004: 100, emphasis added). Social phenomena are seen as the outcome of a multitude of molecular, semi-ordered interactions between large populations of elements. Individual users become part of a vast network culture – of 'the space-time of the swerve', which may lead to emergence (ibid.: 117). These systems only allow for soft control (as in cellular automata models); it is from this perspective that Terranova's definition of network ('the least structured organization that can be said to have any structure at all' (ibid.: 118)) makes sense. The open network, such as the Internet, 'is a global and large realization of the liquid state that pushes to the limits the capacity of control of mechanisms effectively to mould the rules and select the aims' (ibid.: 118). This network culture emphasizes distributed/autonomous forms of organization rather than direct control. In short:

> The biological turn is, as we have seen, not only a new approach to computation, but it also aspires to offer a social technology of control able to

> explain and replicate not only the collective behavior of distributed networks such as the internet, but also the complex and unpredictable patterns of contemporary informational capitalism ... The biological turn thus seems to extend from computing itself towards a more general conceptual approach to understanding the dynamic behavior of the internet, network culture, milieus of innovation and contemporary 'deregulated markets' – that is of all social, technical and economic structures that are characterized by a distributed and dynamic interaction of large numbers of entities with no central controller in charge. (Terranova 2004: 121)

This applies to many social phenomena, from markets to social movements, which can be studied under the rubric of social emergence. It marks a sharp contrast to concepts of control based on Taylorism, classic cybernetics and governmentality, even if these are by no means irrelevant. Similarly to de Landa, Terranova sees pros and cons in this situation; on the downside the multitude/mass cannot be made to unite under any common cause, since the space of a network culture is that of permanent dissonance; yet the benefits in terms of opportunities for self-organization and experimentation based on horizontal and diffuse communication (again, as in the case of many social movements) are real. In the best of cases, the simultaneous tendencies to diverge and separate, on the one hand, and converge and join, on the other, shown by networked movements might lead to 'a common passion giving rise to a *distributed movement* able to displace the limits and terms within which the political constitution of the future is played out' (Terranova 2004: 156). The logic of distributed networks thus amounts to a different logic of the political. In his study of the anti-corporate globalization movement in Catalunya, Juris (2005, 2008) puts it in terms of the intersection between network technologies, organizing forms and political norms that accompany the cultural logic of networking.

There is one final aspect of network theories that is important to discuss. Many recent theories of networks assume that networks are about information above all else. In this modernist view, information is often seen as disembodied, and one could argue that there are many embodied aspects to both knowledge and networks. Information is nevertheless a central component of networks, and it is important to specify why and how. This takes us to the foundational work on information and communications during the early cybernetics period. As Terranova (2004) has argued, this relation was firmly established with the theories of information of the 1940s and 1950s based on thermodynamics and statistical mechanics, particularly in the work of Claude Shannon. Since then, there has been a tendency to reduce information to its technical aspects, overlooking the fact that information always involves practices, bodies and interfaces, particular constructions of the real and, in general, 'a set of relays between the technical and the social' (ibid.: 25), to which one could add the biological (body, nature). For Terranova, there is an entire cultural politics of information associated with distributed networks. This requires a critical examination of information and communication technologies (ICTs) that focuses on how they involve 'questioning the relationship between the probable, the possible, and the real. It entails the opening up of the virtuality of the world by positing not simply different, but radically other codes and channels for expressing and giving expression to an undetermined potential for change' (ibid.: 26).[10] Like to de Landa and Marston and collaborators, Terranova envisions a cultural politics of the virtual, understood as the opening up of the real to the action of forces that may actualize the virtual in different ways.

The relation between networks and information has become central to many network theories. Building on the work of Marilyn Strathern, Riles (2001) has made an eloquent case for the anthropological study of networks as self-producing entities that operate on the basis of knowledge and information. This opens up a serious epistemological problematic, which is at the centre of much work in the anthropology of science and technology (e.g. Marcus 1999; Fischer 2003; Osterweil 2005b): how does one study and describe situations in which the objects/subjects are thoroughly constituted by the same knowledge practices of which the ethnographer herself is also a product? The anthropological solution so far has been: 1) recognize that there is no radical outside from which to conduct a completely detached observation; and 2) a deepened reflexivity that attempts to keep up with a constantly reformulating 'site' or 'object' of study. These conditions have led to various proposals, from 'enacting the network' in the ethnographic description (Riles), to focusing on emergent forms of life (Fischer), to heightened reflexivity coupled with a relational ontology, and novel techniques to render modern artefacts ethnographic (these entail using actor–network approaches, distributed or multi-sited notions of fieldwork, etc.; e.g. Fortun 2001, 2003). The relation between network ethnography and a politics of the virtual and of multiplicities is yet to be explored.

To sum up, a number of theories of networks of the past two decades have tried to make sense of the contemporary logic of the social and the political. The trends based on flat alternatives, self-organization and complexity articulate a network concept from the perspective of new logics operating at the levels of ontology, the social and the political. Flat alternatives make visible design principles based on open architectures, allowing for interconnection of autonomous networks, and the potential for expansive inter-networking, enabled by decentralization, resilience and autonomy. However, places and embodiment have by no means ceased to be important – and let us not forget that there are many embodied aspects to activism, it is not only about information or technology, no matter how networked. This is why struggles over ICTs and the world they contribute to creating become crucial (e.g. Harcourt 1999; Bell and Kennedy 2000); they involve experimentation with appropriations of the fluid architecture of networks, new forms of collaboration, and so forth – in short, network politics are linked to emerging cultural and material assemblages. However, to the extent that information networks and ICTs are also part of the infrastructure of imperial globality, there are very real risks.[11] These risks have everything to do with inextricable and complex relationships between capitalism, the state and efforts at resistance.

Capitalism, the State and Social Movements

There have been two common notions for thinking about the relation between social movements, on the one hand, and the state and capitalism, on the other: co-optation and the dialectic. It is often said, with good reason, that the neoliberal state is more adept than ever at co-opting the demands of movements. As such, the state may appear as the champion of rights, equality, multiculturalism and even local autonomies and sustainability. Contemporary Latin America is full of examples of this neoliberal co-optation. The Marxist tradition has customarily linked the analysis of the contradictions of capitalism to social movements' ability to face them, and perhaps supersede them, through a dialectical framework. Dialectics and co-optation have been at the heart of the assessment of the effectivity of movements by scholars, and of how movements themselves think about the best strategies to follow.

These notions continue to be useful, up to a certain point. A Deleuzian reading of social movements, however, affords new elements that resonate more closely with the character of today's movements. It could be thought, on first reading, that Deleuze and Guattari's account of the relation between territorialization an de-territorialization, between the smooth and the striated, between the arborescent and the rhizomatic, between nomad and state knowledges, between apparatuses of capture and war machines or between minoritarian and majoritarian or molecular and molar logics is a restatement of the dialectic. This is not so, precisely because these are embedded in a series of other concepts that complicate any straightforward interpretation. This is particularly the case with the concepts of becoming, multiplicity and the virtual, but also with the notions of manifold, assemblages, plane of immanence, intensities, and so forth. Deleuze and Guattari actually reserve respectful but unequivocal words for the dialectic; with its ultimate wish of seeking to extract a truth value out of opposing opinions, they say, the dialectic 'reduces philosophy to interminable discussion' (1994: 99), so that even 'beneath the highest ambitions of the dialectic, and irrespective of the genius of the great dialecticians [Hegel, Marx], we fall back into the most abject conditions' (ibid.: 80; see also Deleuze and Guattari (1987: 483) for a critique of the dialectic from the perspective of Riemann's geometry).

The dynamic of the social described by Deleuze and Guattari thus has different overtones, and we can only refer here to some aspects of this dynamic that are of particular relevance to social movements. In their view, social formations such as capitalism and the modern state do not proceed through homogenization (this is already widely accepted) or totalization (less accepted), but through their consolidation of diverse forms; they require 'a certain peripheral polymorphy' (Deleuze and Guattari 1987: 436). This means that there can never be pure opposition or resistance; however, any overcoding by any apparatus of capture such as capitalism or the state, simultaneously frees up decoded flows that escape from it (e.g. the drive to reduce all aspects of social life to the market by neoliberal globalization releases many other forms of organizing and desiring economic and social life, including some 'perverse' ones). In response, of course, the apparatus attempts to organize and 'co-opt' the flows decoded by social movements and to re-subjectify the subjects who try to release themselves from its grip. The area of property rights provides a recent example: to the assertion of collective rights by emergent collective subjects – e.g. indigenous or ethnic minority groups in areas such as rain forests – the state responds by seeking to make this collective subject disappear either discursively, by proposing what appears as an even more commonsensical strategy that articulates well with neoliberal proposals, or literally, through intimidation, assassination or massacres, as in the case of countries like Colombia.

This is why in some cases (USA, Colombia) the state and capitalism could be said to be taking an ultra-neoliberal form: each state agency, it would seem, is busy preparing intelligent counter-strategies to the demands of social movements in order to recode the decoded flows by movements, thus turning every potential defeat into a victory for itself. It is in this way that the ultra-neoliberal State moves its project of radically transforming society one more step forward. We already mentioned that capitalism and the state also develop self-organizing and meshwork-type mechanisms and the ability to operate on the basis of the smooth space. Yet this capacity always runs against limits (Escobar 2004; Marston et al. 2005), and this can give an advantage to movements. Nevertheless, it is important to make this state/capitalism strategy newly visible in terms of those manoeuvres that have caused the greatest deterritorialization of the gains obtained by movements in recent decades.[12] It is here that the framework developed by Deleuze and Guattari can provide some useful elements to social movements as they confront the systematicity of the state project in some world regions. What languages and strategies of liberation, revolution, decoloniality could be invented to tackle this situation and to strengthen the strategies of existence of communities? How do movements themselves come to discover and articulate for themselves and others what strategies and approaches are more likely to work?

Deleuze and Guattari are not completely consistent in their treatment of these dynamics. At some points, they confer upon capitalism and the state a great capacity for integrating the decoded flows – for instance, in their discussion of capitalism as 'a worldwide (or immanent) axiomatic', 'a worldwide enterprise of subjectification' capable of integrating any non-capitalist form, or, appealing to differential calculus, in terms of 'a whole *integral of decoded flows*, a whole *generalized conjunction*' (Deleuze and Guattari 1987: 452–57 ff.).[13] At other points, however, they seem to bracket this possibility, as in certain views of deterritorialization, retroactive smoothing by nomad-like agents or philosophy's ability to turn capitalism against itself (e.g. Deleuze and Guattari 1994: 101). In general, the latter position seems more consistent with the conceptual and political intent of their work. This is clear in a number of discussions of minoritarian logics and of the smooth and the striated. The following quote is telling in this regard:

> The response of the State, or of the axiomatic, may obviously be to accord the minorities regional or federal or statutory autonomy, in short, to add axioms.

> But this is not the problem: this operation consists only in translating the minorities into denumerable sets or subsets, which would enter as elements into the majority, which could be counted among the majority ... [On the contrary,] [w]hat is proper to the minority is to assert a power of the nondenumerable, even if that minority is composed of a single member. That is the formula for multiplicities. Minority as a universal figure, or becoming-everybody/everything (*devenir tout le monde*). Woman: we all have to become that, whether we are male or female. Non-white: we all have to become that, whether we are white, yellow, or black ... The power of the minorities is not measured by their capacity to enter and make themselves felt within the majority system, nor even to reverse the necessary tautological criterion of the majority, but to bring to bear the force of the nondenumerable sets, however small they may be, against the denumerable sets, even if they are infinite, reversed, or changed, even if they imply new axioms or, beyond that, a new axiomatic. (Deleuze and Guattari 1987: 470–71)

Minoritarian movements in many parts of the world seem to be aware of this predicament – again, one could point to black and indigenous struggles in Latin America at present.[14] However, they are often at a loss when faced with concrete situations in which the state seeks to append their struggles to state axioms. It is here that the old dilemma around co-optation resurfaces. Should movements pursue a radical minoritarian logic (e.g. go deeper into their territories, disengage from the regional economies, hold on even more firmly to certain anti-capitalist or a-modern cultural practices, etc.), in which case they become the target of even more virulent repression, or should they rather engage with the state, in which case they might fall into the reformist trap? In the Deleuzian conception, this is not really a dilemma, even if the consequences for strategy are no less dire; for, as they add:

> Once again, this is not to say that the struggle on the level of the axiom [the system's own codes] is without importance; on the contrary, it is determining ... But there is also always a sign to indicate that these struggles are the index of another, coexistent combat. However modest the demand, it always constitutes a point that the axiomatic cannot tolerate: when people demand to formulate their problems themselves, and to determine at least the particular conditions under which they can receive a more general solution (hold to the *Particular* as an innovative form) ... In short, the struggle around axioms is most important when it manifests, itself opens, the gap between two types of propositions, propositions of flow and propositions of axioms. (Deleuze and Guattari 1987: 471)

Or, one could say thinking about contemporary movements, struggles around axioms and struggles around flows – struggles around so-called universal human rights and struggles in terms of minoritarian/non-modernist conceptions and practices of rights, respectively, to give a prominent example; or to paraphrase, 'other rights, and rights otherwise'.

What does this mean in concrete terms for various movements at present? 'The issue is not at all anarchy versus organization, nor even centralism versus decentralization, but a calculus or conception of the problems of nondenumerable sets, against the axiomatic of denumerable sets' (Deleuze and Guattari 1987: 471). Some movements have had clarity about this for some time; one thinks, for instance, about the Zapatista take on autonomy, their clear distancing from the idea of taking state power, their emphasis on language and affect, their vision of 'a world where many worlds can fit' – itself a vernacular notion of multiplicity.[15] Deleuze and Guattari state that we must construct '*revolutionary connections*' rather than '*conjugations of the axiomatic*' (ibid.: 473). Yet movements often get trapped in debates about strategy that are about conceptual or ideological debates (e.g. the debate about 'Fix it or nix it' in the global justice movement, referring to the organizations of world capitalism like the WTO, the World Bank and the IMF – can they be reformed or should they be abolished?). 'Such a calculus', Deleuze and Guattari add, 'may have its own compositions, organizations, even centralizations; nevertheless, it proceeds not via the States or the axiomatic process but via a pure becoming of minorities' (ibid.: 471). How can movements operate on the plane of capital and the state and at the same time escape it? Engage with the axiomatic and challenge it from a minoritarian perspective? Disrupt the model of capitalism – smash it even, as Deleuze and Guattari intuit – without being paralysed by capital's recodings? Use the tools of modernity but within a logic of becoming (becoming-molecular/rhizome/smooth space)? Become dispersed without excessive fragmentation and retaining the capacity to mount significant resistance to the new strategies of 'machinic enslavement'?

We find additional clues in the discussion of the smooth and the striated. It is important to bear in mind that, again unlike the model of the dialectic, the two kinds of spaces always exist in mixture, being continuously transformed into each other (for a characterization of each type of space, see Deleuze and Guattari 1987: 474–500). A first set of elements comes from thinking about how the smooth operates. A patchwork that can be put together in multiple ways, the smooth constructs, occupies territories without striating them in the standard ways, without 'metricizing' them, without bringing them into the real through logocentric rules and norms. Rather, the smooth space (e.g. an indigenous collective territory, a rain forest) is constructed by local operations and local knowledges that are place-specific rather than universally valid.

What is important to bear in mind is that in each instance 'the simple opposition "smooth–striated" gives rise to far more difficult complications, alternations, and superpositions' (Deleuze and Guattari 1987: 481). Activists following these notions might learn to 'voyage smoothly' as they encounter the superpositions of the smooth and the striated. This gets more complicated, as Deleuze and Guattari assert that both the smooth and the striated can indeed operate as multiplicities. We mentioned above capital's capacity to operate as a smooth space.[16] Social movements need to be attuned to this state of affairs, they need to be ready for this moment so that their imaginaries and projects – even their dreams – can become particularly effective, deterritorializing, disturbing. This 'itinerant geometry' might enable movements to travel back and forth between the smooth and the striated, allowing for translations between the two spaces, which, while necessarily metricizing the smooth space movements want to hold on to, it does by 'giving it a [new] milieu of propagation, extension, refraction, renewal, and impulse without which it would perhaps die of its own accord' (ibid.: 486). This could mean that while extracting real concessions from capitalism and the state, say, in terms of defence of particular ecosystems or better working conditions for people, social movements still maintain alive the promise of smooth multiplicities; in short, 'all progress is made by and in striated space, but all becoming occurs in smooth space' (ibid.: 486).

As many in the global justice movement know, this is a politics 'without guarantees' (Hall 1996; Grossberg 2006). In the words of Deleuze and Guattari; 'Of course, smooth spaces are not in themselves liberatory. But the struggle is changed or displaced in them, and life reconstitutes its stakes, confronts new obstacles, invents new paces, switches adversaries'; all of these are aims to which movements can aspire, while being mindful of Deleuze and Guattari's ultimate admonition: 'Never believe that smooth space will suffice to save us' (1987: 500). But this warning could also be read from an exactly opposite direction, and we want to end with a brief discussion of this possibility, namely, the politics of the virtual. In our view, one defining feature of movements today is their appeal to, and engagement with, the virtual. Movements exist not only, or chiefly, as empirical and straightforward objects 'out there' playing out an already determined political role, but also, in their various instantiations, as a potentiality of how politics could be, and as a sphere of action in which people can both dream of a better world and contribute to enacting it, through experimenting with alternative social forms, which might not change the 'actual' world but which make visible the possibilities of new arrangements or imaginaries of the social. It could be said that this has always been the case, at least with some movements of the past, but we want to suggest that there are significant differences today. The first, and perhaps most important, is that the production of knowledge and discursive imaginaries has become increasingly central to many movements. In some cases, as with the Italian '*movimento dei movimenti*',[17] this takes the form of a vast and varied theoretical production. Movements in these instances are best thought about as collectively produced spaces where the discursive work of imagining and creating a politics otherwise is carried out. It is in these spaces that new imaginaries and ideas about how to re/assemble the social are not only hatched but experimented with, critiqued, elaborated upon, and so forth.[18]

In its contemporary phase, Italian activism is characterized by processes that foreground both homegrown theoretical and practical innovations and imaginaries and theories coming from as far away as Chiapas, specifically from the Zapatista movement. This reveals a novel epistemological and geopolitical situation, which suggests that a large part of what activists do is to theorize and experiment with different political practices geared towards developing more adequate forms and frameworks of politics and social change. Moreover, there was something in the cultural politics of the Zapatistas – including their imaginaries and discourses committed to difference, and inflected by post-structuralist critiques of power, which are also deeply committed to and informed by indigenous cosmology – that resonated with many Italian activists and organizations. In addition, these often involve events, gestures and practices that are spectacular or 'mythopoetic', rather than overtly political, in terms of their effects.[19] These characteristics, we want to suggest, point to the need for sustained ethnographic research that, rather than presuming the terrain of the political and the logical objectives of movements, is attuned to the ways these practices seem to be as much about the virtual as about the actual. An investigation that viewed a movement as a complex cultural-political field, in which the production, transformation and circulation of social and political theories and analyses occupy a prominent place, would not only give us better understandings about the nature of social movement practice today: it could serve as a concrete site for engaging the politics of the virtual.[20] That is to say, many movement practices can be viewed as part of an extended theoretical or experimental moment in which the terrain is micro-political and the object is to test out or make visible the possibilities of new arrangements or imaginaries of the social. 'Success', then, is achieved not by creating immediate or actual transformations in the present, but rather by impacting upon people's imaginations and desires – making the imagination of 'other worlds' possible.

The Italian case might be somewhat exemplary in the extent of theoretical production and the number of sites and processes of collective (self-)reflection.[21] But active and sophisticated knowledge production has in effect become increasingly essential to movement functioning. Some movements have even become adept at doing their own brand of activist research, including producing situated forms of knowledge or expertise that contend with 'scientific' knowledge and contest the assumptions of privilege and superiority granted to more traditional sites of knowledge production (Casas-Cortés et al. 2008). Altogether, this has been described as both a 'knowledge turn' in social movement practice and a needed turn in social movement research.[22] While the use of knowledge and theory in and of itself does not necessarily push movements into a politics of the virtual, it could be argued that the particular ways in which this knowledge turn is happening today do indeed portend such a politics; this is so to the extent that the knowledge practices have a focus on non-directional processes such as '*encuentros*' and discussions (i.e. World Social Forum) that are meant to discover political modalities that are less concerned with creating 'actual' change and are more explicitly focused on the micro-political.

It could be also that the excesses of capitalism and neoliberal globalization, coupled with these ever more reflexive movements, are pushing activists into the domain of the virtual. As the awareness that there is not much to be gained by joining the system or even extracting concessions from it grows, activists' politics become and more linked to imagining more creative and radical alternatives. With its founding slogan, 'Another world is possible', the World Social Forum process that started in Porto Alegre, Brazil, in 2001 was an early statement of this orientation. It fostered a space that an important intellectual associated with the Forum, Boaventura de Sousa Santos, has described in terms of a politics of absences and emergences. Santos starts with the notion that 'what does not exist is, in fact, actively produced as non-existent, that is, as non-credible alternatives to what exists'. From here it follows that 'the objective of the sociology of absences is to transform impossible into possible objects, absent into present objects' (Santos 2004: 238). This may happen, as Santos concludes, through a sociology of emergences that 'aims to identify and enlarge the signs of possible future experiences, under the guise of tendencies and latencies, that are actively ignored by hegemonic rationality and knowledge' (ibid.: 241). These notions constitute a particular articulation of the politics of the virtual and its actualization, which Santos, following Ernst Bloch, refers to as the 'not yet' aspect of social life.

These tendencies also resonate with attempts to re-imagine the world's geographies of power and knowledge by social movements that operate from the epistemic borders of the modern colonial world system – again, such as indigenous and ethnic movements in the global South. Besides questioning Western discourses, these movements build on the epistemic potential of local histories embedded in, or arising from, the economic, ecological and cultural practices of difference they embody vis-à-vis hegemonic forms of modernity. We may see these struggles as operating on the basis of a politics of flows first and foremost, although always challenging the axiomatics of state and capitalism (e.g. around questions of so-called free-trade agreements, development, biodiversity, food production, or what have you). These struggles should thus be seen as meaningful sources for political action and for alternative world constructions. However, the local histories from which decolonializing projects emerge have remained largely invisible in Eurocentric theory precisely because they have been actively produced as non-existent or as non-credible alternatives to what exist, to put it in Santos' terms (see also Mignolo 2000). Thus these movements can be said to aim at a sociology of emergences that enables the identification and enlargement of the range of knowledges that could be considered credible alternatives.[23]

Faced with the persistence of these movements and events, one cannot help but think back to Deleuze's conception of the work of the philosopher as being 'to locate those areas of the world where the virtual is still expressed, and *use the unactualized tendencies and capacities one discovers there as sources of insight into the nature of virtual multiplicities*' (de Landa 2002: 67, emphasis added). We would argue that this is exactly what many of today's movements do: some quite clearly offering examples of less actualized tendencies, while others actually doing the work of trying to discover them.

Conclusion

The interest in flat alternatives might well be a sign of the times. 'We are tired of trees' – famously denounced Deleuze and Guattari, two of the prophets of this movement in modern social theory: 'We should stop believing in trees, roots and radicles. They've made us suffer too much. All of arborescent culture is founded on them, from biology to linguistics' (1987: 15). This means that we need to move away from ways of thinking based on binaries, totalities, generative structures, pre-assumed unities, rigid laws, logocentric rationalities, conscious production, ideology, genetic determination, simple dialectics and macro-politics, and embrace instead multiplicities, lines of flight, indetermination, tracings, movements of deterritorialization and processes of reterritorialization, becoming, in-betweenness, morphogenesis, chaosmosis, rhizomes, micropolitics and intensive differences and assemblages. If the dominant institutions of modernity have tended to operate on the basis of the first set of concepts, it would make sense now to build a politics of world-making based on the second set (Gibson-Graham 2006). From biology to informatics, from geography to social movements, from the wise elders of an alternative West (some of the complexity theorists) to many indigenous peoples and activists, at present this is a strong message that can at least be plausibly heard.

Not that this message solves all the problems for theory or political action, although it perhaps renews our sense of hope, to get back to our reflection on the relation between social reality, social theory and the articulation of political purpose and sense of hope. While some, perhaps many, of today's movements seem intuitively or explicitly aiming at a practice informed by flat alternatives, relational approaches and self-organization, it remains to be seen how they will fare in terms of the effectiveness of their action. Most observers would state that the experience of many movements is ambiguous at best in terms of this criterion. There is the need for more empirical, on-the-ground research and activist research on particular experiences (including the kind of time series used in some fields to ascertain longer-term dynamics, for instance, in the globally oriented movements). Flat alternatives contribute to putting issues of power and difference on the table in a unique way. If actual economic, ecological and cultural differences can be seen as instances of intensive differences, and if these can be seen as enactments of a much larger field of virtuality, then the spectrum of strategies, visions, dreams and actions is much larger than conventional views of the world – mainstream or Left – might suggest. The challenge is to translate these insights into political strategies that incorporate multiple modes of knowing and doing in novel ways, while resisting the modern drive to 'organize' (the people) in logocentric, rational and reductionist ways.

We shall end with a recent discussion by biologist Brian Goodwin, a complexity theorist, about these trends. 'We need to hold a vision of what is just dawning,' he said, by which he meant that emergent, self-referential networks are indicative of certain dynamics, signalling an unprecedented epoch and culture for which we need a new vocabulary. By a new culture he means something far deeper than any rationalistic understanding of networks might suggest. For one thing, these dynamics are not something that we invent but which we experience. If the shadow of modernity is death – its greatest fear – the message of biological worlds (from neurons to rivers, from atoms to lightning, from species to ecosystems and evolution) is that of self-organization, self-similarity, multiplicities. If language and meaning, as some of these biologists have begun to suggest, is a property of all living beings and not only of humans – that is, if the world is one of pan-sentience – can activists and others learn to become 'readers of the book of life' and have this reading illuminate their reveries and strategies? How do we learn to live with/in both places and networks creatively? To resort to Maturana and Varela, the lesson of this deeply relational biology is that '*we have only the world that we bring forth with others, and only love helps us bring it forth*' (1987: 248, emphasis in original). For this, it is perfectly fine, of course, to use our rational minds, but it certainly also means embracing ways of knowing other than the rational and the analytical. Rather than naive romanticism, these latter would have to be included, in Goodwin's view, in any new foundation of realism and responsibility. This resonates well, we believe, with Deleuzian intersections.[24]

Notes

1. The category 'flat', as used here, is completely different from the concept of 'flat files' in mathematics, or from the use given to it by Thomas Friedman in his 2005 book, *The World is Flat*. It should be pointed out that flat alternatives and theories of complexity and self-organization have not emerged in a vacuum; the history of some of their most important predecessors and antecedents is rarely told, since they pertain to traditions of thought that often lie outside the scope of the social sciences. These include information theories in the 1940s and 1950s, including cybernetics; systems theories since the 1950s; early theories of self-organization; and the phenomenological biology of Maturana and Varela. See the chapter on Networks in Escobar (2008) for further detail. More recently, the sources of flat alternatives include some strands of thought in geography, cognitive science and informatics and computing; complexity theories in biology; network theories in the physical, natural and social sciences; and Deleuze and Guattari's 'neo-realism'. We also see Foucault's work partially within this frame – e.g. Foucault's theory of the archaeology of knowledge may be seen as a theory of autopoiesis and self-organization of knowledge; his concept of 'eventalization' resembles recent proposals in assemblage theory; and his conception of power anticipated developments in actor–network theory. Useful introductions to complexity and self-organization in the life sciences include Prigogine and Stengers (1984), Prigogine and Nicolis (1989), Solé and Goodwin (2000) and Camazine et al. (2001). See the useful Primer on Complexity at the end of the volume edited by Haila and Dyke (2006).
2. Deleuze deploys a difficult mathematical language, which de Landa explains (not altogether clearly for the uninitiated), particularly the concepts of multiplicity, as a form of organization 'which has no need whatsoever of unity in order to form a system' (de Landa 2002: 13; Deleuze and Guattari explain this concept at length especially in *A Thousand Plateaus*); manifolds, as the space of possible states of a system, regulated by the system's degrees of freedom; dynamical processes, in terms of trajectories in a space, recurrent behaviour and processes of differentiation; singularities, which act as attractors around which a number of trajectories within the same sphere of influence converge (basin of attraction), possibly leading to a steady state (structural stability); and so forth. De Landa also summons complexity concepts to explain the Deleuzian world. Multiplicities are concrete universals, they are divergent, and cannot be thought about in terms of three-dimensional metric Euclidean space but of nth-dimensional (non-metric) topological spaces, although the former is produced through differentiations in the latter. This happens through concrete physical processes of differentiation of an undifferentiated continuous intensive space into extensive structures (i.e. discontinuous, divisible structures with metric properties) through processes that include phase transitions, symmetry-breaking, etc. Multiplicities are thus immanent to material processes.

How does the actualization of the virtual happen? Deleuze makes an ontological distinction between actual trajectories and vector fields (inherent tendencies to behave in certain ways). Actual trajectories converge around a basin of attraction with a certain structural stability. This is to say, concrete realizations of a multiplicity are more accurately actualizations of a vector field – actualizations of a larger field of virtuality. This is not opposed to the real but to the actual – that is, *the virtual is another structural part of reality*. Multiplicities imply virtuality. This approach requires understanding the individuation of possible histories. This is complicated because the actualization of vector fields is rarely a linear process; on the contrary, it is shaped by non-linear dynamics; trajectories may emerge out of an attractor even by accident or external shocks; they are always the result of a contingent history. Alternatives that are pursued at a given point (especially at bifurcation) may depend on chance fluctuations in the environment (a point underscored by complexity theorists, e.g. Prigogine and Nicolis 1989; Solé and Goodwin 2000), in a conjunction of chance and necessity. What matters in the investigation is to remain close to the specific individuation going on – that is, to the formation of spatio-temporal structures, boundaries, etc.

3. The concealment of the virtual takes particular forms within modernity. Escobar (2008) has linked this idea to the existence of alternatives to modernity, and we shall link it to social movements in the third part of the chapter. The task de Landa envisions could also be carried out by working backwards from the concrete actualizations towards the virtual and by considering the population of multiplicities that exist in the plane of consistency. Alternative possibilities need to be shown as 'historical results of actual causes possessing no causal power of their own' (de Landa 2002: 75). Information can play a key role in these processes, for instance, in systems or networks poised at the edge of a threshold (see also, for example, Kauffman 1995).
4. While Deleuze and Guattari did not develop a systematic theory of assemblages, there are many discussions throughout their work that provide elements for such a theory – e.g. from their discussion of assemblages in *A Thousand Plateaus* to their notion of multiplicity and their account of the formation of concepts in *What is Philosophy?* A concept itself could be seen as an assemblage that reaches a degree of 'endoconsistency' (Deleuze and Guattari 1994: 25). What de Landa adds to Deleuze and Guattari's framework is a series of elements that he brings from fields that, while not entirely absent from Deleuze and Guattari, are not as salient as in de Landa; these include evolutionary theory, theories of complexity (particularly emergence and self-organization) and systems theories. The fact that in his most recent work de Landa focuses on modern sociological theory also enables him to develop further certain elements of assemblage theory. The extent to which de Landa succeeds or not in laying the foundations for a new sociology remains to be seen. Brown (this volume) is right in stating that, by refusing to engage with the main works of contemporary social theory, de Landa also gives up the possibility of constructing a plane of consistency for the social, which he sees as essential for a systematic and Deleuzian social science. In our opinion, however, such a social science has to go beyond the plane identified by Donzelot and Rose ('the rise of the social', which was central to the modern experience) to really encompass the multiplicities that exist, and might arise, from non-modern and non-Western cases.
5. The above explanation is of necessity schematic, and is only intended to highlight a few features of the theory. Let us mention a few other aspects of de Landa's assemblage theory. First, assemblage theory emphasizes the exteriority of relations, Secondly, it postulates two dimensions of analysis: (1) the role played by the components, from the purely material to the purely expressive; and (2) processes of territorialization and deterritorialization, which either stabilize or destabilize the identity of an assemblage (internal consistency and sharpness of boundaries). Thirdly, it pays attention to several other mechanisms, particular those of coding and decoding (e.g. by genes and language). Assemblage theory also seeks to account for the multi-scaled character of social reality, and provides adjustments to this end. First, it recognizes the need to explain the historical production of the assemblage, but without placing emphasis only on the moment of birth (e.g. as in the origin of a given collectivity or social movement) or on the original emergence of its identity at the expense of the processes that maintain this identity through time. Secondly, assemblages are produced by recurrent processes; given a

population of assemblages at one scale, these processes can generate larger-scale assemblages using members of existing populations as components. Thirdly, assemblages are complex entities that cannot be treated as simple individuals. Here de Landa introduces other (non-metric, topological) concepts from Deleuze, particularly those of possibility space or phase space (from physical chemistry), and attractors that might be shared by many systems; and the concept of the diagram as that which structures the space of possibilities of a particular assemblage.

Finally, there is the question of how assemblages operate at larger time scales – they often endure longer than their components and change at a slower rate. Does it take longer to effect change in organizations than in people, for example? At this level, it is important to identify: (1) collective unintended consequences – slow cumulative processes – that result from repeated interactions; and (2) products of deliberate planning. The first item is more common in long-term historical change. In the second case, enduring change happens as a result of mobilization of internal resources (from material resources to, say, solidarity). In general, the larger the social entity targeted for change the larger the amount of resources that must be mobilized. This implies that the spatial scale has temporal consequences since the necessary means for change may have to be accumulated over time. Said differently, the larger the spatial scale of the change desired, the more extensive the alliances among those involved need to be, and the more enduring their commitment to change. There is no simple correlation, however, between larger spatial extension and long temporal duration. In the case of assemblages that do not have a well-defined identity, such as dispersed, low-density networks, this dynamic is a strength and a weakness at the same time. On the one hand:

> low density networks, with more numerous weak links, are for this reason capable of providing their component members with novel information about fleeting opportunities. On the other hand, dispersed networks are less capable of supplying other resources, like trust in a crisis, the resources that define the strength of strong links. They are also less capable of providing constraints, such as enforcement of local norms. The resulting low degree of solidarity, if not compensated for in other ways, implies that as a whole, dispersed communities are harder to mobilize politically and less likely to act as causal agents in their interaction with other communities. (de Landa 2006: 35)

De Landa applies this theory systematically to the worlds of persons, organizations and governments. Its applicability to social movements should be apparent.

6. Castells' characterization of networks is suggestive:

> A network is a set of interconnected nodes ... Networks are open structures, able to expand without limits, integrating new nodes as long as they are able to communicate within the network ... Networks are appropriate instruments for a capitalist economy based on innovation, globalization, and decentralized concentration; for work, workers and firms based on flexibility, and adaptability; for a culture of endless deconstruction and reconstruction ... Switches connecting the network ... are the privileged instruments of power ... Since networks are multiple, the interoperating codes and switches between networks become the fundamental sources in shaping, guiding, and misguiding societies. (Castells 1996: 469–71)

The consequences of this conceptualization assume a dystopian dimension:

> Dominant functions are organized in networks pertaining to the space of flows that links them up around the world, while fragmenting subordinate functions, and people, in the multiple space of places, made of locales increasingly segregated and disconnected from each other. ... *Not that people, locales, or activities disappear. But their structural meaning does, subsumed in the unseen logic of the meta-network where value is produced, cultural codes are created, and power is decided.* (ibid.: 476–77, emphasis added)

7. A well-known aspect of ANT is that this process greatly depends on materials that are not only human. Technologies of all kinds are important in the generation of powerful networks. For Latour (1993),

moderns have been able to construct more powerful networks precisely to the extent that they have been able to enlist non-human elements – technologies, scientific knowledge, etc. – in the creation of longer and more connected networks. A question that arises of interest to social movements is: how does one compare networks? ANT's conclusion in this respect, it seems to us, is epistemologically weak: that one can only talk about longer and more powerful networks (invariably, those of the moderns) in terms of the methods and materials they utilize to generate themselves. Strathern (1996) presents a corrective to this view in that 'pre-moderns' (Latour's term) have a greater capacity for constructing hybrids and networks than one might suspect, by enrolling into it equally unsuspected entities or materials – from clans and animals to the ancestors; pre-moderns might also be more adept at 'cutting the network' than moderns, whose greed and sense of property (e.g. intellectual property) might compel them to arrive at premature closure in some cases (e.g. a patent, which forecloses the inventiveness of the previously existing network that produced it as a possibility). Castells and Latour have a tendency for limitless expansion; this all-embracing logic is in keeping with a particular style of theorizing. Social movements are seen as reactive mobilizations, which lead to the production of 'secluded identities'; they emerge from 'historically exhausted social forms' and, while they affect the network society, they cannot guide the reconstruction of the social order (e.g. Castells 1997: 104–9). Contrary to this view, it is important to see networks as a source for the production of culture, power, information, and the like. This requires a theory of the virtual, which Castells and ANT lack, and to which we shall return in the last section.

8. Cyberspace is seen by some as embodying a new model of life and world-making. Variously called by enthusiasts a knowledge space, a space of collective intelligence and a 'noosphere' (a sphere of collective thought, after Teilhard de Chardin), cyberspace, in these views, constitutes a signifying space of subject–subject interaction for the negotiation of visions and meanings. The resulting systems of networked intelligence could make up an inter-networked society of *intelligent communities*, centred on the democratic production of culture and subjectivity. Pierre Lévy (e.g. 1997) has most powerfully articulated this thesis in recent years. The liberation theologian Leonardo Boff's recent work on *religación* (2000) – a 'reconnecting' of humans with nature, each other, the earth, the cosmos, God – could also be interpreted in this light (he appeals explicitly to complexity). Discussions of the impact of ICTs on daily life abound, including those examining 'cybercultures' (e.g. Harcourt 1999; Bell and Kennedy 2000; Burbano and Barragán 2002). As Terranova (2004: 75–97) warns, the 'distributed mode of production' is not free from capitalist exploitation; there are clear interfaces between capital and the digital economy. Cyber-cultural politics can be most effective if it fulfils two conditions: awareness of the dominant worlds that are being created by the same technologies on which the progressive networks rely; and an ongoing tacking back and forth between cyber-politics and place-based politics, or political activism in the physical locations where networkers or net-weavers live. This is precisely the politics that some of today's movements are attempting to develop in creatively combining strategies for action at various scales. See Harcourt (1999); Escobar (2004); King (2006).
9. As an advocate of this research said in a comprehensive introduction to the subject, 'networks will dominate the new century to a much greater degree than most people are yet ready to acknowledge … Network thinking is poised to invade all domains of human activity and most fields of human enquiry' (Barabási 2002: 7, 222). The scientists' most striking claim is that there are some basic laws governing all networks. Networks are highly interconnected, so that huge networks constitute 'small worlds' in the sense that all elements in the network are only a few links away from all others, due to the presence of clusters, hubs and connectors. Not everything goes in networks, since some sites and hubs are much more connected than others, so that there are hierarchies of interconnection. Often, the network topology is determined by a few large hubs, as in the case of the World Wide Web, where links such as Google, Yahoo or amazon.com have a much greater weight in defining the Web's architecture than millions of much smaller nodes. These hubs determine preferential attachments; something similar happens in global movement networks, in which Zapatista and a few other key nodes (including the World Social Forum) are crucial to the structuring of the overall network. In sum, even if self-organized, networks of this type follow certain rules, which scientists refer to as 'power laws' (e.g. Barabási 2002; Duncan 2003; King 2006).

10. Notably, while she is concerned primarily with cyber-culture, she is also aware of the cultural overlaps and relations between cyber-activism and other activisms associated with the alter-globalization movement. See Terranova (200); Chesters and Welsh (2006).
11. Nobody has diagnosed the dangers of the worlds enabled by new information and communication technologies like Paul Virilio (see, for example, 1997, 1999).
12. For instance, the gains in terms of recognition of difference, rights, collective territories, identities, and so forth, by indigenous and black groups or the collective territories granted to these groups in the 1990s in many Latin American countries, which are now being dismantled by the state.
13. Hence Gibson-Graham's (1996) critique of the lingering capitalocentrism in Deleuze and Guattari, because of these concepts.
14. Deleuze and Guattari explicitly define the minoritarian (and majoritarian) as different from numeric minorities (or majorities) (1987: 469–73). It is not a matter of numbers, but of logics, the logic of denumerable sets versus that of non-denumerable sets. That is why not all self-identified social movements can be considered minoritarian or as enacting minoritarian politics. Discovering what is effectively minoritarian is an important task for movements.
15. Whether or not these conceptions are based on having engaged with Deleuze and Guattari's own work or are arrived at by other means would be an interesting empirical question.
16. In terms of capital, the limits can be thought about in terms of the dynamic between striated and smooth capitals (e.g. Deleuze and Guattari 1987: 492).
17. *'Movement of movements'* has become another way to refer to the 'anti-globalization' or alter-globalization movements associated with protests at Seattle, Genoa, Prague, but also the World Social Forum and the Zapatistas.
18. For a more elaborated explanation of this theoretical production, see Osterweil (2006).
19. For a description of the use of mythopoesis in Italian movements see Wu Ming I (2001).
20. These contentions about the nature of contemporary Italian activism– the centrality of theoretical and reflexive practices and products, and the active experimentation and exploration with and of different arrangements of the social – inform the doctoral dissertation project being currently undertaken by one of the authors (for more, see Osterweil 2005b, 2006).
21. In a sense this has always been the case in the Italian Left. (e.g. Gramsci 1971; see also Borio et. al. 2002; Wright 2002).
22. The 'knowledge turn' in social movement studies can be located in multiple sites and from various sources that cannot be summarized here. A main site for the development of social movement research based on this is the interdisciplinary Social Movements Working Group at the University of North Carolina at Chapel Hill (www.unc.edu/smwg), with which both authors are associated. Works on the knowledge turn produced by members of this group include Casas-Cortés (2005); Castas-Cortés et. al. (2006); Osterweil with Chesters (2007). Furthermore, there are a growing number of activist research texts published and written by activists making this claim (see Malo 2004; Graeber and Shukaitis forthcoming). Other important works that examine social movements and knowledge or theory production include: Eyerman and Jamison (1991); Barker and Cox (2001); Conway (2004, 2006); Chesters and Welsh (2006).
23. The hypothesis that the practices of difference that might exist at the borders of the modern colonial world system could be anchoring important 'decolonial' projects at the epistemic, social, cultural, economic and ecological levels is being advanced by a group of Latin American researchers, working principally in, or in conversations with, struggles in the Andean countries. See Escobar (2003) for a presentation and discussion of this group (referred to as the 'modernity/coloniality/decoloniality group'). The connection between this group's work and Deleuze is yet to be made.
24. These remarks come from a lecture by Goodwin at Schumacher College in Devon, England, 20 February 2006. See also Goodwin (2007). Goodwin finds great hope in a holistic science that integrates mainstream science with the science of qualities, forms and intensities that he sees present in the work of Goethe and also in many indigenous traditions. These sciences incorporate experience, feelings and

intuition as modes of knowledge. In sum, Goodwin, among others, is trying to articulate anew the role of experience, feelings, intuition and embodied knowledge as epistemological and ontological questions. For him, the issue is how to rethink our place within the flows of creative emergence on the planet on the basis of a deeper understanding of living process that moves back and forth between the life of form and forms of life. This has tremendous implications for ecological design. The analysis of biological life in terms of meaning has been developed by Markos (2002) by building bridges between hermeneutics and biology.

Bibliography

Alvarez, S.E. (1998). 'Latin American Feminisms "Go Global": Trends in the 1990s and Challenges for the New Millennium', in S.E. Alvarez, E. Dagnino and A. Escobar (eds) *Cultures of Politics/Politics of Cultures. Revisioning Latin American Social Movements*. Boulder, Westview Press, pp. 293–324.

———.(forthcoming). *Contentious Feminisms: Critical Readings of Social Movements, NGOs, and Transnational Organizing in Latin America*. Durham, Duke University Press.

Alvarez, S.E., E. Dagnino and A. Escobar. (eds). (1998). *Cultures of Politics/Politics of Culture: Revisioning Latin American Social Movements*. Boulder, Westview Press.

Barabási, A.-L. (2002). *Linked: The New Science of Networks*. Cambridge, Perseus Publishing.

Barker, C. and L. Cox. (2001). '"What Have the Romans Ever Done for Us?" Academic and Activist Forms of Movement Theorizing'. Available at

Barker, C. and L. Cox. (2001). '"What Have the Romans Ever Done for Us?" Academic and Activist Forms of Movement Theorizing'. Available at http://www.iol.ie/~mazzoldi/toolsforchange/afpp/afpp8.html

Bell, D. and B. Kennedy. (eds). (2000). *The Cybercultures Reader*. London, Routledge.

Boff, L. (2000). *El cuidado esencial*. Madrid, Editorial Trotta.

Borio, G., F. Pozzi and G. Roggero. (2002). *Futuro Anteriore: Dai 'Quaderni Rossi' ai movimenti globali: richezze e limite dell'operaismo italiano*. Rome, Derive Approdi.

Burbano, A. and H. Barragán. (eds). (2002). *hipercubo/ok/. arte, ciencia y tecnología en contextos próximos*, Bogotá, Universidad de los Andes / Goethe Institut.

Camazine, S., J.-L. Deneubourg, N.R. Franks, J. Sneyd, G. Theraulaz and E. Bonabeau. (2001). *Self-Organization in Biological Systems*. Princeton, Princeton University Press.

Casas-Cortés, M.I. (2005). 'Reclaiming Knowledges/Reclamando Conocimientos: Movimientos Sociales y la producción de saberes', *Latin American Studies Association Forum* 36 (1), 14–17.

Casas-Cortés, M.I., M. Osterweil and D. Powell. (2006). *Blurring Boundaries: Recognizing Knowledge-Practices in the Study of Social Movements*, Anthropological Quarterly 81 (1), 17–58.

Castells, M. (1996). *The Rise of the Network Society*. Oxford, Blackwell.

———.(1997). *The Power of Identity*. Oxford, Blackwell.

Chesters, G. (2003). 'Shape Shifting: Civil Society, Complexity and Social Movements', unpublished manuscript, Centre for Local Policy Studies, Edge Hill University College, Lancashire, UK.

Chesters, G. and I. Welsh. (2006). *Complexity and Social Movements: Multitudes at the Edge of Chaos*. London, Routledge.

Conway, J. (2004). *Identity, Place, Knowledge: Social Movements Contesting Globalization*. Halifax, Fernwood Publishing.

———.(2006). *Praxis and Politics: Knowledge Production in Social Movements*. New York and London, Routledge

de Landa, M. (1997). *A Thousand Years of Nonlinear History*. New York, Zone Books.

———.(2002). *Intensive Science and Virtual Philosophy*, New York, Continuum.

———. (2003). '1000 Years of War. *CTHEORY* Interview with Manuel De Landa'. Available at www.ctheory.net/text_file?pick=383

———.(2006). *A New Philosophy of Society: Assemblage Theory and Social Complexity*. London, Continuum Press.

———.(n. d.). 'Meshworks, Hierarchies and Interfaces'. Available at http://www.t0.or.at/De Landa/

Deleuze, G. and F. Guattari. (1987). *A Thousand Plateaus: Capitalism and Schizophrenia.* Minneapolis, University of Minnesota Press.

———.(1994). *What is Philosophy?.* New York, Columbia University Press.

Diani, M. and D. McAdam. (eds). (2003). *Social Movements and Networks.* Oxford, Oxford University Press.

Duncan, W. (2003). *Six Degrees. The Science of a Connected Age.* New York, W.W. Norton.

Escobar, A. (2000). 'Notes on Networks and Anti-Globalization Social Movements', presented at AAA Annual Meeting, San Francisco, 15–19 November. Available at www.unc.edu/~aescobar/

———.(2003). '"World and Knowledges Otherwise": the Latin American Modernity/Coloniality Research Program', *Cuadernos del CEDLA* 16, 31–67.

———.(2004). 'Other Worlds Are (Already) Possible: Self-Organisation, Complexity, and Post-Capitalist Cultures', in J. Sen, A. Anad, A. Escobar and P. Waterman (eds) *The World Social Forum. Challenging Empires.* Delhi, Viveka, pp. 349–58.

———.(2008). *Territories of Difference: Place. Movements. Life, Redes.* Durham, Duke University Press.

Eyerman, R. and A. Jamison. (1991). *Social Movements: A Cognitive Approach.* University Park, Pennsylvania State University Press.

Fischer, M.J. (2003). *Emergent Forms of Life and the Anthropological Voice.* Durham, Duke University Press.

Fortun, K. (2001). *Advocacy After Bhopal: Environmentalism, Disaster and New Global Orders.* Chicago, University of Chicago Press.

———.(2003). 'Ethnography In/Of/As Open Systems', *Reviews in Anthropology* 32, 171–90.

Friedman, T. (2005). *The World is Flat.* New York, Farrar, Strauss and Giroux.

Gibson-Graham, J.K. (1996). *The End of Capitalism (As We Knew It).* Oxford, Blackwell.

———.(2006). *A Postcapitalist Politics.* Minneapolis, University of Minnesota Press.

Goodwin, B. (2007). *Nature's Due: Healing Our Fragmented Culture.* Edinburgh, Floris Books.

Graeber, D. and S. Shukaitis. (eds). (forthcoming). *Constituent Imagination: Militant Investigation // Collective Theorization.* Oakland and Edinburgh, AK Press.

Gramsci, A. (1971). *Selections from the Prison Notebooks of Antonio Gramsci.* New York, International Publishers.

Grossberg, L. (2006). 'Does Cultural Studies Have Futures?? Should It? (Or What's the Matter with New York?): Cultural Studies, Contexts, and Conjunctures', *Cultural Studies* 20(1), 1–32.

Haila, Y, and C. Dyke. (eds). (2006). *How Nature Speaks: The Dynamics of the Human Ecological Condition,* Durham, Duke University Press.

Hall, S. (1996). 'Introduction: Who Needs "Identity"?', in S. Hall and P. du Gay (eds) *Questions of Cultural Identity.* London, Sage, pp. 1–17.

Harcourt, W. (ed.). (1999). *Women@Internet: Creating New Cultures in Cyberspace.* London, Zed Books.

Harcourt, W. and A. Escobar. (eds). (2005). *Women and the Politics of Place.* Bloomfield, Kumarian Press.

Juris, J. (2005). 'The New Digital Media and Activist Networking within Anti-Corporate Globalization Movements', *The Annals of the American Academy of Political and Social Science* 597 (1), 189–208.

———.(2008). *Networking Futures: The Movements Against Corporate Globalization.* Durham, Duke University press.

Kauffman, S. (1995). *At Home in the Universe: The Search for Laws of Self-Organization and Complexity.* Oxford, Oxford University Press.

Keck, M. and K. Sikkink. (1998). *Activists Beyond Borders: Advocacy Networks in International Politics.* Ithaca, Cornell University Press.

King, M. (2005). 'Looking Out from Within, Confronting the Dilemmas of Activist Research', presented at the AAA Annual Meeting, Washington, DC. 1–5 December.

———.(2006). 'Emergent Socialities: Networks of Biodiversity and Anti-globalization'. PhD Dissertation, Department of Anthropology, University of Massachusetts, Amherst.

Latour, B. (1993). *We Have Never Been Modern.* Cambridge, MA, Harvard University Press.

———.(1997). 'The Trouble with Actor–Network Theory'. Available at http://www.ensmp.fr/~latour/poparticles/poparticle/p067.html

Law, J. (2000) [1992]. 'Notes on the Theory of the Actor Network'. Available at http://tina.lancs.ac.uk/sociology/soc054jl.html

Lévy, P. (1997). *Collective Intelligence: Mankind's Emerging World in Cyberspace.* New York, Plenum Trade.
Leyva Solano, X. (2002). 'Neo-Zapatismo: Networks of Power and War', PhD Dissertation, Department of Anthropology, University of Manchester.
———. (2003). 'Concerning the Hows and Whys in the Ethnography of Social Movement Networks', presented at the Department of Anthropology, University of North Carolina, Chapel Hill, 17 March.
Malo, M. (ed.). (2004). *Nociones Comunes: Experiencias y ensayos entre investigación y militancia.* Madrid, Traficantes de Sueños.
Marcus, G. (ed.). (1999). *Critical Anthropology Now.* Santa Fe, School of American Research.
Markos, A. (2002). *Readers of the Book of Life: Contextualizing Developmental Evolutionary Biology.* Oxford, Oxford University Press.
Marston, S., J.P. Jones III and K. Woodward. (2005). 'Human Geography without Scale', *Transactions of the Institute of British Geography* (NS) 30, 16–432.
Maturana, H. and F. Varela. (1980). *Autopoiesis and Cognition,* Boston, Reidel Publishing.
Maturana, H. and F. Varela. (1987). *The Tree of Knowledge.* Boston, Shambhala.
Mignolo, W. (2000). *Local Histories/Global Designs,* Princeton, Princeton University Press.
Mignolo, W. (2001). 'Local Histories and Global Designs: An Interview with Walter Mignolo'. *Discourse* 22 (3), 7–33.
Olesen, T. (2005). *International Zapatismo: the Construction of Solidarity in the Age of Globalization.* London, Zed Books.
Osterweil, M. (2004). 'A Cultural-political Approach to Reinventing the Political', *International Social Science Journal,* Special Issue: Explosions in Open Space: The World Social Forum and Cultures of Politics 182, December.
———.(2005a). 'Place-based Globalism: Locating Women in the Alternative Globalization Movement', in W. Harcourt and A. Escobar (eds) *Women and the Politics of Place.* Bloomfield, CT, Kumarian Press, pp. 174–89.
———.(2005b). 'Social Movements as Knowledge Practice Formations: Towards an "Ethno-Carto-Graphy" of Italy's Movimento Dei Movimenti', PhD Dissertation Prospectus, Department of Anthropology, University of North Carolina, Chapel Hill.
———.(2006). 'Theoretical-practice: Il movimento dei movimenti and (Re)Inventing the Political', paper presented at Cortona Colloquium: Cultural Conflicts, Social Movements and New Rights: A European Challenge, 20–22 October, Cortona, Italy.
Osterweil, M with G. Chesters. (2007). 'Global Uprisings: Towards a Politics of the Artisan', in D. Graeber, and S. Shukatis (eds) *Constituent Imagination: Militant Investigation // Collective Theorization.* Oakland and Edinburgh, AK Press, pp. 253–63.
Peltonen, L. (2006). 'Fluids on the Move: an Analogical Account of Environmental Mobilization', in Y. Haila and C. Dyke (eds) *How Nature Speaks. The Dynamics of the Human Ecological Condition.* Durham, Duke University Press, pp. 150–76.
Prigogine, I. and G. Nicolis. (1989). *Exploring Complexity.* New York, W.H. Freeman.
Prigogine, I. and I. Stengers. (1984). *Order Out of Chaos: Man's New Dialogue With Nature.* New York, Bantam Books.
Riles, A. (2001). *The Network Inside Out.* Ann Arbor, University of Michigan Press.
Santos, B. de S. (2002). *Towards a New Legal Common Sense.* London, Butterworth.
———.(2004). 'The World Social Forum: Towards a Counter-Hegemonic Globalisation (Part I)', in J. Sen, A. Anand, A. Escobar and P. Waterman (eds) *The World Social Forum. Challenging Empires.* Delhi, Viveka, pp. 235–45.
Smith, J., C. Chatfield and R. Pagnucco. (eds). (1997). *Transnational Social Movements and Global Politic.* Syracuse, Syracuse University Press.
Solé, R. and B. Goodwin. (2000). *Signs of Life: How Complexity Pervades Biology.* New York, Basic Books.
Strathern, M. (1996). 'Cutting the Network', *Journal of the Royal Anthropological Institute* (ns) 2, 17–535.
Summer, K. and H. Halpin. (2005). 'The End of the World as We Know it', in D. Harvie, K. Milburn, B. Trott and D. Watts (eds) *Shut Them Down! The G8, Gleneagles 2005 and the Movement of Movements.* Brooklyn, Autonomedia, pp. 351–60.

Terranova, T. (2001). 'Demonstrating the Globe: Virtual Action in the Network Society', in D. Holmes (ed.) *Virtual Globalization: Virtual Spaces/Tourist Spaces.* London, Routledge, pp. 95–113.

———. (2004). *Network Culture,* London, Pluto Press.

Virilio, P. (1997). *The Open Sky.* New York, Verso.

———. (1999). *Politics of the Very Worst.* New York, Semiotext(e).

Wright, S. (2002). *Storming Heaven: Class Composition and Struggle in Italian Autonomist Marxism.* London, Pluto Press.

Wu Ming I. (2001). 'Tute Bianche: La Prassi della Mitopoiesi in Tempi di Catastrofe'. Available at http://www.wumingfoundation.com/italiano/outtakes/monaco.html. October, 2001.

Chapter 10

Intensive Filiation and Demonic Alliance

Eduardo Viveiros de Castro

Life is robbery. (Whitehead 1978: 105)[1]

Deleuze and Anthropology

For my generation, the name of Gilles Deleuze immediately suggests the change in thought that marked the years *circa* 1968, during which some key elements of our present cultural apperception were invented.[2] The meaning, the consequences and the very reality of this change have given rise to a controversy that still rages. Just like the postmodernity of which it is one of the (anti-) foundational dates, '1968' seems never to end. For some, it does not stop having never occurred; for others, it does not cease to have not begun yet. I place myself among the latter; and thus I would say the same about the influence of Deleuze and his long-time associate Félix Guattari, the authors of the most radically consistent (conceptually) and most consistently radical (politically) *oeuvre* created in philosophy during the second half of the last century. Its presence in certain contemporary disciplines or investigation fields is indeed much less evident than one might expect, being felt mostly (when at all) through its effects on the general cultural matrix. This seems to be the case, for example, of social anthropology.

From an anthropological point of view, the novelty which was (and remains) Deleuze's philosophy was immediately perceived by the countercultural and counterpolitical movements that proliferated in the West during the last three decades, most notably feminism (some versions of it, at least), as well as some currents in experimental art (*art de vivre* included). Not much later, it was incorporated in the toolkit of certain auto-anthropological projects such as science and technology studies or the sundry disquisitions on the dynamics of late capitalism. On the other hand, the attempts at a more direct engagement between anthropology at large, or allo-anthropology,[3] and Deleuzian concepts have been surprisingly rare. I deem this dearth surprising for two reasons. First, because *Anti-Oedipus* and *A Thousand Plateaus*, the two volumes of *Capitalism and Schizophrenia*, support many of their arguments by a vast bibliography on non-Western peoples – from the Guayaki to the Nuer to the Mongolians – developing thereby some hypotheses of great analytical potential for anthropology. Secondly, because the work carried out by some of the most innovative anthropologists in the last two decades – I am thinking, for example, of Roy Wagner,

Marilyn Strathern and Bruno Latour – shows evident connections with Deleuze's ideas. In the case of Wagner, such connections seem to be purely virtual, a serendipitous convergence manifesting an 'aparallel evolution' (in Deleuze's sense) or an independent 'invention' (in Wagner's sense); but this does not make the connections any the less real or less surprising. As concerns Strathern, the connections are certainly partial (*et pour cause*) and mediated (through Donna Haraway, for example); but the great Cambridge anthropologist shares with Deleuze a dense web of conceptually charged notions, such as multiplicity, perspective, dividual, fractality. In many respects, of the three anthropologists cited, Strathern is the one whose work shows the greatest 'molecular' affinity to Deleuzian ideas. In the case of Bruno Latour, the connections are actual and explicit, forming one of the major articulations of Latourian anthropology. At the same time, there are important aspects of Latour's work which remain foreign to the spirit of Deleuzian philosophy.

It is certainly no accident that the three above-mentioned anthropologists are among the few who could be called post-structuralists (instead of, say, post-modernists) with some propriety. They have managed creatively to take on board the insights of structuralism and move ahead, instead of, like so many of their colleagues, bending backwards and embracing conservative theoretical projects, such as the sentimental pseudo-immanentism of lived worlds and embodied practices, or the macho-positivist truculence of TOEs such as evolutionary psychology or so-called 'political economy'. By the same token, Deleuze's thought can be seen as a form of extreme deterritorialization of structuralism, a movement or a style from which he extracted – some would say, into which he introduced – its most radically original insights (Deleuze 2002 [1972]) so as to move forward, with their help, in other, sometimes very different, directions.[4] After all, what is a multiplicity, if not a structure at long last freed from all complicity with transcendence?

In this chapter, one of the things I shall do is remark upon a few evocative parallels between Deleuzian conceptual motifs and the work of Wagner, Strathern and Latour. There would be no point in undertaking anything resembling a 'systematic comparison', which would risk reifying a dynamic, rhizomatic, multiple connection of ideas into an arborescent tracing of origins and influences; besides, this sort of endeavour almost inevitably retroprojects a conception of 'similarities and differences' as causal-like properties that exist out there (or in there – the authors' minds, perhaps), rather than as pragmatico-theoretical effects of the comparison itself and, as such, partial, interested and instrumental relations.

This is not to deny that there exists something like a wider cultural matrix, an intellectual configuration of which Deleuze and Guattari, as well as Wagner, Strathern and Latour, are named singularities or points of inflection with different chronological and disciplinary coordinates. Deleuze's first 'own' book (*Différence et Répétition*), let it be recalled, was presented as the expression of a certain Zeitgeist of which the author intended to draw all the philosophical consequences. Hence the sense of strange familiarity ('Perhaps the sense of *déjà vu* is also a sense of habitation within a cultural matrix' – Strathern 1991a: xxv, italics in original) that any anthropologist, after a few years, say, of exclusive immersion in the literature of his or her own discipline cannot but feel when reading or rereading Deleuze and Guattari's books: a curious sensation of temporally inverted déjà vu, not unrelated to the Deleuzian phenomenon of a 'dark precursor'. Indeed, quite a few of the descriptive techniques and theoretical perspectives in anthropology that have only recently begun to lose their scandalous overtones are powerfully, if subterraneously, connected to Deleuzo-Guattarian texts of twenty or thirty years ago.

In order to assess with reasonable precision the anthropological value of those texts it would be necessary to describe the constellation of forces in which anthropology is implicated today, something that far exceeds the limits of my competence. If we decide to remain generic, however, it is easy to point out that Deleuze played a major, perhaps decisive, role in the sedimentation of a certain pervasive contemporary conceptual aesthetics. This aesthetics can be described with the aid of the binary vocabulary of structuralism, all the more so since it is a response to the latter – or, more precisely, a reproblematization internal to it.

For some time now, the human sciences have displayed a shift in interest towards semiotic processes such as metonymy, indexicality and literality – three modes of rejecting metaphor and representation (metaphor as the essence of representation), privileging pragmatics over semantics and valorizing paratactic coordination over syntactic subordination. The 'linguistic turn' that formed the virtual point of convergence for such widely diverse temperaments, projects and systems over the last century now appears to be turning in other directions, away from linguistics and (to some extent) language as an anthropological macro-paradigm: in fact, the emphases indicated above show how the lines of escape from language as a model have been glimpsed within the model of language itself. Put otherwise, the ancient premise of the ontological discontinuity between language and the world, which assured the reality of the former and the intelligibility of the latter (and vice versa) and that served as ground and pretext for so many other discontinuities and exclusions – those between myth and philosophy, magic and science, primitive and civilized, for example – seems to be in the throes of metaphysical obsolescence. Indeed, it is primarily in this sense that we are ceasing to be – or, better, that we remain never having been – modern.

On the 'world' side (a side that no longer has another side, given that the world itself is now only made of sides), the corresponding shift in emphasis has been towards the fractional and differential instead of the whole and the combinatory, flat multiplicities in lieu of hierarchical totalities, the transcategorial connection of heterogeneous elements instead of the correspondence between internally homogeneous series, the wave-like or topological continuity of forces rather than the particle-like or geometric discontinuity of forms.

In sum, '[t]his is what we are getting at: a generalized chromaticism' (Deleuze and Guattari 2004 [1988]: 108). The molar discontinuity between the two internally homogeneous series of signifier and signified, on the one hand – series that are themselves in structural discontinuity – and the phenomenologically continuous series of the real, on the other, has gradually been diffracted into molecular or fractal discontinuities; into trans-serial self-similarities that potentialize difference and reveal it as continuous variation – or rather, that project continuity as intrinsically differential and heterogenic (implying a distinction between the ideas of the continuous and the undifferentiated). A 'flat' ontology (de Landa 2002) and a corresponding 'symmetrical' epistemology (Latour 1993); the collapse, in fact, of the distinction between epistemology (language) and ontology (world), and the gradual emergence of a 'practical ontology' (Jensen 2004), in which knowing is no longer a way of representing the (un)known but of interacting with it; that is, a way of creating rather than of contemplating, reflecting or communicating (Deleuze and Guattari 1991). The task of knowledge ceases to be the unification of the diverse through representation, becoming instead the 'multiplication of the number of agencies that people the world' (Latour 1996). A new image of thought. Nomadology. Multinaturalism. (The phrasing is entirely Deleuzian here, of course; but other conceptual resources could be mustered to more or less the same effect, from Niels Bohr to Derrida and beyond.)

In what follows, I sketch out a map of a very limited sector of this contemporary conceptual aesthetics. By way of an example more than anything else, I suggest two possible directions among many others for the intensification of a dialogue between Deleuzian philosophy and current anthropology. In the first part below, I draw a few parallels between Deleuzian concepts and certain influential motifs in contemporary anthropology; in the second part, I concentrate on a specific incidence of classical social anthropology – kinship theory – upon the Deleuzo-Guattarian conception of the primitive territorial machine, or presignifiying semiotics.

I
An Anti-sociology of Multiplicities

In *Anti-Oedipus*, as is well known, Deleuze and Guattari overthrow the temple of psychoanalysis by imploding its central pillar – the reactionary conception of desire as lack – and replace it with a theory of desiring machines as sheer positive productivity that must be coded by the *socius*, the social production machine. This theory runs into a vast panorama of universal history, painted in the book's central chapter in a quaintly archaic style, which may make the anthropological reader wince. Not only does it employ the venerable savagery/barbarism/civilization triad, but all ethnographic references are treated in a seemingly cavalier way that the same reader might be tempted to call 'uncontrolled comparison'. Yet, if that reader stops to think for a moment, it is probable that she will reach the conclusion that the traditional three-stage topos is submitted there to a far from traditional interpretation, and that the impression of comparative erraticism derives from the fact that the controls used by the authors are other than the usual ones – of a differentiating rather than a collectivizing type, as Wagner (1981) would put it. *Anti-Oedipus* is indeed the result of a 'prodigious effort to think differently' (Donzelot 1977: 28); its purpose is not merely to denounce the repressive paralogisms of psychoanalysis but to establish a true 'anti-sociology' (ibid.: 37).[5] An obviational project like this should certainly appeal to contemporary anthropology; or at least to that anthropology bent on exploring the vast territory that lies beyond the jurisdiction of the three infernal dichotomies that contain the discipline within an iron ring: nature and culture, individual and society, traditional and modern.[6]

A Thousand Plateaus distances itself from *Anti-Oedipus*' psychoanalytical concerns. The project to write a 'universal history of contingency' (Deleuze and Guattari 2003: 290) is carried on in a decidedly non-linear fashion through the crossing of different intensity 'plateaus' (a notion inspired by Bateson) corresponding to diverse material-semiotic formations and peopled by a disconcerting quantity of new concepts. The book puts forward and illustrates a theory of multiplicities, the Deleuzian theme that has perhaps achieved greatest repercussions in contemporary anthropology.

Multiplicity is the meta-concept that defines a new type of entity; the well-known (by name at least) 'rhizome' is its concrete image.[7] The sources of the Deleuzian idea of multiplicity lie in Riemann's geometry and Bergson's philosophy (Deleuze 1966: chapter 2; Deleuze and Guattari 2004: 532–38), and its creation aims at dethroning the classical metaphysical notions of essence and type (de Landa 2002).[8] It is the main tool of a 'prodigious effort' to imagine thought as an activity other than that of identifying (recognition) and classifying (categorization), and to determine what is there for thought to think as intensive difference rather than as extensive substance. The political-philosophical intentions of this decision are clear: it is a matter of severing the link between the concept and the state. Multiplicities are the tools of a fractal, anexact counter-geometry, a minor mathematics that aims at 'inverting Platonism'. Thinking through multiplicities is thinking against the state.[9]

The notions of type and entity are, in fact, inadequate to define rhizomatic multiplicities. If there is 'no entity without identity', as Quine famously quipped, one must conclude that multiplicities do not qualify for that enviable status. A rhizome does not behave as an entity, nor does it instantiate a type; it is an acentric reticular system of $n - 1$ dimensions, constituted by intensive relations ('becomings') between heterogeneous singularities that correspond to events, or extra-substantive individuations ('hecceities'). Hence, a rhizomatic multiplicity is not truly one *being* but an assemblage of becomings: a 'difference engine' or, rather, the intensive diagram of its functioning. Bruno Latour, who in his recent book on ANT indicates how much it owes to the rhizome concept, is particularly emphatic: a network is not a thing because any thing can be described as a network (Latour 2005: 129–31). A network is a perspective, a way of inscribing and describing 'the registered movement of a thing as it associates with many other elements' (Jensen 2003: 227). Yet this perspective is internal or immanent; the different associations of the 'thing' make it differ from itself – 'it is the thing itself that has been allowed to be deployed as multiple' (Latour 2005:

116). In short, there are no viewpoints on things – it is the things and the beings that are viewpoints themselves (Deleuze 1968: 79, 1969: 203). If no entity without identity, then no multiplicity without perspective.

A rhizome is not truly *one* being, either. Multiplicities are constituted by the absence of any extrinsic coordination imposed by a supplementary dimension ($n + 1$: n plus its 'context', for example); congenitally devoid of unity, they differ constantly from themselves. They evince an immanent organization 'belonging to the many as such, and which has no need whatsoever of unity in order to form a system' (Deleuze 1968: 236).[10] Multiplicities are, in short, tautegorically anterior to their own 'contexts'; like Roy Wagner's (1986) symbols that stand for themselves, they possess their own internal measure. This turns them into systems whose complexity is 'transversal', that is, resistant to hierarchy or to any other type of transcendent unification – a complexity of alliance rather than descent, to anticipate an argument. Emerging when and where open intensive lines (lines of force, not lines of contour; see Deleuze and Guattari 2004: 549) connect heterogeneous elements, rhizomes project a radically flat ontology, which ignores the distinctions between 'part' and 'whole'.[11]

From the viewpoint of the periodizations we regularly borrow from other disciplines, it could be said that the ontology of pure difference is 'neo-baroque' (as Kwa 2002 persuasively argued), thus escaping the canonical alternation to which the history of anthropology is usually reduced, that is, between 'classical' mechanist atomism (with the associated individual/society dichotomy) and 'romantic' organicist holism (with its powerful nature/culture dialectics). At another temporal scale, these new ontologies must be classified, of course, as 'post-structuralist'.

Wagner's fractal person, Strathern's partial connections, Callon and Latour's socio-technical networks are some well-known anthropological examples of flat multiplicities. 'A fractal person is never a unit standing in relation to an aggregate, or an aggregate standing in relation to a unit, but always *an entity with relationship integrally implied*' (Wagner 1991: 163, emphasis added).[12] The mutual implication of the concepts of multiplicity, intensity and implication is in fact a point elaborated at length by Deleuze (1968: chapter VI). François Zourabichvili, the most perceptive commentator on this philosopher, observes that 'implication is the fundamental logical movement in Deleuze's philosophy' (2004a: 82); elsewhere, he underlines that Deleuzian pluralism supposes a 'primacy of relations'.[13]

A primacy of relations: every anthropologist should feel at home here. Not 'every' relation will do, though – not every anthropological home truth would do, either. Multiplicity is a system defined by a modality of relational synthesis different from a connection or conjunction of terms. Deleuze calls it disjunctive synthesis or inclusive disjunction, a relational mode that does not have similarity or identity as its cause (formal or final), but divergence or distance; another name for this relational mode is 'becoming'. Disjunctive synthesis or becoming is 'the main operator of Deleuze's philosophy' (Zourabichvili 2003: 81), as it is the movement of difference as such – the centrifugal movement through which difference escapes the powerful circular attractor of dialectical contradiction and sublation. A difference that is positive rather than oppositive, an indiscernibility of the heterogeneous rather than a conciliation of contraries, disjunctive synthesis takes disjunction as 'the very nature of relation' (Zourabichvili 2004a: 99) and relation as a movement of 'reciprocal asymmetric implication' (Zourabichvili 2003: 79) between the terms or perspectives connected by the synthesis, which is not resolved either into equivalence or into a superior identity: 'Deleuze's most profound insight is perhaps this: that difference is also communication and contagion between heterogeneities; in other words, that a divergence never arises without reciprocal contamination of points of view ... To connect is always to communicate across a distance, through the very heterogeneity of the terms' (Zourabichvili 2004a: 99).

Coming back to the parallels with contemporary anthropological theory, it is worth recalling that the theme of separation-as-relation is emblematic of Strathernian anthropology. The conception of relations as 'comprising disjunction and connection *together*' (Strathern 1995: 165, emphasis added) is the basis of the theory of differential relations, the idea that '[r]elations make a difference between persons' (Strahern 1999: 126, see also, naturally, 1988: chapter 8 and 1996: 525). To cut a long argument short, let us say that the celebrated 'system M' (Gell 1999), the description of Melanesian sociality both as an exchange of perspectives and a process of relational implication-explication, is the prototypical allo-anthropological theory of disjunctive synthesis. From the auto-anthropological standpoint, in turn, it is possible to observe that the subtractive rather than additive multiplicity of rhizomes turns the latter into non-merological, post-plural (Strathern 1992a) 'objects', capable of tracing a line of flight from the dilemma of the one and the many that Strathern insightfully identifies as anthropology's characteristic analytical trap: '[A]nthropologists by and large have been encouraged to think [that] the alternative to one is many. Consequently, we either deal with ones, namely single societies or attributes, or else with a multiplicity of ones ... A world obsessed with ones and the multiplications and divisions of ones creates problems for the conceptualization of relationships' (Strathern 1991a: 52–53).

To compare multiplicities is different from making particularities converge around generalities, as in the usual case of anthropological comparisons that seek out the substantial similarities underlying accidental differences.[14] Yet it is also different from establishing correlational invariants through formal analogies between extensive differences (oppositions), as in the case of structuralist comparisons, in which 'it is not the resemblances, but the differences that resemble one another' (Lévi-Strauss 1962a: 111). To compare multiplicities – which are comparing devices in their own right – is to determine their characteristic ways of diverging, their distances both internal and external; here, every comparative analysis is necessarily a separative synthesis. When it comes to multiplicities, it is a case less of (extensive) relations that vary than of (intensive) variations that relate: it is the differences that differ.[15] As the strange molecular sociologist Gabriel Tarde wrote more than a century ago: 'The truth is that differences go differing, and changes go changing, and that, as they take themselves thus as their own finality, change and difference bear out their necessary and absolute character' (1999: 69).[16]

Intensive difference, difference of perspectives, difference of differences. Nietzsche observed that health's viewpoint on illness differs from illness' viewpoint on health.[17] It was perhaps this observation that inspired Roy Wagner to say about his early relations with the Daribi: 'their misunderstanding of me was not the same as my misunderstanding of them' (Wagner 1981: 20) – my candidate for the best anthropological definition of 'culture' ever proposed. Since the difference is never the same, the way is not the same in both directions.[18] The comparison of multiplicities – comparison as production of multiplicity (or 'invention of culture') – is always a disjunctive synthesis, like the relations it relates.

Partial Dualities

Deleuzian texts revel in conceptual dyads: difference and repetition; intensive and extensive; nomadic and sedentary; virtual and actual; flows and quanta; code and axiomatic; deterritorialization and reterritorialization; minor and major; molecular and molar; supple and rigid; smooth and striated – the list is long and colourful. Owing to this stylistic 'signature', Deleuze has sometimes been classified as a dualist philosopher (Jameson 1997) – a hasty interpretation, to say the least, of the morphogenesis of this philosopher's conceptual system.[19]

It is interesting to notice how the expositive pace of the two *Capitalism and Schizophrenia* books, in which dualities particularly abound, is time and again interrupted by provisos, qualifications, distinctions, involutions, subdivisions and other argumentative displacements of the dual (or other) distinctions that had just been proposed by the authors themselves. Such methodical interruptions are exactly this, a question of method and not a moment of regret after the binary sin; they are perfectly determined moments of conceptual construction.[20] Neither principle nor result, the Deleuzian dyads – one might wish to call them 'conceptual duplexes', after Strathern (2005) – are means to arrive elsewhere. The exemplary case here is, once again, the distinction between root-tree and canal-rhizome:

> The important point is that the root-tree and the canal-rhizome are not two opposed models; the first operates as a transcendent model and tracing, even if it engenders its own escapes; the second operates as an immanent process that overturns the model and outlines a map, even if it constitutes its own hierarchies, even if it gives rise to a despotic channel. It is not a question of this or that place on earth, or of a given moment in history, still less of this or that category of thought. It is a question of a model that is perpetually in construction or collapsing, and of a process that is perpetually prolonging itself, breaking off and standing up again. No, this is not a new or different

> dualism ... We invoke one dualism only in order to challenge another. We employ a dualism of models only in order to arrive at a process that challenges all models. Each time, mental correctives are necessary to undo the dualisms we had no wish to construct but through which we must pass. Arrive at the magic formula we all seek – PLURALISM = MONISM – via all the dualisms that are the enemy, an entirely necessary enemy, the furniture we are constantly rearranging. (Deleuze and Guattari 2004: 22–23)

Along with this brushing off of the readings that reduce their philosophy to another great divide theory,[21] the authors illustrate two characteristic procedures. First, the treatment of concepts in a 'minor' or pragmatic key, as tools or vehicles rather than as ultimate objects, meanings or destinations; the philosopher as *penseur sauvage* – whence the authors' warily pragmatic attitude towards the dualistic propensities of inertial thinking. In *Anti-Oedipus*, they expound a monist conception of desiring production; in *A Thousand Plateaus*, they develop a 'post-plural' theory of multiplicities – two pointedly non-dualistic enterprises. Yet they do not suppose that dualisms are a surmountable obstacle through the sheer power of wishful unthinking, like those who fancy that it is enough to call someone else a dualist to stop being such themselves. Dualisms are real and not imaginary; they are not a mere ideological mirage but the modus operandi of an implacable abstract machine of overcoding. It is necessary to undo dualisms because first of all they were made. Moreover, it is possible to undo them for the same reason: for the authors do not think that dualisms are the event horizon of Western metaphysics, the absolute boundary that can only be exposed – deconstructed – but never crossed by the prisoners in the cave. There are many other possible abstract machines.

This takes us to the second procedure. Deleuzian dualities are constructed and transformed according to a recurrent pattern, which determines them as minimal multiplicities – partial dualities, one might say. Every conceptual distinction begins by the establishment of an extensive actual pole and an intensive virtual one. The subsequent analysis consists in showing how the duality changes its nature as it is taken from the standpoint of one and then the other pole. From the standpoint of the extensive (arborescent, molar, rigid, striated, etc.) pole, the relation that distinguishes it from the second pole is typically an opposition: an exclusive disjunction and a limitative synthesis, that is, an extensive, molar and actual relation itself. From the standpoint of the other (rhizomatic, molecular, supple, smooth) pole, however, there is no opposition but intensive difference, implication or disjunctive inclusion of the extensive pole in the intensive or virtual pole; the duality posed by the first pole reveals itself as the molar echo of a molecular multiplicity at the other pole.[22]

The two poles or aspects are said to be always present and active in every phenomenon or process. Their relation is typically one of 'reciprocal presupposition', a notion many times advanced in *A Thousand Plateaus* (Deleuze and Guattari 2004: 49–50, 73, 97, 235, 554) in lieu of the classic schematisms of causality (linear or dialectical), micro-macro reduction (ontological or epistemological) and expressivity (hylomorphic or signifying). From an anthropological standpoint, it is tempting to relate reciprocal presupposition to the Wagnerian double semiotics of invention and convention, in which each mode of symbolization precipitates or 'counter-invents' the other, according to a figure-ground reversal scheme (Wagner 1981: chapter 3, 1986);[23] or even, to the behaviour of certain central analytical duplexes in *The Gender of the Gift* (Strathern 1988), such as those that preside over the economy of gender or the logic of exchange in Melanesia, in which a pole – cross-sex/same-sex, mediated/unmediated exchange – is always described as a version or transformation of the other, 'each provid[ing] the context and grounding for the other', as Strathern summarized in a quite different (precisely!) context (1991a: 72).[24]

The crucial point here is that reciprocal presupposition determines the two poles of any duality as being equally necessary, since they are mutually conditioning, but does not thereby make them into symmetrical or equivalent poles. Inter-presupposition is a asymmetric relation: 'the way is not the same in both directions'. Hence, as they distinguish the rhizomatic maps from arborescent tracings, Deleuze and Guattari observe that the maps are constantly being totalized, unified and stabilized by the tracings, which are, in turn, subject to all sorts of anarchic deformations induced by rhizomatic processes. Yet, at the end of the day, 'the tracing should always be put back on the map. This operation and the previous one are not at all symmetrical' (Deleuze and Guattari 2004: 14).[25] They are not symmetrical because the latter operation, tracing or *calque*, works contrary to the process of desire (and 'becoming is the process of desire' (ibid.: 334)) whereas the other forwards it. Tracing is dangerous because it 'injects redundancies' into the map, organizing and neutralizing the rhizomatic multiplicity: 'What the tracing reproduces of the map or rhizome are only the impasses, blockages, incipient taproots, or points of structuration … Once a rhizome has been obstructed, it's all over, no desire stirs; for it is always by rhizome that desire moves and produces' (ibid.: 15). This asymmetrical relation between processes and models in reciprocal presupposition (in which the rhizome is process, while the tree is model) reminds one very much of the distinction between difference and negation developed in *Differénce et Répétition* (Deleuze 1968: 302ff.): negation is real but its reality is purely negative; it is only inverted, extended, limited and reduced difference. Thus, although Deleuze and Guattari warn more than once that it is not the case of establishing an axiological contrast between the rhizome and the tree, the molecular and the molar and so on (Deleuze and Guattari 2004: 22, 237), the fact remains that there is always a tendency and a counter-tendency, two entirely different movements: the actualization and the counter-effectuation (or 'crystallization') of the virtual (Deleuze and Guattari 1991: 147–52). The first movement consists in the decay of differences of potential or intensity as these explicate themselves into extension and body forth empirical matters of fact. The second is the creator or 'implicator' of difference as such; it is a movement of return or reverse causality (Deleuze and Guattari 2004: 476), a 'creative involution' (ibid.: 203), but this does not prevent it from being strictly contemporaneous with the first as its transcendental and therefore non-annullable condition. This last movement is the event or the becoming, a pure reserve of intensity – the part, in everything that happens, that escapes its own actualization.

Once again, it seems natural to approximate this asymmetry of inter-implicated processes to certain aspects of Wagnerian semiotics (Wagner 1981: 51–53, 116, 121–22). The 'dialectical' or obviational nature of the relation between the two modes of symbolization belongs as such to one of the modes, the invention-differentiation mode, whereas the contrast between the two modes is, by itself, the result of the other mode's operation, the conventionalization-collectivization one. Moreover, although the two modes operate simultaneously and reciprocally in every act of symbolization (they operate one upon the other, since there is nothing 'outside' them), there is 'all the difference in the world' (ibid.: 51) between those cultures whose controlling context – in the terms of *A Thousand Plateaus*, the dominant form of territorialization – is the conventional mode and those in which the control is the differentiating mode. If the contrast between the modes is not axiological in itself, the culture that favours conventional and collectivizing symbolization – the culture that generated the theory of culture as 'collective representation' – is firmly territorialized on tracing mechanisms, thereby blocking or repressing the dialectics of invention; it must, in the final analysis, 'be put back in the map'. This, according to Wagner, is what anthropologists 'do', or, rather, counter-do.

Similarly, the contrast advanced in *The Gender of the Gift* between gift-based and commodity-based socialities is explicitly assumed as internal to the commodity pole (Strathern 1988: 16, 136, 343), but at the same time it is as if the commodity form were a unilateral transformation of the gift instead of the contrary, in so far as the analysis of a gift-based sociality forces the anthropologist to recognize the contingency of the cultural presuppositions of anthropology itself and to displace its own commodity-based metaphors (ibid.: 309). The point of view of the gift with respect to the commodity is not the same that the point of view of the commodity with respect to the gift: reciprocal asymmetric implication.[26]

II

If there is effectively one implicative asymmetry that can be said to be primary within the Deleuzian conceptual system, it resides in the distinction between the intensive and the extensive. The second part of this chapter discusses the relevance of this distinction to the reinterpretation made in *Capitalism and Schizophrenia* of two key concepts of classical anthropological kinship theory: alliance and filiation. The choice might be seen as requiring justification. I would argue, then, that the treatment given by Deleuze and Guattari to these two notions manifests with especial clarity an important theoretical displacement that occurs between *Anti-Oedipus* and *A Thousand Plateaus*; and, secondly, that it suggests the possibility of a transformation of the anthropology of kinship so as to align it with the 'non-humanist' developments that take place today in other fields of investigation (Jensen 2004). The issue is: how to convert conceptually the notions of alliance and filiation, traditionally seen as the basic social coordinates of hominization, in so far as the latter is effected in and through kinship, into an opening to the extra-human; in other words, how to transform these intra-anthropological operators into cross-ontological ones. If humanity is no longer an essence, what is to be made of kinship?

After having played a quasi-totemic function in anthropology from the 1950s to the 1970s, when they signified two diametrically opposed conceptions of kinship (Dumont 1971), the notions of alliance and filiation, following the general fate of the Morganian paradigm to which they belonged, suddenly lost their synoptic value, receding into the more modest role of analytical conventions, when they did not pass (away) from use to mention.[27] The following pages propose a reflective interruption of this movement, suggesting that some parts of classical kinship theory can be recovered and put back into use. It is certainly not a case of turning back to the *status quo ante* and plunging back into the analytical formalisms of prescriptive alliance, and even less into the substantialist metaphysics of descent group theory, but of imagining the lineaments of a rhizomatic conception of kinship capable of extrapolating all the possible consequences from the premise that 'persons have relations integral to them' (Strathern 1992b: 101). If the theory of descent groups had as its abstract archetype the ideas of substance and identity (the group as a metaphysical Individual), and the theory of marriage alliance the notions of opposition and totalization (society as a dialectical Totality), the perspective suggested here poaches on Deleuzian philosophy in search of some elements for a theory of kinship as difference and multiplicity (the relation as inclusive disjunction).

Against Exchange

Anthropological literature is given pride of place in *Capitalism and Schizophrenia*. From Bachofen and Morgan to Lévi-Strauss and Leach, the first book of the diptych rewrites from scratch the anthropology of kinship. Its main interlocutor and controversial target is Lévi-Strauss' structuralism, for whom and largely against whom are mobilized a number of references, from Malinowski's functionalism to Fortes' structural functionalism, from Griaule and Dieterlen's Dogon experiment to Meillassoux and Terray's ethno-Marxism, from Nuer relational segmentarity to Ndembu social dramaturgy.

The general subversionary and liberating effects of *Anti-Oedipus* in its time cannot be overstated. But it should be regarded as an epoch-making book also from the restricted viewpoint of kinship theory. In its refusal to take the family as the primary referent of desire, defining the latter as immediately social, *Anti-Oedipus* articulated a general philosophical justification (since it applies just as well to the so-called descriptive systems) of the anti-extensionist and anti-genealogist stance defended by many anthropologists then. The argument is still important today, since the popularity of genealogistic or 'genetic' interpretations of human relationality has been on the rise as a result of the universal percolation of neoliberal ontologies. Similarly, extensionist interpretations of semantics remain embedded in the many anthropological theories that still use the explanatory framework of metaphorical projection to account for personification modes deemed 'illegal' in our cosmology.

The thesis of the immediate identity between desiring production and social production fits in with the wider issue of literality in Deleuze's philosophy, or, rather, with his refusal of any distinction between metaphorical and non-metaphorical discourse (Zourabichvili 2004b). In this sense, less than supporting a 'categorial' reading of kinship semantics, in the terms of the classic 'genealogy vs. category' debate, what is at stake in *Anti-Oedipus* is rather a contrast between constitutive-intensive and regulative-extensive conceptions of relationality. A connection to Wagner comes to mind once again. This anthropologist's remarks on the tautological character of the incest prohibition (Wagner 1972) – it is impossible to separate relationships, categories and kinship roles, since these aspects are inter-defined – coincide in an intriguing way with Deleuze and Guattari's arguments over the impossibility of incest:

> [I]f relationship is part of the definition of a category ... then a statement of incest prohibition vis-à-vis categories is the purest and most trivial of tautologies. (Wagner 1972: 603)

> [T]he possibility of incest would require both persons and names – son, sister, mother, brother, father. Now in the incestuous act we can have persons at our disposal, but they lose their names inasmuch as these names are inseparable from the prohibition that proscribes them as partners; or else the names subsist, and designate nothing more than prepersonal intensive states that could just as well 'extend' to other persons ... one can never enjoy the person and the name at the same time. (Deleuze and Guattari 1983: 161–62)[28]

The structuralist conception of kinship is grounded, as is well-known, in the transcendental deduction of the incest prohibition as the condition of sociogenesis (Lévi-Strauss 1969). The authors of *Anti-Oedipus* turn down this conception, in accord with the argument that it is an anthropological generalization of Oedipal thinking. Their criticism of the Freudian reduction of desiring production to Oedipus is thus extended to what could be called the 'Maussian reduction' of social production by Lévi-Strauss, the more eminent contemporary propagator of 'exchangist notions of society' (Deleuze and Guattari 1983: 142, 185). Deleuze and Guattari compare disadvantageously *The Gift* with Nietzsche's *On the Genealogy of Morals*; the latter, they suggest, should be the anthropologist's true bedside book (Deleuze and Guattari 1983: 190; see also the introduction to this volume).[29]

The anti-Oedipal reconstruction of kinship theory made in the first volume of *Capitalism and Schizophrenia* is, however, interestingly partial or incomplete. It remains riveted to a 'humanist' or anthropocentric conception of sociality, and is haunted by the empirico-metaphysical problem of hominization. The blind spots of this focus only show, of course, from the radically an-Oedipal vantage point of *A Thousand Plateaus*, published one decade later. The first book was intended to be a critique of both psychoanalysis and Oedipus; the vocabulary is almost parodically Kantian: transcendental illusions, illegimate usages of the syntheses of the unconscious, the four paralogisms of Oedipus, and so on. It could thus be argued – somewhat mischievously – that, by imagining itself as a sort of critique of psychoanalytic reason, *Anti-Oedipus* remains, in a fundamental philosophical sense, an Oedipal book, and, worse, dialectically so.[30]

It is this limitation that might explain the systematic interpretation of alliance as a sort of superconductor for the Oedipal triangle, an argument that poses parenthood as prior to conjugality (the first 'extends into' the latter) and alliance as something merely instrumental to the deployment of filiation (Deleuze and Guattari 1983: 71). In other words, the criticism of all 'exchangist conceptions' expressed by *Anti-Oedipus* is grounded on a counter-theory of Oedipus in which filiation and production, rather than alliance and exchange, are primary. In this and other senses, *Anti-Oedipus* is an anti-structuralist book. Yet, if its authors keep their distance from the Lévi-Straussian *conception* of human kinship, it is necessary first that they accept some of the terms in which the anthropological *theory* of kinship was formulated by Lévi-Strauss. And it is exactly this that changes, from *Anti-Oedipus* to *A Thousand Plateaus*.

Intensive Filiation

Against the theme of exchange as the socio-institutive synthesis of opposing (contradictory) interests, *Anti-Oedipus* puts forth the postulate that the social machine works paramountly to code the flows of desire. Social production is desiring production in a coded state. Deleuze and Guattari propose an inscriptive conception of society (1983: 184) – the task of the *socius* is the marking of bodies (memory creation) – on the one hand, and a productionist cosmology, on the other: 'everything is production' (ibid.: 4). In good *Grundrisse* style, production, distribution and consumption are defined as different moments of production seen as a universal process. Inscription is a moment of this production, the moment of the recording or coding of production, which counter-effects a fetishized *socius* as the form of the natural or divine given, a magical surface of inscription and an element of anti-production (the Body without Organs).

In the third chapter of *Anti-Oedipus*, the authors engage in a detailed exposition of the primitive territorial machine and analyse its characteristic 'declension of alliance and filiation' (Deleuze and Guattari 1983: 171). The basic hypothesis and crucial analytical decision consist in positing that filiation is doubly inflected by the primitive machine: first, as a generic and intensive state of proto-kinship; and, secondly, as a particular and extensive state in complementary opposition to alliance, the latter appearing exclusively on this extensive plane. What is more, alliance appears in order to (as it were) accomplish the task of extending and coding kinship, that is, of actualizing it. Deleuze and Guattari postulate thus the existence of a filiation prior to incest as prohibition; a nocturnal and biocosmic, disjunctive and ambiguous filiation, a germinal implex or influx (ibid.: 162) that is the first character of inscription marked on the full, unengendered body of the earth: 'a pure force of filiation or genealogy, Numen' (ibid.: 154).[31]

This analysis here relies centrally on an interpretation of the mythical narratives collected by Marcel Griaule and his team, especially that of the well-known Dogon origin myth in *Le Renard pâle* (Griaule and Dieterlen 1965): Amma the cosmic egg, the placentary Earth, Yurugu the incestuous trickster, the hermaphroditic successive Nommo twin pairs and so on.[32] The narrative works as a kind of anti-myth of Oedipus for Deleuze and Guattari.[33]

It is hardly surprising that this reference myth – a cosmogonic story quite widespread in West Africa (Adler and Cartry 1971: 15) – determines filiation as the original element, and alliance as a supervening dimension whose function is to differentiate lineage affiliations. We are here – as it were? – within a classical 'Africanist' universe of discourse (Fortes 1969, 1983). What is intense and primordial are the ambiguous, involved, implicated and (pre-)incestuous filiative lineages, which abandon their inclusive and illimitable regime as (being the subject of a 'nocturnal and biocosmic' memory) they must 'suffer repression' by alliance in order to be explicated or actualized in the physical space of society (Deleuze and Guattari 1983: 155).

It is as if the system of the Dogon, who are the synecdochic savages at this point of *Anti-Oedipus,* was descent-theoretical at the virtual or intensive level and alliance-theoretical at the actual or extensive level. Thus, the authors take totally on board Leach's criticism of Fortes concerning 'complementary filiation'. They also reach the conclusion – from a famous Lévi-Straussian demonstration of the logic of cross-cousin marriage (1969: 129–33) – that '[a]t no time ... does alliance derive from filiation', and that 'in this system in extension there is no primary filiation, nor is there a first generation or an initial exchange, but there are always and already alliances' (Deleuze and Guattari 1983: 157).[34] In the extensive order, filiation assumes a secondary 'administrative and hierarchic' trait, whereas alliance, which comes first in this order, is 'political and economic' (ibid.: 146). The affine as a socio-political persona is there from the start ('always and already') to prevent any Oedipal closing to the *socius* of families, that is, to make sure that familial relations are radically coextensive with the wider social field (ibid.: 166). Yet there is something before the start: in the metaphysical order of genesis – from the mythical standpoint, precisely (ibid.: 155) – alliance comes second. 'The system in extension is born of the intensive conditions that make it possible, but it reacts on them, cancels them, represses them, and allows them no more than a mythical expression' (ibid.: 160).[35] Post-prohibition kinship is, therefore, conceived in terms of a reciprocal pressuposition between alliance and filiation that is actually (politico-economically) ruled by the former and virtually (mythically) by the latter.

The intensive level of myth is thus peopled by (pre-)incestuous filiations that ignore alliance. It is impossible not to remember here the famous last paragraph of *The Elementary Structures of Kinship*, in which it is written that 'mankind has always dreamed of seizing and fixing that fleeting moment when it was permissible to believe that the law of exchange could be evaded' (Lévi-Strauss 1969: 497). Compare, however, this somewhat Freudian conclusion with that other oft-quoted passage in which the author defines myth as 'a story from the time when humans and animals did not distinguish themselves from one another' (Lévi-Strauss and Eribon 1988: 193), adding that the one thing mankind has never reconciled itself to is the lack of communication with the other species that people the earth. Not the same nostalgia as the former one, then; in a sense, its very opposite.

Reformulating the problem in the terms of the Deleuzian general conceptual economy, it seems to me that the crucial step in this analysis of the Dogon myth is the determination of intensive filiation as an operator of disjunctive synthesis – the Nommo who is/are one and two, male and female, human and snake; the Fox who is son, brother and husband of the Earth – whereas alliance is the operator of conjunctive synthesis or pairing:

> Such is alliance, the second characteristic of inscription: alliance imposes on the productive connections the extensive form of a pairing of persons, compatible with the disjunctions of inscription, but inversely reacts on inscription by determining an exclusive and restrictive use of these same disjunctions. It is therefore inevitable that alliance be mythically represented as supervening at a certain moment in the filiative lines (although in another sense it is already there from time immemorial). (Deleuze and Guattari 1983: 155)

As we know, disjunctive synthesis is the diagnostic relational regime of multiplicities. On the same page of the passage just cited, we can read that the problem is not one of going from filiations to alliances, but of moving from 'an intensive energetic order to an extensive system'. In this sense, '[n]othing is changed by the fact that the primary energy of the intensive order ... is an energy of filiation, and does not as yet comprise any distinctions of persons, nor ... of sexes, but only prepersonal variations in intensity' (ibid.: 155).

Here, one might wish to add that, if this intensive order does not know distinctions either of person or of gender, it does not know either any distinction of species, particularly that between humans and non-humans. In myth all actants are deployed on a single interactional field, which is at one and the same time ontologically heterogeneous and sociologically continuous. There are no humans 'there'; or, rather, everything is 'human' there. And, of course, there where everything is human, the human is something else *entirely* (Viveiros de Castro 2004: 16).

We are now in possession of all the elements of the problem: if 'nothing is changed by the fact that the primary energy be an energy of filiation', could it be possible to imagine an intensive order in which the primary energy be an energy of alliance? Is it really necessary that alliance work always to ordain, discern, separate and police a prior, pre-incestuous order of filiation? Or could it conceivably be an intense, an-Oedipal alliance that comprises only 'prepersonal variations in intensity'? In a few words, the problem is that of imagining a concept of alliance as a disjunctive synthesis.

In order to do it, we would probably have to take a greater distance from the Lévi-Straussian kinship cosmology than *Anti-Oedipus* does, at the same time that the exchange concept must be submitted to a properly Deleuzian, or 'perverse', interpretation.[36] Minimally, it means abandoning the description of the kinship atom in terms of an exclusive alternative – this woman as either my sister or my wife, that man as either my brother or my brother-in-law – and rephrase it in terms of an inclusive or non-restrictive disjunction: 'either … or … or'. The difference between sister and wife, brother and brother-in-law must be taken as an internal difference, 'indecomposable and unequal to itself'. Just like with the schizophrenic and the male/female, dead/alive disjunctions which s/he confronts, we might say that this woman is indeed my sister or my wife, but she 'belongs precisely to both sides', sister on the side of sisters (and brothers) and wife on the side of wives (and husbands) – not both at once to me, 'but each of the two as the terminal point of a distance over which [s]he glides' (Deleuze and Guattari 1983: 76).

The point can be rephrased in a language that every anthropologist will recognize (Strathern 1988, 2001). My sister is my sister in so far as she is the wife of another, and vice versa. It is the cross-sex relationship of my sister/wife to myself that generates my same-sex relationship with my brother-in-law. Hence, cross-sex relations not only engender same-sex relations but also communicate their own internal differential potential to the latter. Two brothers-in-law are related in the same way as the cross-sex dyads presupposed by their relation (brother/sister, husband/wife): not despite their difference but because of it. One of the brothers-in-law sees the conjugal face of his sister in her husband; the other sees his wife's sororal side in her brother. One sees the other as determined by the opposite-sex link that differentiates both of them: each sees him/herself as 'same-sex' in so far as the other is seen 'as' cross-sex, and reciprocally. The two sides of the relating term thus create a division that 'always and already' is internal to the related terms. All become double, the relater and the related reveal themselves to be interchangeable without thereby becoming redundant; each vertex of the affinity triangle (two triangles, actually, one for each sex as 'relator') includes the other two vertices as versions of itself.[37] This complex duplication is explicitly described by Deleuze and Guattari in a commentary on the analogy between homosexuality and vegetal reproduction in Proust's *Sodom and Gomorrah*. Something like an atom of gender is suggested:

> [T]he vegetal theme … brings us yet another message and another code: everyone is bisexual, everyone has two sexes, but partitioned, noncommunicating: the man is merely the one in whom the male part, and the woman the one in

> whom the female part, dominates statistically. So that at the level of elementary combinations, at least two men and two women must be made to intervene to constitute the multiplicity in which transverse communications are established ... the male part of a man can communicate with the female part of a woman, but also with the male part of a woman, or with the female part of another man, or yet again with the male part of the other man, etc. (Deleuze and Guattari 1983: 69–70)

Two men and two women at least. Make it a 'sibling-exchange' matrimonial arrangement between two pairs of opposite-sex siblings (or two bisexual dividuals), and there you have it: an extensive, structural version of the intensive rhizomatic multiplicity that is gender. But then, as a matter of course, the tracing must be put back on the map, and 'everything must be interpreted in intensity' (ibid.: 158). This is the job the little 'etc.' at the end of the passage above may be conceived as doing.

Demonic Alliance

The possibility of an intensive interpretation of alliance is effectively consolidated in *A Thousand Plateaus*. Although many things change from *Anti-Oedipus* to *A Thousand Plateaus*, the single most important change from the limited viewpoint of this chapter is introduced in the tenth plateau, '1730: Becoming-intense, becoming-animal, becoming-imperceptible.' It is there that the concept of becoming is developed, a development that carries away all the Deleuzo-Guattarian conceptuality in an astonishing other-becoming.[38]

The chapter begins with an exposition of the contrast established by Lévi-Strauss (1962b) between serial-sacrificial logic and structural-totemic logic: the imaginary identification between human and animal, on the one hand, and the symbolic correlation between social differences and natural differences, on the other. Between these two analogical models, the series and the structure, Deleuze and Guattari introduce the Bergsonian theme of becoming, a type of relation irreducible to serial resemblances as well as to structural correspondences. The concept of becoming designates a relation that has no right of citizenship within classical structuralism's theoretical frame, which tends to treat relations as molar logical objects, apprehended essentially in extension, as oppositions, contradictions and mediations. Becoming is a real relation, molecular and intensive, belonging to a truly relational ontology beyond the pale of the more limited epistemological relationality of structuralism.[39] The disjunctive synthesis of becoming is not possible in the terms of the formal combinatorial rules that engender structures; it is born in the far-from-equilibrium fields of real multiplicities (de Landa 2002: 75). 'Becoming and multiplicity are the same thing' (Deleuze and Guattari 2004: 275).

If serial resemblances are imaginary and structural correspondences are symbolic,[40] becomings are real. Neither metaphor nor metamorphosis, becoming is a movement that deterritorializes both terms of the relation, extracting them from their prior definitional contexts to associate them through a new, necessarily partial connection. The verb to become, in this conceptual sense, does not denote a predicative judgement or a transitive operation: to be implicated in a becoming-jaguar is not the same as 'becoming a jaguar'. It is the becoming itself that is feline, not its 'subject'. For as soon as a man becomes a jaguar, the jaguar is no longer there. '[I]n his study of myths, Lévi-Strauss is always encountering these rapid acts by which a human becomes animal at the same time as the animal becomes . . . (Becomes what? Human, or something else?)' (Deleuze and Guattari 2004: 262).[41] To become, the authors proceed, is a verb with a consistency all its own; becoming is not imitating, appearing, being, corresponding. And – surprise – 'neither is it producing, producing a filiation or producing through filiation' (ibid.: 263). Neither filiation nor production. We are not in *Anti-Oedipus* any more.

'Intensive thinking in general is about production,' remarks de Landa (2003: 15). Perhaps it is a little more complicated than that. Becoming does play the same axial, cosmological role in *A Thousand Plateaus* that production does in *Anti-Oedipus*. Not exactly because 'everything is becoming' – that would be a conceptual solecism – or because there are no other important notions in the book (war machines, segmentarity, ritornello, regime of signs, assemblage: an embarrassment of riches), but because *A Thousand Plateaus*'s anti-representational concept par excellence, in the sense of being the device that pre-empts the work of representation, is the concept of becoming, just as production was *Anti-Oedipus*'s major anti-representational concept. Two distinct movements, therefore, whose relation remains to be more precisely determined: production and becoming. Both involve nature; both are intensive and pre-representative; in a sense, they are the same: becoming is the process of desire, desire is the production of the real, becoming and multiplicity are the same thing, becoming is a rhizome, and the rhizome is the process of production of the unconscious. But in another sense – 'sense' also in the sense covered by the French *sens*: direction – they are definitely not the same: between production and becoming, 'the way is not the same in both directions', *les deux sens*. Production is a process in which the identity of man and nature is realized, and nature reveals itself as a historical process of production ('the human essence of nature and the natural essence of man become one within nature in the form of production' (Deleuze and Guattari 1983: 4)). Becoming, on the contrary, is an unnatural participation (Deleuze and Guattari 2004: 265) between man and nature, an instantaneous ('these rapid acts') or *non-processual* movement of capture, symbiosis, transversal connection between heterogeneities. 'That is the only way Nature operates – against itself. This is a far cry from filiative production or hereditary reproduction' (ibid.: 267). Becoming is counterproductive.

'The Universe does not function by filiation' (ibid.). One could hardly be more direct. The universe, one might note, in all its states, both virtual-intensive and actual-extensive. And, if it does not work by filiation (rather than not by something else), I am tempted to conclude that it can only work by alliance. In the book's first plateau, as a matter of fact, we had already learnt that 'the tree is filiation, but the rhizome is alliance, uniquely alliance' (ibid.: 27). And now we learn that 'becoming is not an evolution, at least not an evolution by descent and filiation. Becoming produces nothing by filiation; all filiation is imaginary. Becoming is always of a different order than filiation. It concerns alliance ... Becoming is a rhizome' (Deleuze and Guattari 2004: 263)

What has changed, from the affirmative analysis of that intensive, ambiguous, nocturnal filiation of the Dogon myth in *Anti-Oedipus* to the flat denial of any significant role played by this relational principle in *A Thousand Plateaus*? How did filiation turn from intensive to imaginary? I think that this change reflects the displacement of the analytical focus from an intraspecific horizon to an interspecific one; from a human economy of desire – world-historical, racial, socio-political rather than familial or Oedipal desire, but human nonetheless – to a trans-specific economy of affects, which ignores the natural order of genera and species and its limitative syntheses, disjunctively including 'us' (all and any of 'us') in the plane of immanence. From the standpoint of the human economy of desire, extensive alliance works to limit intensive and molecular filiation, actualizing it under the molar form of the descent group. But, from the standpoint of the cosmic economy of affects (of desire as an unhuman force), it is filiation that comes now to limit, with its imaginary identifications, an alliance as real as it is unnatural between radically heterogeneous beings: 'If evolution includes any veritable becomings, it is in the domain of symbioses that bring into play beings of totally different scales and kingdoms' (Deleuze and Guattari 2004: 263). Then follows Deleuze's favourite example of the wasp and the orchid, an assemblage 'from which no wasp-orchid can ever descend' – and without which, one might add, no wasp and no orchid as we know them could ever leave descendants. The wasp-orchid assemblage is the origin myth, as it were, of both the wasp and the orchid tribes: their shared deterritorialization, their common anti-memory (ibid.: 324).

The deterritorialisation of sexuality begun in *Anti-Oedipus* is now completed; the binary organization of the sexes, bisexuality included (cf. the atom of gender), gives way to *n* sexes that connect to *n* species at the molecular level: 'sexuality proceeds by way of the becoming-woman of man and the becoming-animal of the human: an emission of particles' (ibid.: 307).[42] If every animal implicated in a becoming-animal is a molecular multiplicity ('every animal is fundamentally a band, a pack' (ibid.: 264)), it is because it defines a multiple, transversal, extra-filiative and non-reproductive sociality that drags human sociality along in a universal metonymic flow: 'We oppose epidemic to filiation, contagion to heredity, peopling by contagion to sexual reproduction, sexual production... Unnatural participations or nuptials are the true Nature spanning the kingdoms of nature' (ibid.: 266).

Alliance, then. But, then again, not any kind of alliance. As we have seen, *Capitalism and Schizophrenia*'s first volume postulated two kinds of filiations: one, intensive and germinal; the other, extensive and somatic, posed by and counterposed to alliance seen as an extensive principle playing the role of 'repressing representation' of the representative of desire, the germinal influx.[43] Now we can see two alliances appearing:

that discussed in *Anti-Oedipus*, internal to the *socius* and even to the male gender (primary, collective homosexuality: affinal alliance as the archetypal same-sex relation *sensu* Strathern); and another alliance, intrinsic to becoming, as irreducible to imaginary production and metamorphosis (mythical genealogy, filiation to the animal) as to exchange and symbolic classifications (exogamy, totemism).

Every becoming is an alliance, which does not mean, I repeat, that every alliance is a becoming. There is extensive alliance, which is cultural and socio-political, and intensive alliance, which is unnatural and cosmopolitical. If the former distinguishes filiations, the latter confuses species, or, rather, counter-effectuates by implicative synthesis the continuous differences that 'had been' actualized by the limitative synthesis of discontinuous speciation. When a shaman activates a becoming-jaguar, he does not 'produce' a jaguar, he does not 'affiliate' himself to the descent of jaguars either; he makes an alliance: 'One can say rather that a zone of indistinction, of indiscernibility, of ambiguity, establishes itself between two terms, as if they had attained the point immediately preceding their respective differentiation: not a similitude, but a slippage, an extreme vicinity, an absolute contiguity; not a natural filiation, but a counter-natural alliance' (Deleuze 1993: 100).

One should note that this concise description of becoming cuts right through the middle of the molar contrast between filiation, metonymic continuity and serial resemblance, on the one hand, and alliance, metaphorical discontinuity and oppositive difference, on the other – a contrast characteristic of structuralist theories of kinship. That 'absolute contiguity' established by counter-natural alliance, of a differential, tangential kind, differs absolutely from the absolute 'discontiguity' between filiative lineages established by cultural alliance (exogamy). But it is irreducible as well, it goes without saying, to any identification or imaginary indifferentiation between 'the two terms'. Therefore, it is not the case of opposing natural filiation to cultural alliance, as in the classical alliance-theoretical models. The counter-naturalness of intensive alliance is countercultural or counter-social as well, in so far as human sociality is necessarily counter-intensive, being generated through the extensivization of 'the primary energy of the intensive order'. We are speaking of an included middle, a new alliance, a relation that separates.[44]

It is not necessary to leave Africanist ground to find this other alliance. In the section entitled 'Memories of a Sorcerer II', in the tenth Plateau, the authors evoke the demonic were-animals of the 'sacred deflowerer type' studied by Pierre Gordon, and the were-hyenas of some Sudanese traditions described by Calame-Griaule. The latter furnish the occasion for a decisive commentary:

> [T]he hyena-man lives on the fringes of the village, or between two villages, and can keep a lookout on both directions. A hero, or even two heroes with a fiancée in each other's village, triumphs over the man-animal. It is as though it were necessary to distinguish two very different states of alliance: a demonic alliance that imposes itself from without, and imposes its laws upon all the filiations

> (a forced alliance with the monster, with the man-animal), and a consensual alliance, which is on the contrary in conformity with the law of filiations and is established after the men of the villages have defeated the monster and have organized their own relations. This sheds new light on the question of incest. For it is not enough to say that the prohibition against incest results from the positive requirements of alliance in general. There is instead a kind of alliance that is so foreign and so hostile to filiation that it necessarily takes the position of incest (the man-animal always has a relation to incest). The second kind of alliance prohibits incest because it can subordinate itself to the rights of filiation only by lodging itself, precisely, between two distinct filiations. Incest appears twice, once as a monstrous power of alliance when alliance overturns filiation, and again as a prohibited power of filiation when filiation subordinates alliance and must distribute it among distinct lineages. (Deleuze and Guattari 2004: 597–98)

'This sheds new light on the question of incest.' The authors seem to be referring to the theory of *The Elementary Structures*, but I believe this observation is also apposite to the manner the issue is dealt with in *Anti-Oedipus*. Now, it is alliance that has a double incidence, not only to regulate 'sexuality as a process of filiation', but also as 'a power of alliance, inspiring illicit unions or abominable loves'. Its purpose is not that of regulating, but of 'preventing procreation': an anti-filiative alliance (Deleuze and Guattari 2004: 271). And, more importantly still, even exchangist, repressive, filiation-producing alliance begins to show some hidden powers – as if it had been conceptually contaminated by that other, intense alliance. 'It is true that alliance and filiation come to be regulated by laws of marriage, but even then alliance retains a dangerous and contagious power. Leach was able to demonstrate . . . ' (ibid.: 272).[45] It is interesting to notice how the word 'power' (orig. '*puissance*', not '*pouvoir*') begins persistently to qualify and determine the notion of 'alliance' from this point onwards. The concept of alliance ceases designating an institution – a structure – and becomes a power, a potential – a becoming. From alliance as form to alliance as force, bypassing filiation as substance. We are no longer in the structural-mythical element of totemism, but neither are we in the serial-mystical element of sacrifice; we are in the real-magical element of becoming (Goldman 2005; Viveiros de Castro 2009).

We are not in the element of the contract, either – of exchange or interest '*à l'anglaise*' (Deleuze and Guattari 1983: 190). 'Desire knows nothing of exchange, it knows only theft and gift' (ibid.: 186). But, then again, there may be exchange and exchange. There is an exchange that is certainly not 'exchangist' in the commodity-economic sense, since it belongs to the 'theft and gift' category: debt or gift exchange precisely, a movement of double capture in which people shift (counter-alienate) invisible perspectives by moving visible personified (inalienable) things.[46] Gifts may be reciprocal; but this does not make them any the less violent: the whole point of the gift act is to make the partner act, to extract an act from the other, to provoke a response. And in this sense there is no social action that is not an exchange of 'gifts', in so far as all action is 'social' by being an action upon another action, a reaction to a reaction. Reciprocity means only recursivity; no society intended. Not to mention altruism.[47] Life is robbery.

Amazonian Alliance

The distinction between the two alliances proposed in *A Thousand Plateaus* seems to impose itself, from an ethno-theoretical point of view, when we move over from a West African(ist) to an Amazonian(ist) landscape. It corresponds closely to the distinction made by the ethnographers of this region between an intensive or potential affinity, of which one could certainly state that it is 'ambiguous, disjunctive, nocturnal, demonic', and an extensive or effective affinity, subordinated to consanguinity (Viveiros de Castro 2001, 2009). In the context of local and cognatic 'prescriptive endogamy' prevailing in many Amazonian societies, affinity as a particular relationship is masked or neutralized by consanguinity (or filiation). Terminological affines are seen as types of cognates (namely, close cross-kin), actual affines are attitudinally consanguinized, specifically affinal terms are avoided in favour of their cross-kin alternatives or of teknonyms expressing co-consanguinity, spouses are conceived as becoming consubstantial by way of sex and commensality, and so forth. We can then say that affinity as a particular relation is eclipsed by consanguinity as part of the process of making kinship. As Rivière observed (1984: 70), 'within the ideal settlement affinity does not exist'. But this may be taken as implying that, if affinity does not exist within the ideal community, it must then exist somewhere else. Within real settlements, to be sure; but, above all, outside the ideal settlement, that is, in the ideal outside of the settlement, as 'ideal' (intensive) affinity. For as the perspective shifts from local relationships to wider contexts the value distribution is inverted, and affinity becomes the overall mode of sociality.

Marriage alliance is, then, both locally concentrated and masked in Amazonian socialities. Supra-local relations, on the other hand, are a variable mixture: statistically residual but politically strategic intermarriage; formal friendship and trade-partnership links; intercommunal ritual and feasting; physical and spiritual, actual or latent predatory warfare. This relational complex straddles species boundaries: animals, plants, spirits and divinities are equally engaged in such synthetic-disjunctive relations with humans. All these relationships, whatever their components, manifest the same general set of values and dispositions, as witness the common idiom in which they are expressed, that of affinity. Guests and friends as much as foreigners and enemies, political allies or clients as much as trade partners or ritual associates, animals as much as spirits, all these kinds of beings bathe, so to speak, in affinity. They are conceived either as generic affines or as marked versions – sometimes inversions – of affines.[48] The Other is first and foremost an affine.[49]

This affinity undoubtedly belongs to the 'second kind of alliance' mentioned by Deleuze and Guattari. It is hostile to filiation, since it appears above all when and where marriage is not an option (or at least not a preference), and its productivity is not of a kinship-procreative kind; it is, rather, part of a war-machine which is anterior and exterior to kinship as such. This is an alliance against filiation, not in the sense of being a repressing representation of a prior intensive filiation, but in that of being what prevents filiation from functioning as the seed of transcendence (the descent group, the origin, the ancestor). Every filiation is imaginary, say the authors of *A Thousand Plateaus.* We might add: every filiation projects a state, it is state filiation. Amazonian intensive alliance is an alliance against the state (Clastres 1977).

The 'affinity first' configuration – that is, a configuration where the 'second kind of alliance' is the condition of possibility of extensive kinship (of marriage alliance as well as parental filiation) – although perhaps more easily apprehended in the cognatic-endogamous morphologies found among northern Amazonian peoples like the Trio, the Piaroa, the Jívaro or the Yanomami, is not restricted to them. It is present, under different guises, everywhere in Amazonia and central Brazil, including those societies that feature descent constructs, house-like corporations, local exogamy rules, Crow-Omaha terminologies, semi-complex alliance systems and whatnot. I believe it is one of the telltale signs of the unity of Amazonian cultures, and probably beyond; for we might be touching here 'the bedrock' (Lévi-Strauss 1991: 295) of Amerindian cosmology as a coherent historical entity.

Consider the continent-wide mythological complex analysed in the *Mythologiques,* whose subject matter is the origin of human culture. If we compare the Amerindian myths with our own mythology of cultural origins, the first difference that strikes the eye is the dominance of relations of affinity in the former and that of parenthood in the latter. The central figures of the Amerindian myths of culture are canonically related as affines. One conspicuous character of this mythology, to take an example, is the cannibal father-in-law, the non-human master of things cultural, who subjects his son-in-law to a series of ordeals before the latter succeds in returning to his fellow-humans with the precious bounty of culture. The content of this archi-myth (Lévi-Strauss 1971: 503ff.) is not too different from the Promethean plot: there is heaven and earth, and a hero stranded in between the two, civilizational fire, the 'gift' of women and human mortality. But the protagonists of the Amerindian myth are wife-givers and wife-takers, not parental and filial figures like those haunting our mythologies, be they Greek, Near Eastern, West African or Freudian. Not to put too fine a point on it, in the Old World humans had to steal 'fire' from a divine father, while Amerindians had to steal it from an animal wife-giver, or received it as a marriage counter-gift from an animal wife-taker.

Mythology may be said to constitute the discourse of the given (see Wagner 1978). It originally grants what must thenceforth be taken for granted, the primordial conditions from/against which humans must define (construct) themselves: it sets the terms of the ontological debt. If such is the case, then in Amerindian thought the debt of the Given does not concern filiation and parenthood, but marriage and affinity; the Other, as we saw, is first and foremost an affine. Please note that I am not referring to the trivial fact that the myths treat affinal relations as 'already there' – they do the same to consanguineal relations, or they may imagine worlds in which pre-humans did not care for the incest taboo, etc.[50] – but to the fact that affinity constitutes the frame, the sociological armature in which the mythic message is couched. Such a frame or setting is always peopled by more than one kind of people; in particular, it is filled with animal affines. It is absolutely essential that they be animal, or, more generally, non-human (future non-humans, that is; in myth everyone is partially human, present humans included, though the way is not the same in both directions).

It is this alliance with the non-human that defines the 'intensive conditions of the system' in Amazonia. At this level there is not, strictly speaking, a distinction – necessarily extensive – between alliance and filiation. Or, rather, if there are two alliances, then there are two filiations as well. If all production is filiative, not all filiation is (re-)productive; if there are reproductive and administrative (representative and state-like) filiations, there are contagious and monstrous filiations as well, those resulting from unnatural alliances or becomings. That is why incest has an intrinsic 'affinity' to trans-specific unions: hyper-exogamy and hyper-endogamy flow into each other in the intensive world of myth, the conditioning world of fluent difference that accompanies the actual world as its virtual counterpart.[51]

Production and Exchange

A steadfast refusal of any 'exchangist' conception of the *socius* pervades the two *Capitalism and Schizophrenia* books. In *Anti-Oedipus*, exchange is rejected in favour of production as the general 'action' concept, and circulation (to which Maussian exchange is univocally assimilated[52]) is subordinated to inscription. In *A Thousand Plateaus*, as we have seen, production makes room for a different non-representational relation, becoming. If production is inherently filiative, becoming shows an affinity with alliance. What happens, then, to the anti-exchangist stance when we pass from production to becoming?

It is well known though sometimes conveniently forgotten that the concept of production in *Anti-Oedipus* is not exactly identical to the homonym Marxist concept. Deleuze and Guattari's desiring production or functioning production cannot be confused with the Hegelian/Marxist, Oedipal-oriented production, dominated by the ideas of need and lack (Deleuze and Guattari 1983: 26ff.).[53] The latter concept of production has wide currency in anthropological circles, and it is for its sake and its convenient appurtenances (domination, false consciousness, etc.) that, as a rule, 'exchangist' positions are criticized within the discipline. Besides, *Anti-Oedipus*' desiring production is not that easily distinguishable from a process of generalized circulation; as Lyotard (1977: 15) provocatively argued, '[t]his configuration of *Kapital*, the circulation of flows, is imposed by the predominance of the point of view of circulation over that of production'.

If it is possible, and indeed necessary, to make this disambiguation between necessitated production and desiring production, between labour-production and machinic production, it can be argued by analogy that it is equally pertinent to distinguish between alliance as structure and alliance as becoming. The distinction undermines the contractualist concept of alliance, playing with a deliberately equivocal homonymy between the intensive alliance of Amazonian socio-cosmology and the extensive alliance of classical kinship theory. Naturally, in both cases the equivocation is a little more than that, since there is a filiation, albeit monstrous rather than reproductive, between the respective homonym concepts. *Anti-Oedipus*'s production owes much to the production of political economy, even as it subverts it. Similarly, the potential Amazonian alliance exists in filigree or in counter-light (virtually, as it were) in the Lévi-Straussian kinship-theoretical corpus, whose anti-Oedipal and, therefore, (self-)subversive potential must be brought about.

The problem is ultimately that of imagining a non-contractualist *and* non-dialectical concept of exchange: neither rational interest nor the a priori synthesis of the gift, neither unconscious teleology nor the work of the signifier, neither inclusive fitness nor desire for the Other's desire, neither 'contract' nor 'conflict' – but a mode of becoming-other.[54] Alliance is the mode of becoming-other proper to kinship. The machinic and rhizomatic laterality of alliance is ultimately much closer to Deleuzian philosophy than the organic and arborescent hierarchic verticality of filiation. Therefore, the question is that of freeing alliance from the managerial control of (and by) filiation, thus liberating its 'monstrous', that is, creative, potencies. As for exchange, it is now tolerably clear that this manifestation of relationality has never been a true conceptual antagonist of production, notwithstanding the prevailing dogma. On the contrary, today we are in a position to see that exchange has always been treated by anthropology as the most eminent form of production: production of society, precisely. The question then is not one of revealing the naked truth of production behind the hypocritical veil of exchange, but rather that of liberating the latter concept from any equivocal function within the repressive machine of filiative and subjectifying production, giving it back to its proper element, the element of becoming: exchange, or the circulation of perspectives: exchange of exchange, that is, change.

Notes

1. And 'the robber requires justification'. The quote was plundered from Isabelle Stengers' brilliant study of Whitehead (Stengers 2002: 349).
2. The reader certainly knows to what a small fraction of humankind the pronoun 'our' (cultural apperception) refers.
3. 'Allo-anthropology', 'hetero-anthropology' – or perhaps 'cross-anthropology'. The reader is invited to take the redundancy as a provocation.
4. If not the century itself, as Foucault famously predicted, it is at least structuralism that seems to be turning Deleuzian of late (which is not the same as saying that Deleuze is turning structuralist). See,

for example, the recent re-examination of Saussure's ideas by Maniglier (2006), which modify dramatically the traditional image of the ancestor.

5. In *Anti-Oedipus*, 'the reversal of psychoanalysis [is] the primary condition for a shake-up of a completely different scope ... on the scale of the whole of the human sciences, there is an attempt at subversion on the general order of what Laing and Cooper had carried out solely on the terrain of psychiatry' (Donzelot 1977: 27).
6. 'Infernal dichotomies' in an analogous, if not identical, sense to that of the 'infernal alternatives' exposed by Pignarre and Stengers (2005: chapter 4).
7. I say meta-concept because every concept is a multiplicity in its own right, though not every multiplicity is conceptual (Deleuze and Guattari 1991: 21ff.).
8. De Landa (2002: 9–10, 38–40 and *passim*) is a detailed exposition of the mathematical origins and implications of the Deleuzian concept of multiplicity (also evoked in Plotnitsky 2003). Zourabichvili (2003: 51–54), in turn, is the best overview of the concept's properly philosophical connections and its place in Deleuze's work.
9. In remembrance of Pierre Clastres (1977). Clastres was (and remains) one of the rare French anthropologists who knew how to make something out of *Anti-Oedipus*'s ideas, besides being one of the inspirations for the theory of the war machine developed on Plateaus 12 and 13 in *A Thousand Plateaus*.
10. A multiplicity or a rhizome is a *system*, one must notice, and not a sum of 'fragments'. It is simply another concept of system, which differs from the arborescent system as an immanent process differs from a transcendent model (Deleuze and Guattari 2004: 22). We are not talking postmodernism here.
11. Or, rather, where the whole is only one part among parts. 'We believe only in totalities that are peripheral. The whole not only coexists with all the parts; it is contiguous to them, it exists as a product that is produced apart from them and yet at the same time is related to them' (Deleuze and Guattari 1983: 42, 43–44). About the heterogeneity of the elements connected in a rhizome, it is important to notice that it does not concern a previous ontological condition, or essence of the terms (what counts as heterogeneous, in this sense, depends on the observer's 'cultural predispositions' – Strathern 1996: 525), but an effect of its capture by a multiplicity, which renders heterogeneous the terms that it connects by making them operate as singularities, 'representatives of themselves'.
12. The 'one man and many men' of Strathern (1991b) is not without evoking an n – 1 multiplicity: '(m)any-minus-one men'. If I understand her correctly, in Melanesian aesthetics the unity of a plurality comes from itself (it is a 'moment' of itself), not from an exterior/superior principle. Alternatively, one could imagine a '1 – n multiplicity', since, to take another famous Melanesian example, 'the greatest possible number that the Iqwaye counting system can and does reach is one' (Mimica 1992: 95).
13. 'Pluralism' as opposed to dualism – see Deleuze and Guattari (2004: 23).
14. For an anthropological comparison of multiplicities, see Strathern (2005: especially 161): '[C]ontrasting types of multiplicity come into view. If we talk of multiple origins in relation to Euro-American works, then multiplicity comes from the way persons are added to one another's enterprises. If we talk of multiple origins in relation to their Melanesian counterparts, then multiplicity comes from the way people divide themselves from one another.' So perhaps we could see in the former (Euro-American) multiplicity a molar and arborescent, or 'false', multiplicity, while the latter would be of the molecular and intensive type, 'composed of particles that do not divide without changing in nature' (Deleuze and Guattari 2004: 37). Strathern goes on to remark, in the passage just cited: 'I am not comparing like with like. Quite so.' I take this as implying a definition – comparing the incomparable – of anthropology.
15. This could be taken as an acceptable gloss of the canonical formula of myth of Lévi-Strauss (1958 [1955]). The recurrent presence of this synoptic figure in his *Mythologiques* and beyond bears out the fact that, for structuralism, not *all* the 'differences that make a difference' are reducible to the reversible proportionalities of the totemic scheme.
16. The ideas of Tarde, an ultra-Leibnizian thinker who was the greatest metaphysical adversary of Durkheim (a strictly Kantian soul), were rescued from the land of lost theories by Deleuze (1968:

104–5, n. 1; Deleuze and Guattari 2004: 240–41). Recently they have been taken up again by B. Latour and M. Lazzarato, among others. Chunglin Kwa, in an article already mentioned here, observes a 'fundamental difference between the romantic conception of society as an organism and the baroque conception of an organism as a society' (2002: 26). This is a perfect description of the difference between the sociologies of Durkheim and Tarde. Against the *sui generis* character of the social facts proposed by the former, the 'universal sociological viewpoint' of the latter states that 'every thing is a society, every phenomenon is a social fact' (Tarde 1999: 67, 58).

17. See Deleuze (1969: 202–3). Similarly, in the dialectics of the master and the slave, it is the slave who is dialectician and not the master (Deleuze 1962: 11).
18. Once again Zourabichvili: 'A meditation on Nietzschean perspectivism puts one in a position to see the positive consistence of the [Deleuzian] concept of disjunction: a distance between viewpoints which is at the same time undecomposable and unequal to itself, since the way is not the same in both directions' (2003: 79, my translation, emphasis removed).
19. For a subtler interpretation of Deleuze as a philosopher of 'immediate or non-dialectical duality', see Lawlor (2003).
20. Thus, with the duality between arborescence and rhizome ('have we not ... reverted to a simple dualism?' Deleuze and Guattari 2004: 14), two schemes that do not cease to interfere with each other. Thus, with the two types of multiplicity, molar and molecular, which operate always at the same time and in the same assemblage – there is no dualism of multiplicities but only 'multiplicities of multiplicities' (ibid.: 38). Thus, with the distinction between form of expression and form of content, in which there is neither parallelism nor representation but 'a manner in which expressions are inserted into contents ... in which signs are at work in things themselves just as things extend into or are deployed through signs' (ibid.: 96). Thus, with the opposition between segmentary and centralized, which must be replaced by a distinction between two different but inseparable segmentations, the supple and the rigid: 'they overlap, they are entangled' (ibid.: 231, 234). Thus, lastly, with smooth (nomadic, war-machinic) and striated (sedentary, state-like) spaces, whose difference is said to be complex, both because 'the successive terms of the oppositions fail to coincide entirely' – that is, smooth versus striated is not exactly the same thing as nomadic versus sedentary, etc. – and because 'the two spaces in fact exist only in mixture' (ibid.: 524). To summarize, soon after distinguishing two poles, processes or tendencies, the Deuleuzian analysis, on the one hand, unfolds the polarity into further polarities, asymmetrically embedded in the first (thus bringing about a 'mixture' de jure), and, on the other, it indicates the de facto mixture of the initial poles. And the typical conclusion is: 'All of this happens at the same time' (ibid.: 246).
21. Anthropologists are in general much given to this type of off-the-shelf deconstruction. See Rival (1998) and Rumsey (2001) for two apposite examples: both authors protest against a supposed great divide between the West = arborescence and the Rest = rhizome. These two critics show a certain naivety as they imagine a certain naivety on the part of the criticized, who knew perfectly well what they were (not) doing: '[W]e are on the wrong track with all these geographic distributions. An impasse. So much the better' (Deleuze and Guattari 2004: 22).
22. This pattern appears early in the Deleuzian corpus: see his comments on the Bergsonian division between duration and space, which cannot be simply defined as a difference in nature: the division is rather between duration, which supports and conveys all the differences in nature, and space which presents only differences in degree. 'There is thus no difference in nature between the two halves of the division: the difference in nature is wholly on one side' (Deleuze 1966: 23, my translation). It is as if each pole would 'apprehend' its relation to the other according to its own nature; or, said otherwise, as if the relation between the poles belonged necessarily and alternatively to the regime of either pole, either the regime of contradiction or that of the line of flight (Deleuze and Guattari 2004: 238), and these two regimes organize their mutual differences very differently. In any case, that relation cannot be traced from the outside, from a third encompassing pole. Perspectivism (duality as multiplicity) is what dialectics (duality as unity) must negate in order to impose itself as universal law.

23. Wagner qualifies the reciprocal co-production between cultural convention and invention as 'dialectical' (1981: 52; the term is widely used in Wagner 1986), which may confuse a Deleuzian reader. Yet the characterization of this dialectics, besides being *explicitly* non-Hegelian, makes it very evocative of the Deleuzian reciprocal presupposition and disjunctive synthesis: 'a tension or dialogue-like alternation between two conceptions or viewpoints that are simultaneously contradictory and supportive of each other' (Wagner 1981: 52). 'Dialectics' without resolution or conciliation, in short.
24. In the Melanesian gender-kinship model, 'each relation can come only from the other' and 'conjugal and parent–child relations are metaphors for one another, and hence a source of internal reflection' (Strathern 2001: 240). In the same paper we find the remark: 'cross-sex relations both alternate with same-sex relations, and contain an inherent premise of alternation within' (ibid.: 227). A fine example of reciprocal asymmetric presupposition.
25. In their approach to the smooth/striated contrast, the authors make the same methodological point: although in reciprocal presupposition, 'the two spaces do not communicate with each other in the same way ... the principles of mixture ... are not at all symmetrical'; the passage from the smooth to the striated and vice versa are 'entirely different movements' (Deleuze and Guattari 2004: 524).
26. The same strategy of evoking one dualism only in order to challenge another is employed, for example, by Latour in his counter-critical booklet on 'factishes': 'The double repertoire of the Moderns does not reside in their distinction of facts from fetishes, but, rather, in the ... subtler distinction between the theoretical separation of facts from fetishes, on the one hand, and an entirely different practice, on the other hand' (Latour 1996: 42–43, my translation). And: 'The choice proposed by the Moderns is not one between realism and constructivism; it is rather an alternative between this choice itself and a form of practical existence which cannot understand either the terms or the importance of such a choice.' This is an apt illustration of the Deleuzian concept of disjunctive synthesis: the meta-relation between exclusive ('the choice') and inclusive disjunctions is itself an exclusive disjunction, from the standpoint of the former ('You must choose!'), and an inclusive one, from the standpoint of the latter ('What are you talking about?').
27. The effective antagonism between the two kinship conceptions is somewhat debatable (Schneider 1965, 1984).
28. Also: 'In reality, global persons – even the very form of persons – do not exist prior to the prohibitions that weigh on them and constitute them' (Deleuze and Guattari 1983: 70). See Adler and Cartry (1971: 7) for one probable source of the argument.
29. The difference made in *Anti-Oedipus* between Mauss and Nietzsche is perhaps a little overstated. First, 'exchange' versus 'debt' is not as clear-cut a conceptual distinction as the authors (and, earlier, Deleuze 1962: 155) make it appear to be. Secondly, the Nietzschean theory of the proto-historical repression of a 'biological memory' necessary to create a 'social memory' is also not that obviously antipodal to the nature/culture anthropogenetic paradigm shared by Maussian or structuralist theories of exchange. It is only when Deleuze and Guattari determine becoming as an anti-memory, in their second book (Deleuze and Guattari 2004: 324), that the terms of the problem could be said to change radically. The Mauss/Nietzsche contrast drawn in *Anti-Oedipus* has a very complex polemical background, which we cannot enter into here. It involves Hegel, Kojève, Bataille, the Collège de Sociologie and, more proximally, Lévi-Strauss, Lacan and Baudrillard, among others. The 'generalized economy' derived by Bataille from a Nietzschean reading of *The Gift* is practically not mentioned (but see Deleuze and Guattari 1983: 190). Deleuze and Guattari's utter contempt for the Bataillean category of transgression (the remark comes from Lyotard) may be involved here. In the essay on Klossowski included in *Logic of Sense*, however, Deleuze establishes an insightful contrast between exchange, generality (equivalence) and false repetition, on the one hand, and gift-giving, singularity (difference) and true repetition, on the other hand; the contrast, which anticipates much of the theses of *Anti-Oedipus*, is partially and somewhat ambiguously connected, via Klossowksi, to Bataille. Deleuze writes that Théodore, the hero of one of Klossowski's novels, 'knows that true repetition lies in the gift, in an economy of the gift which is the opposite of a mercantile economy of exchange (... homage to Georges Bataille)' (Deleuze 1969: 334; my translation, suspension points in the original).

30. See Deleuze and Guattari (1983: 74–75, 109–15). The problem is indicated by the authors themselves in the preface to the Italian edition of *A Thousand Plateaus*: '*Anti-Oedipus* had a Kantian ambition' (Deleuze and Guattari 2003: 289, my translation). It is not preposterous to conjecture that the authors of *A Thousand Plateaus* would agree, at variance with the authors of *Anti-Oedipus*, that *any* philosophico-anthropological disquisition on the distinctiveness of the human species, regardless of what is to serve as a sign or cause of its election (or curse) – neoteny, language, toolmaking, an immortal soul, the incest taboo, a higher-order intentionality and whatnot – is irremediably committed to Oedipus. The purpose of contemporary anthropology can no longer be that of finding what makes humans 'different' from 'the rest of nature'. (Humans as the West and non-humans as the Rest? Quite so.) In so far as nature is concerned, I am afraid it is a case of 'same difference'. Anthropologists would be better employed in studying the different differences humans are capable of making; the divide between humans and non-humans is not the only one possible, and certainly not a particularly great one.
31. One should notice, then, that the Deleuzian attack on any form of genealogical extensionism does not imply an 'anti-biological' stance: 'It serves no purpose to recall that genealogical filiation is social rather than biological, for it is necessarily biosocial inasmuch as it is inscribed on the cosmic egg of the full body of the earth' (Deleuze and Guattari 1983: 154). Needless to say, this cosmic egg is quite a different animal from the selfish gene.
32. The already-mentioned article by Adler and Cartry (1971) on the Dogon origin myth is the main stimulus behind the strategic importance of these materials in *Anti-Oedipus*, being cited at crucial junctures of the analysis. The two anthropologists, along with A. Zempléni, read and 'corrected' the third chapter of *Anti-Oedipus* (see Nadaud 2004: 20–21, citing Guattari's correspondence with Deleuze). At the same time, Deleuze and Guattari's general anti-Oedipal thesis was quite obviously a determining influence on Adler and Cartry's text (see Adler and Cartry 1971: 37 n.1).
33. Or perhaps 'the myth of anti-Oedipus'. The contrast between expressive and productive conceptions of the unconscious had led the authors to put the flippant question in *Anti-Oedipus* (Deleuze and Guattari 1983: 57): 'Why return to myth?', referring to the psychoanalytic hijacking of the Oedipus story. A hundred pages later, in the discussion of the Dogon materials, the question receives, as it were, an unexpected answer. The reference to the pale fox myth in support of the authors' arguments is justified by the tantalizing remark: 'Only myth can determine the intensive conditions of the system (the system of production included) in conformity with indigenous thought and practice ... [myth] does not express but conditions' (ibid.: 157; see also Adler and Cartry 1971: 16). These two contrasting usages of the concept of myth in *Anti-Oedipus* need a much closer examination than I shall be able to provide here.
34. See also Deleuze and Guattari (1983: 187) for a typically structuralist remark: 'When considering kinship structures, it is difficult not to proceed as though the alliances derived from the lines of filiation and their relationships, although the lateral alliances and the blocks of debt condition the extended filiations in the system in extension, and not the opposite.'
35. What would be a 'mythical expression', though, since myth 'does not express but conditions'? Here is where the two usages of 'myth' (see note 33) meet, somewhat uncomfortably.
36. The Lévi-Straussian theory of marriage exchange remains, after all, a much better anthropological proposition than the jural metaphysics of group descent it pre-empted. In a sense, *The Elementary Structures of Kinship* was the first *Anti-Oedipus*, a radical break with the family-centred, parenthood-dominated image of kinship; or, to put it differently, the complex relationship of *Anti-Oedipus* to *The Elementary Structures* is anamorphically analogous to the relationship of the latter to *Totem and Taboo.*
37. See Wagner's analysis (1977) of matrimonial exchange for the case of the Melanesian Daribi: the wife-giving patrilineal clan sees the women it provides as an efferent flow of its own male substance; but the wife-taking clan sees the afferent flow as constituted by female substance. At the moment the matrimonial prestations start moving in the opposite direction, perspectives are inverted. The author concludes: 'What might be described as exchange or reciprocity is in fact a[n] ... intermeshing of two views of a single thing' (ibid.: 628). See Gell (1999: 67–68) for a detailed exposition of this idea as

developed by Strathern, the anthropologist who made the decisive move of *defining* every exchange as an exchange of perspectives, thereby completely changing the rules of the whole conceptual language game of 'exchange' within anthropology.

38. Becoming is, of course, a central Deleuzian concern ever since his essays on Bergson and Nietzsche, not to mention the *Logic of Sense* (Deleuze 1969). Starting with the joint essay on Kafka (Deleuze and Guattari 1986), however, 'becoming' acquires a specific conceptual consistency that will be fully deployed in *A Thousand Plateaus*: becoming is that which escapes both mimesis (imitation, reproduction) and 'memesis' (memory, history).
39. Note, however, that besides the importance he attributes to the intensive (morphodynamical: Petitot 1988) 'canonical formula' in his mythological work (Lévi-Strauss 1985), the founder of anthropological structuralism explicitly recognizes the limitations of the relational vocabulary of extensional logics to account for the transformations that take place in/between myths (Lévi Strauss 1971: 567–68). Moreover, if *The Elementary Structures of Kinship* can be said to deal with objects resembling, well, structures, the mythic analyses consolidated in the *Mythologiques* tetralogy – which develops a whole 'theory of primitive codes' (Deleuze and Guattari 1983: 185) – and the sequels are far more evocative of rhizomatic mappings than they are of structural tracings. The relations that constitute Amerindian mythic narratives, rather than forming combinatorial totalities in discrete arborescent logical distribution, in concomitant variation with and dialectical relation to socio-ethnographic *realia*, appear to instantiate to the point of explicitness the very rhizomatic principles of 'connection and heterogeneity', 'multiplicity', 'asignifying rupture' and 'cartography' that Deleuze and Guattari counterpose to structural models (Deleuze and Guattari 2004: 7–13). There is a whole case to be built for this affinity between late, i.e. non-sociologistic, Lévi-Straussian structuralism and Deleuzian differential analytics; in Maniglier (2000, 2005) one can glean some essential elements for the task.
40. In this sense, both resemblances and correlations are Oedipal, 'mythic' in the negative sense; see Deleuze and Guattari (1983: 83).
41. The Amerindian myths studied by Lévi-Strauss, very far from displaying any 'indifferentiation' or originary identification between humans and non-humans, as it is usually formulated, are defined by a regime of infinite difference, albeit (or because) internal to each persona or agent, in contrast to the finite and external differences constituting the species and qualities of our contemporary world (Viveiros de Castro 2001). This explains the regime of qualitative multiplicity proper to myth: the question of knowing whether the mythic jaguar, to stay with this animal, is a block of human affects in the shape of a jaguar or a block of feline affects in the shape of a human is in any rigorous sense undecidable, since mythic transformations describe an intensive superposition of heterogeneous states, not an extensive transposition of two homogeneous states.
42. Emphasis removed. The '*n* sexes' should be written '*n* – 1 sexes' – unity being represented by the male sex, which must be subtracted in order for sexuality to reach the state of multiplicity – just as the *n* species of animal-becomings are really *n* – 1 species (1 = humankind).
43. Incest (or what amounts to the same thing, its prohibition) is, in turn, only the retroactive effect of repressive alliance on repressed germinal filiation (Deleuze and Guattari 1983: 164–66).
44. 'A becoming is neither one nor two, *nor the relation of the two:* it is the in-between, the border or line of flight or descent *running perpendicular to both*' (Deleuze and Guattari 2004: 323, emphasis added).
45. The reference, of course, is to 'Rethinking Anthropology', in which Leach establishes the well-known connection between affinity and 'metaphysical influence' (1961: 20). See Viveiros de Castro (2009) for a recent comment on this correlation.
46. 'The gift economy, then, is a *debt economy*. The aim of a transactor in such an economy is *to acquire as many gift-debtors as he possibly can,* and not to maximize profit, as it is in a commodity economy. What a gift transactor *desires* is the personal relationships that the exchange of gifts creates, and not the things themselves' (Gregory 1982: 19, emphasis added). It might be amusing to see what would happen if we gave the verb 'desire' in this passage a strict Deleuzo-Guattarian sense. On exchange and perspective, see

Strathern (1988: 230, 271, 327, 1991a: *passim*, 1992a: 96–100, 1999: 249–56); Munn (1992: 16). On double capture, see Deleuze and Parnet (1996: 7–9); Stengers (1996: 64 n. 11).

47. 'Language can work against the user of it ... Sociality is frequently understood as implying sociability, reciprocity as altruism and relationship as solidarity' (Strathern 1999: 18). 'An action upon an action' is Foucault's definition of power, which is close to the Nietzschean ontology of force (a force always refers to another force, not to an 'exterior' object; Deleuze 1962: 7 ff.). And 'a reaction to a reaction' is how Bateson (1958) defined schismogenesis.
48. As an example of inversion, see the Araweté spouse-swapping ritual friends, who are 'anti-affines' without thereby being 'consanguines' (Viveiros de Castro 1992: 167–78).
49. It should be stressed that this affinization of 'others' occurs in spite of the fact that the vast majority of actual matrimonial alliances take place within the local group. And, at any rate, such alliances cannot but accumulate in the local group, since their concentration defines what a 'local group' is. By this last remark I mean to imply that the situation does not change much when we consider those regimes that feature village or descent group exogamy, such as those prevailing among the Tukanoan and Arawakan peoples of north-west Amazonia. Potential affinity and its cosmological attendants continue to mark the generic relations with non-allied groups, enemies, animals and spirits.
50. Amerindian myths, of course, also feature 'Oedipal' motives, parental figures and filio-parental conflicts. Lévi-Strauss even alluded, not without irony, to a Jivaroan *Totem and Taboo* (1985: chapter XVI). But it is quite obvious that for him the mythology of the continent and, above all, the myths dealing with the origin of culture revolve around affinity and exchange, not parenthood and procreation.
51. There are mythological complexes in Amazonia that feature a pre-cosmological scenario very similar to the 'intensive filiation' discerned by Deleuze and Guattari in the Dogon case; north-western Amazonian origin myths, in particular, must be mentioned here (Hugh-Jones 1979, 1993). These myths, however, articulate the same meta-schema of potential affinity as constituting the ontological base state (Andrello 2006).
52. Thus *Anti-Oedipus* repeats the historical-materialist cliché of the 1970s to the effect that Maussian structuralist ethnology is 'burdened' by 'the reduction of social reproduction to the sphere of circulation' (Deleuze and Guattari 1983: 188).
53. See Donzelot (1977), Lyotard (1977) and Zourabichvili (2003: 48–51) for three excellent assessments of *Anti-Oedipus*'s relation to Marxist conceptuality.
54. If '[t]he expression "difference of intensity" is a tautology' (Deleuze 1968: 287, my translation), 'becoming-other' is another, or perhaps the same, tautology.

Bibliography

Adler, A. and M. Cartry. (1971). 'La transgression et sa dérision', *L'Homme* 11 (3), 5–63.
Andrello, G. (2006). *Cidade do índio. Transformações e cotidiano em Iauaretê.* São Paulo, Edunesp/ISA/NuTI.
Bateson, G. (1958). *Naven: A Survey of the Problems Suggested by a Composite Picture of the Culture of a New Guinea Tribe Drawn from Three Points of View.* Stanford, CA, Stanford University Press.
Clastres, P. (1977). 'Archéologie de la violence: la guerre dans les sociétés primitives', *Libre* 1, 137–73.
de Landa, M. (2002). *Intensive Science and Virtual Philosophy.* London, Continuum.
———. (2003). '1000 Years of War: *CTheory* interview with Manuel de Landa', *CTheory*, a127. Available at www.ctheory.net/articles.aspx?id=383
Deleuze, G. (1962). *Nietzsche et la philosophie.* Paris, PUF.
———. (1966). *Le Bergsonisme.* Paris, PUF.
———. (1968). *Différence et répétition.* Paris, PUF.
———. (1969). *Logique du sens.* Paris, Minuit.
———. (1993). *Critique et clinique.* Paris, Minuit.

———. (2002) [1972]. 'A quoi reconnaît-on le structuralisme?', in D. Lapoujade (ed.) *L'Ile déserte et autres textes. Textes et entretiens 1953–1974*. Paris, Minuit, pp. 238–69.

Deleuze, G. and F. Guattari. (1983). *Anti-Oedipus: Capitalism and Schizophrenia*. Minneapolis, University of Minnesota Press.

———. (1986) [1975]. *Kafka: Toward a Minor Literature*. Minneapolis, University of Minnesota Press.

———. (2004) [1988]. *A Thousand Plateaus: Capitalism and Schizophrenia*. London, Continuum.

———. (1991). *Qu'est-ce que la philosophie*. Paris, Minuit.

———. (2003) [1987]. 'Préface pour l'édition italienne de Mille Plateaux', in D. Lapoujade (ed.) *Deux régimes de fous. Textes et entretiens 1975–1995*. Paris, Minuit, pp. 288–90.

Deleuze, G. and C. Parnet (1996) [1977]. *Dialogue*, Paris, Flammarion.

Donzelot, J. (1977). 'An Anti-sociology', *Semiotext(e)* II (3), 27–44.

Dumont, L. (1971). *Introduction à deux théories d'anthropologie sociale. Groupes de filiation et alliance de marriage*. Paris, Mouton.

Fortes, M. (1969). *Kinship and the Social Order: The Legacy of Lewis Henry Morgan*, London, Routledge and Kegan Paul.

———. (1983). *Rules and the Emergence of Society*. London, Royal Anthropological Institute of Great Britain and Ireland.

Gell, A. (1999). 'Strathernograms, or the Semiotics of Mixed Metaphors', in A. Gell, *The Art of Anthropology: Essays and Diagrams*. London, Athlone, pp. 29–75.

Goldman, M. (2005). 'Formas do saber e modos do ser: observações sobre multiplicidade e ontologia no candomblé' *Religião e Sociedade* 25 (2), 102–20.

Gregory, C. (1982). *Gifts and Commodities*. London, Academic Press.

Griaule, M. and G. Dieterlen. (1965). *Le Renard pâle*, Paris, Institut d'Ethnologie.

Hugh-Jones, S. (1979). *The Palm and the Pleiades: Initiation and Cosmology in North-west Amazonia*. Cambridge, Cambridge University Press.

———. (1993). 'Clear Descent or Ambiguous Houses? A Re-examination of Tukanoan Social Organization', *L'Homme* 126–128, 95–120.

Jameson, F. (1997). 'Marxism and Dualism in Deleuze', *The South Atlantic Quarterly* 96 (3), 393–416.

Jensen, C.B. (2003). 'Latour and Pickering: Post-human Perspectives on Science, Becoming, and Normativity', in D. Ihde and E. M. Selinger (eds) *Chasing Technoscience: Matrix for Materiality*. Bloomington, Indiana University Press, pp. 225–40.

———. (2004). 'A Non-humanist Disposition: On Performativity, Practical Ontology, and Intervention', *Configurations* 12, 229–61.

Kwa, C. (2002). 'Romantic and Baroque Conceptions of Complex Wholes in the Sciences', in J. Law and A. Mol (eds) *Complexities: Social Studies of Knowledge Practices*. Durham, Duke University Press, pp. 23–52.

Latour, B. (1993) [1991]. *We Have Never Been Modern*. New York, Harvester-Wheatsheaf.

———. (1996). *Petite réflexion sur le culte moderne des dieux faitiches*. Le Plessis-Robinson, Les Empêcheurs de Penser en Rond.

———. (2005). *Reassembling the Social: An Introduction to Actor-Network Theory*, Oxford, Oxford University Press.

Lawlor, L. (2003). 'The Beginnings of Thought: the Fundamental Experience in Derrida and Deleuze', in P. Patton and J. Protevi (eds) *Between Deleuze and Derrida*. London, Continuum, pp. 67–83.

Leach, E.R. (1961) [1951]. 'Rethinking Anthropology', in E. Leach, *Rethinking Anthropology*. London, Athlone, pp. 1–27.

Lévi-Strauss, C. (1958) [1955]. 'La structure des mythes', in C. Lévi-Strauss *Anthropologie structurale*. Paris, Plon, pp. 227–55.

———. (1962a). *Le Totémisme aujourd'hui*. Paris, Presses Universitaires de France.

———. (1962b). *La Pensée sauvage*. Paris, Plon.

———. (1969) [1967]. *The Elementary Structures of Kinship*. Boston, Beacon Press.

———. (1971). *L'Homme nu. Mythologiques IV*. Paris, Plon.

———. (1985). *La Potière jalouse*. Paris, Plon.

———. (1991). *Histoire de lynx.* Paris, Plon.

Lévi-Strauss, C. and Eribon, D. (1988). *De près et de loin.* Paris, Odile Jacob.

Lyotard, J.-F. (1977). 'Energumen capitalism', *Semiotext(e)* II(3), 11–26.

Maniglier, P. (2000). 'L'humanisme interminable de Lévi-Strauss', *Les Temps Modernes* 609, 216–41.

———. (2005). 'Des us et des signes. Lévi-Strauss: philosophie pratique', *Revue de Métaphysique et de Morale ('Repenser les structures', G.-F. Duportail ed.)* 1.Available at: http://formes-symboliques.org/article.php3?id_article=159

———. (2006). *La Vie énigmatiques des signes. Saussure et la naissance du structuralisme.* Paris, Léo Scheer.

Mimica, J. (1992) [1988]. *Intimations of Infinity: The Cultural Meanings of the Iqwaye Counting and Number System.* Oxford, Berg.

Munn, N. (1992) [1986]. *The Fame of Gawa: A Symbolic Study of Value Transformation in a Massim (Papua New Guinea) Society.* Durham, Duke University Press.

Nadaud, S. (2004). 'Les amours d'une guêpe et d'une orchidée.' Preface to F. Guattari, *Félix Guattari, Écrits pour l'Anti-Oedipe,* Paris, Éditions Lignes et Manifestes, pp. 7–27.

Petitot, J. (1988). 'Approche morphodynamique de la formule canonique du mythe', *L'Homme* 106–107, 24–50.

Pignarre, P. and I. Stengers. (2005). *La Sorcellerie capitaliste. Pratiques de desenvoûtement.* Paris, La Découverte.

Plotnitsky, A. (2003). 'Algebras, Geometries, and Topologies of the Fold: Deleuze, Derrida, and Quasi-Mathematical thinking, with Leibniz and Mallarmé', in P. Patton and J. Protevi (eds) *Between Deleuze and Derrida.* London: Continuum, pp. 98–119.

Rival, L. (1998). 'Trees: From Symbols of Life and Regeneration to Political Artefacts', in L. Rival (ed.) *The Social Life of Trees: Anthropological Perspectives on Tree Symbolism.* Oxford, Berg, pp. 1–36.

Rivière, P. (1984). *Individual and Society in Guiana: A Comparative Study of Amerindian Social Organization.* Cambridge, Cambridge University Press.

Rumsey, A. (2001). 'Tracks, Traces, and Links to Land in Aboriginal Australia, New Guinea, and Beyond', in A. Rumsey and J. F. Weiner (eds) *Emplaced Myth: Space, Narrative, and Knowledge in Aboriginal Australia and Papua New Guinea.* Honolulu, University of Hawai'i Press, pp. 19–43.

Schneider, D.M (1965). 'Some Muddles in the Models: Or, How the System Really Works', in M. Banton (ed.) *The Relevance of Models for Social Anthropology.* London, Tavistock, pp. 25–85.

———. (1984). *A Critique of the Study of Kinship.* Ann Arbor, University of Michigan Press.

Stengers, I. (1996). *Cosmopolitiques, tome 1. La Guerre des sciences.* Paris, La Découverte/Les Empêcheurs de Penser en Rond.

———. (2002). *Penser avec Whitehead.* Paris, Seuil.

Strathern, M. (1988). *The Gender of the Gift: Problems with Women and Problems with Society in Melanesia.* Berkeley, University of California Press.

———. (1991a). *Partial Connections,* Lanham, MD, Rowman and Littlefield.

———. (1991b). 'One Man and Many Men', in M. Godelier and M. Strathern (eds) *Big Men and Great Men: Personification of Power in Melanesia.* Cambridge, Cambridge University Press, pp. 197–214.

———. (1992a). *After Nature: English Kinship in the Late Twentieth Century.* Cambridge, Cambridge University Press.

———. (1992b). 'Parts and Wholes: Refiguring Relationships in a Post-Plural world', in M. Strathern, *Reproducing the Future: Essays on Anthropology, Kinship, and the New Reproductive Technologies.* New York, Routledge, pp. 90–116.

———. (1995). 'The Nice Thing about Culture is That Everyone Has It', in M. Strathern (ed.) *Shifting Contexts: Transformations in Anthropological knowledge.* London and New York, Routledge, pp. 153–76.

———. (1996). 'Cutting the Network', *Journal of the Royal Anhtropological Institute* (NS) 2 (4), 517–35.

———. (1999). *Property, Substance and Effect: Anthropological Essays on Persons and Things.* London, Athlone.

———. (2001). 'Same-sex and Cross-sex Relations: Some Internal Comparisons', in T. Gregor and D. Tuzin (eds) *Gender in Amazonia and Melanesia: An Exploration of the Comparative Method.* Berkeley, University of California Press, pp. 221–44.

———. (2005). *Kinship, Law and the Unexpected: Relatives are Always a Surprise.* New York, Cambridge University Press.

Tarde, G. (1999) [1895]. *Oeuvres de Gabriel Tarde*, volume 1: *Monadologie et sociologie.* Le Plessis-Robinson, Institut Synthélabo.

Viveiros de Castro, E. (1992). *From the Enemy's Point of View: Humanity and Divinity in an Amazonian Society.* Chicago, University of Chicago Press.

———. (2001). 'GUT feelings about Amazonia: Potential Affinity and the Construction of Sociality', in L. Rival and N. Whitehead (eds) *Beyond the Visible and the Material: the Amerindianization of Society in the Work of Peter Rivière.* Oxford, Oxford University Press, pp. 19–43.

———. (2004). 'Perspectival Anthropology and the Method of Controlled Equivocation', *Tipití* 2 (1), 3–22.

———. (2009). 'The Gift and the Given: Three Nano-Essays on Kinship and Magic', in S. Bamford and J. Leach (eds) *Kinship and Beyond: The Genealogical Model Reconsidered.* Oxford, Berghahn Books.

Wagner, R. (1972). 'Incest and Identity: a Critique and Theory on the Subject of Exogamy and Incest Prohibition', *Man* 7 (4), 601–13.

———. (1977). 'Analogic Kinship: A Daribi Example', *American Ethnologist* 4 (4), 623–42.

———. (1978). *Lethal Speech: Daribi Myth as Symbolic Obviation.* Ithaca and London, Cornell University Press.

———. (1981). *The Invention of Culture,* (2nd edn). Chicago, University of Chicago Press.

———. (1986). *Symbols that Stand for Themselves,* Chicago, University of Chicago Press.

———. (1991). 'The Fractal Person', in M. Godelier and M. Strathern (eds) *Big Men and Great Men: Personification of Power in Melanesia.* Cambridge, Cambridge University Press, pp. 159–73.

Whitehead, A.N. (1978). *Process and Reality.* New York, The Free Press.

Zourabichvili, F. (2003). *Le Vocabulaire de Deleuze.* Paris, Ellipses.

———. (2004a). 'Deleuze. Une philosophie de l'événement', in F. Zourabichvili, A. Sauvargnargues and P. Marrati (eds) *La philosophie de Deleuze.* Paris, PUF, pp. 1–116.

———. (2004b). 'Deleuze et la question de la littéralité', unpublished manuscript.

Notes on Contributors

Geoffrey C. Bowker, Carnegie Mellon Professor in Cyberscholarship, University of Pittsburgh. Author of *Science on the Run: Information Management and Industrial Geophysics at Schlumberger, 1920–40*, MIT Press, 1994, *Sorting Things Out: Classification and its Consequences* (with Susan Leigh Star), MIT Press, 2000, *Memory Practices in the Sciences*, MIT Press, 2005.

Steven D. Brown, Professor, School of Management, University of Leicester, UK. Author of *The Social Psychology of Experiences: Studies in Remembering and Forgetting* (with David Middleton), Sage, 2005 and *Psychology Without Foundations* (with Paul Stenner), Sage, 2008.

Arturo Escobar, Distinguished Professor, Department of Anthropology, University of North Carolina at Chapel Hill, US. Author of *Encountering Development: The Making and Unmaking of the Third World*, Princeton, 1995 and numerous other works.

Mariam Fraser, Senior Lecturer, Sociology Department, Goldsmiths, UK. Author of numerous articles and of *Identity without Selfhood: Simone de Beauvoir and Bisexuality*, Cambridge, 1999. Co-editor of *The Body: A Reader* and of *Inventive Lives: Approaches to the New Vitalism*, Sage, 2006.

Casper Bruun Jensen, Associate Professor, Department of Organization, Copenhagen Business School, Denmark and at the IT-University of Copenhagen. Casper has published in *Configurations, Science, Technology and Human Values, Social Studies of Science* and edited a Danish introduction to STS.

Adrian Mackenzie, Reader, Center for Economic and Social Aspects of Genomics (Cesagen), Lancaster University, UK. Author of *Transductions: Bodies and Machines at Speed*, Continuum Press, 2002, *Cutting Code: Software and Sociality*, Peter Lang, 2006 and *Wirelessness: Radical Network Empiricism*, MIT Press, 2010.

Michal Osterweil, Doctoral canditate, Department of Anthropology, University of North Carolina at Chapel Hill, US.

Andrew Pickering, Professor, Department of Sociology and Philosophy, University of Exeter, UK. Author of *Constructing Quarks: A Sociological History of Particle Physics*, 1984, *Science as Practice and Culture* (ed.), 1992, *The Mangle of Practice: Time, Agency and Science*, 1995 all published by University of Chicago Press, and *The Mangle in Practice: Science, Society and Becoming* (edited with Keith Guzik), Duke University Press, 2008.

Kjetil Rödje, Doctoral candidate, School of Communication, Simon Fraser University, Canada.

Erich Schienke, Assistant Professor, Science, Technology, and Society Program, and Research Associate, Rock Ethics Institute, Pennsylvania State University, US. Author of the forthcoming book, *Greening the Dragon: The Rise of Ecological Governance in Contemporary China*, as well as various articles on Climate Ethics and the Ethical Dimensions of Scientific Research.

Isabelle Stengers, Professor, Department of Philosophy, Free University of Brussels, Belgium. Author of multiple books in French, including *Cosmopolitiques* vols. 1–7 (1996–97) and *Penser avec Whitehead: Une libre et sauvage création de concepts* (2002). English translations include *Power and Invention*, University of Minnesota Press, 1997, and *The Invention of Modern Science*, University of Minnesota Press, 2000.

Katie Vann, Assistant Director of Public Engagement-Center for Science, Technology, and Society, Santa Clara University.

Eduardo Viveiros de Castro, Professor, Museu Nacional, Rio de Janeiro, Brazil. Author of many works in Portuguese. *From the Enemy's Point of View: Humanity and Divinity in an Amazonian Society* is translated to English (University of Chicago Press, 1992). *The Anti-Narcissus* is forthcoming from Prickly Pear Press.

Index

E

F

H

I

J

K

L

M

N

Q

R

T